SOIL PHYSICS WITH PYTHON

Soil Physics with Python

Transport in the Soil–Plant–Atmosphere System

Marco Bittelli
University of Bologna, Italy

Gaylon S. Campbell
Decagon Devices Inc., USA

Fausto Tomei
Regional Agency for Environmental Protection, Emilia Romagna, Italy

OXFORD
UNIVERSITY PRESS

OXFORD
UNIVERSITY PRESS

Great Clarendon Street, Oxford, OX2 6DP,
United Kingdom

Oxford University Press is a department of the University of Oxford.
It furthers the University's objective of excellence in research, scholarship,
and education by publishing worldwide. Oxford is a registered trade mark of
Oxford University Press in the UK and in certain other countries

First published 2015
First published in paperback 2020

Impression: 1

Published in the United States of America by Oxford University Press
198 Madison Avenue, New York, NY 10016, United States of America

British Library Cataloguing in Publication Data
Data available

Library of Congress Cataloging in Publication Data
Data available

ISBN 978–0–19–968309–3 (Hbk.)
ISBN 978–0–19–885479–1 (Pbk.)

Printed and bound by
CPI Group (UK) Ltd, Croydon, CR0 4YY

Preface

The first version of this book was published in 1985 under the title *Soil Physics with BASIC*, by Gaylon S. Campbell. It was the first book to present numerical solutions of soil physics problems using computers. Since its publication, developments in computer science have led to significant advances in numerical solutions of such problems.

Software is now available with numerical solutions for water, heat and solute transport in soils. Convenient windows user-interfaces allow users to set up numerical solutions quickly and with various options regarding parameterizations, numerical schemes, domain of simulations and so forth. However, today's software often does not allow users to access and modify the code, relegating them to a more superficial understanding of the problems and their solutions. This book employs the same open source philosophy as the original *Soil Physics with BASIC*. Computer code is written, explained and discussed for each problem, to provide the reader with a full understanding of the solutions. Moreover, the availability of open source code allows the reader to incorporate the code into his or her own model, fostering further development of computer models.

We have selected a different programming language from that used in the original book. The decision to use Python instead of BASIC was motivated by various factors described in more detail in Appendix A. A list of additional programs and modules needed to run the programs is presented in Appendices A and B. In general, Python has a simple syntax, it implements object-oriented programming techniques, comes with powerful mathematical and numerical tools, and is user-friendly. It is freeware and runs on different platforms, including MS Windows, Macintosh and Linux. These features make it an accessible language for both beginner and expert programmers. With respect to the original book, we have included additional features, subjects and examples and we have extended the numerical solutions from 1D, to 2D and 3D. Overall, we aim to build on the numerical solutions provided in Campbell's original book, making available updated and new source code applicable to the rapidly advancing computer models available today.

Acknowledgements

We would like to thank the following institutions for institutional and financial support. Department of Agricultural Sciences, University of Bologna, Italy; Decagon Devices Inc., Pullman, WA, USA and Regional Agency for Environmental Protection, Emilia–Romagna region (ARPA–EMR), Bologna, Italy.

We thank Eva Kroener, who reworked problems and examples and Markus Flury for reviewing the chapter on solute transport.

We thank Sonke Adlung for editorial expertise, Jessica White and Ania Wronski for administrative support and Mac Clarke for copy editing at Oxford University Press. We thank B. Gogulanathan, Project Manager for production support. We also wish to thank our students, whose feedback helped us to improve the book. A thanks for their endearing support to Andrea Vogt, Judy Campbell and Murielle Moise.

Contents

1

Introduction

The energy and mass exchanges between living organisms and their environment are the principal phenomena that determine the existence of life on earth. Plants' ability to take up water and nutrients from the soil, fix carbon from the atmosphere and use solar energy to synthetize sugars through photosynthetic activity is a key process that summarizes the exchange of mass (water, nutrients, carbon) and energy (solar radiation) between soil, plants and the atmosphere. Soil, a thin layer of unconsolidated material on the earth's surface, plays a fundamental role in supporting terrestrial life, and it determines these exchanges of energy and mass between the lithosphere and the atmosphere. The storage and redistribution of water from precipitation is regulated by the soil's ability to store water in its pores and, at the same time, make it available for plants and microorganisms.

The multitude of chemical reactions occurring in soils, often determined by organisms such as bacteria, fungi, nematodes and insects, are of the utmost importance for the decomposition of organic matter into molecules utilized by plants. The diurnal temperature cycle regulates heat exchange between earth and atmosphere, while soil water and its modulation of sensible and latent heat affect the evolution of the atmospheric boundary layer determining weather patterns.

These phenomena are strictly interconnected in a complex array of nonlinear processes, which are often described by coupled physical, chemical and biological laws. The physical aspects of such processes are very relevant for understanding and quantifying these exchanges of mass and energy. An understanding of the principles behind these processes provides an important tool for assessing the physical relationship between an organism and its environment, and therefore their reciprocal effects. The study of soil physics involves the state and movement of energy and material in soil. This includes the structure and texture of the solid matrix, the diffusion of gases, soil temperature and heat flux, retention and flow of water, and the movement of solutes. Consideration of any of these subjects is difficult. However, they are necessary to know how quickly plants are withdrawing water or nutrients from the soil, or how much energy is being supplied or removed at the soil surface. These qualitative representations of the physical state of the soil system have given way to more quantitative expressions, and the dynamics of the system has begun to be considered.

Soil Physics with Python. First Edition. Marco Bittelli, Gaylon S. Campbell and Fausto Tomei.
© Marco Bittelli, Gaylon S. Campbell and Fausto Tomei 2015. Published in 2015 by Oxford University Press.

The first step in the description of transport in soil is to write the differential equations that represent transport and to determine the appropriate driving forces. Writing the correct equations is a necessary first step in solving transport problems, but unless the equations are solved, they have little practical value. Because of this, a lot of effort has been and is being expended on finding methods for solving the differential equations that describe transport in soil. Early work concentrated on analytical solutions, but these are available only for a few very simple systems. On the other hand, numerical solutions have the advantage that, unlike the analytical solutions, they are not restricted to simple system geometry or simple boundary conditions.

Another advantage of using numerical techniques is that the numerical method converts a differential equation into a set of algebraic equations. The algebraic equations are much easier to solve than the differential equations. The availability of microcomputers, combined with the power and simplicity of numerical methods, allows soil scientists, plant scientists, hydrologists and others who need to solve transport problems to understand the procedures and to use the methods for practical purposes.

The intent of this book is to present the equations that describe transport of mass and energy in soil, show how to solve them using numerical methods, and then demonstrate how to apply them to problems. To aid in this, computer programs written in Python are provided in each chapter and problems are given that require the use of these programs. This approach will help the reader to become familiar with the power as well as the simplicity of numerical methods for solving soil physics problems. The purpose of the different parts of the programs is explained in the chapters, and the programs are also downloadable from the book's website and organized by projects as presented in the book.[1]

The appendices of this book provide an overview of basic Python programming and some useful examples of numerical applications in Python. The reader should refer to these appendices for details about specific modules and functions used in the book. Throughout the book, units consistent with the International System (SI) will be used. Units can be reduced to fundamental units of length [L], mass [M] and time [T]. In the metric system, these units are metres [m], kilograms [kg] and seconds [s], defining the mks system. The SI unit of temperature is the kelvin [K], but degrees Celsius [°C] can also be used.

---

[1] The programs presented in this book are downloadable at: http://www.dista.unibo.it/~bittelli/

2

Basic Physical Properties of Soil

One of the most common questions a physicist is asked by non-scientists is the classic question about whether two snowflakes are ever identical. When asked what were the main differences between the macroscopic world and the microscopic world, Enrico Fermi responded that in the macroscopic world, each object is different from one to another, while in the microscopic world (he was referring to atomic and subatomic particles as the microscopic world), we can find particles that are equal to one another (Barone, 2009). He wrote that even if we take two objects made of iron, of the same shape, density and volume, the billions of iron atoms and their different geometrical arrangements would make the two objects different. On the other hand, we could consider two electrons of two different nuclei to be exactly equal—therefore we could consider them as two identical objects. Therefore, according to Fermi's definition, two snowflakes are never identical.

Soil is a heterogeneous, disperse, three-phase system composed of solids, liquids and gases. A system that is, at the same time, heterogeneous and disperse exhibits several interfaces not only between phases, but also between internal parts of a single phase. The interfacial activities in soil are behind many important phenomena such as surface tension, friction, adsorption and pressure changes. The difficulties encountered when attempting to describe macroscopic objects such as soil or its constituents are illustrated in large part by the differences that we experience when we measure a specific property (say density or mineralogical composition) at different spatial scales. Not only are the soil constituents never the same, as pointed out by Fermi, but they may display different features depending on the observation scale. Therefore, we must employ several simplifications to describe this complex system.

The solid phase is made up of mineral particles and organic materials. The liquid phase consists of water and dissolved solutes, and the gas phase is composed of a combination of gases. About 79% is nitrogen, with oxygen and carbon dioxide making up most of the remaining 21%. The geometrical features of the solid particles and their complement, the pore space, are complex. Figure 2.1 shows a microscope image of a soil thin section. The white area is the pore space, while the black and the grey areas are solid particles (minerals and organic matter). In natural conditions, the pore space can be filled with water and gases in different proportions.

Soil Physics with Python. First Edition. Marco Bittelli, Gaylon S. Campbell and Fausto Tomei.
© Marco Bittelli, Gaylon S. Campbell and Fausto Tomei 2015. Published in 2015 by Oxford University Press.

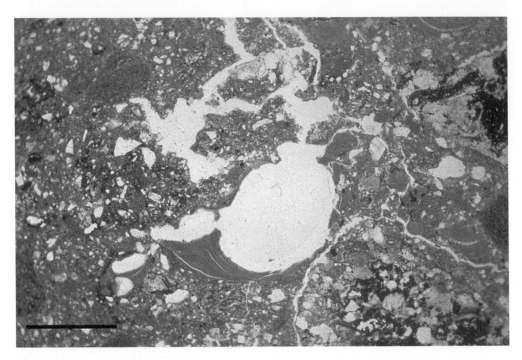

Figure 2.1 *Microscope image of a soil thin section. The black line in the image is 50 μm. Courtesy of Cranfield University, 2014. LandIS Soils Guide. Available at <http://www.landis.org.uk/services/ soilsguide/>.*

2.1   Geometry of the Soil Matrix

The soil matrix is characterized by complicated geometries formed by particles of different sizes and shapes, aggregates of different dimensions and shapes, interconnected pores, isolated pores, conglomerations, and aggregates of different molecules (iron oxides, carbonates, salts) and organic matter. The organic matter can be present as poorly degraded material (leaves, small branches, small dead animals, plant tissues of different forms) or as degraded organic material such as different forms of humus, humic acids and fulvic acids. Microorganisms such as bacteria, fungi and viruses can be present as isolated microorganisms or as colonies of different sizes and shapes. Providing a geometrical description of such a system is a complicated task. However, there are geometrical models that have been proposed over the years and that provide a relatively satisfactory description, at least for the solution of some problems we are faced with in soil physics.

Figure 2.1 shows areas that resemble circles and others that look like long and thin channels. Others resemble triangles, squares, etc. How do we represent such a complicated system in terms of geometrical properties that can be used to derive properties necessary to compute, for instance, water, heat or solute movement? A variety of theories are used to represent the complex soil geometry. Arrangements are also defined as 'structure', a term that refers to the way solid particles are organized in the soil matrix.

2.1.1 Basic Geometry

One model of the soil matrix is obtained by representing the soil particles as spheres, arranged in different geometrical shapes. The ideal and easiest representation (although certainly very far from natural conditions) is to imagine a medium composed of spheres of the same diameter, in a two- or three-dimensional arrangement. Depending on the geometrical arrangements of the spherical particles, various packing geometries are possible, having different coordination numbers and being named according to the packing arrangement (cubic, tetrahedral, etc.). A detailed geometrical and mathematical analysis was given by Deresiewicz (1958), who computed bulk density and porosity for different geometrical arrangements of spherical particles (Fig. 2.2).

The most basic arrangement of spherical particles (solid phase) is the cubic arrangement, for which the porosity can be calculated by considering a sphere of radius r (with volume $V = \frac{4}{3}\pi r^3$) and a cube of side $L = 2r$ (with volume $V = (2r)^3 = 8r^3$). The volume in the pore space of the unit cube is then given by

$$V_{\text{pore}} = 8r^3 - \frac{4}{3}\pi r^3 \qquad (2.1)$$

By dividing the volume of the pore space by the total volume of the unit cube, the porosity is obtained: porosity $= V_{\text{pore}}/8r^3$. The total porosity for this configuration is 0.4764 (or 47.64%). Based on the same approach, Deresiewicz (1958) presented a table with the relationships for other configurations.

Such a simple description cannot represent the soil matrix, however, since the soil has particles of many different sizes and shapes that determine a complicated packing of particles. For instance, spheres of smaller particles would fit between larger spheres

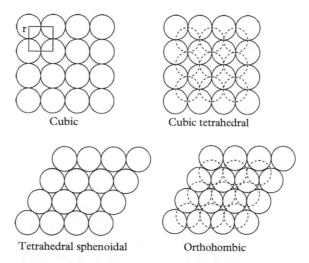

Cubic   Cubic tetrahedral

Tetrahedral sphenoidal   Orthohombic

Figure 2.2 *Models of regular packing of spherical particles.*

Figure 2.3 *Scanning electron microscope picture of a soil sample. Each tick on the label is 1 μm. Courtesy of Annamaria Pisi and Gianfranco Filippini, Microscopy Centre, Department of Agricultural Sciences, University of Bologna.*

and particles of non-spherical shape (clay platelets) would also fill the spaces between larger particles. An example of particle arrangement is provided in the scanning electron microscope picture of a soil sample shown in Fig. 2.3.

Moreover, aggregating and cementing agents create units of secondary particles and structural units, called macro-aggregates. Variously decomposed organic matter, humus, roots, earthworms and microorganisms also contribute to the complex geometrical arrays present in soils, creating a network of geometries varying in space over several orders of magnitude. In general, the structure represents the soil morphology, which refers to the heterogeneous arrangement of solids and pores at any given time. Soil heterogeneity is classified as either deterministic or stochastic. Indeed, many geometries in nature (such as that of soils) are not easily described by simple, classical Euclidean geometries such as spheres, cylinders, lines or triangles.

2.2   Soil Structure

We define as 'structure' the arrangement and organization of solid particles in the soil matrix. Because of the variety of different geometries, soil structure is a very difficult property both to measure and to classify. There are a variety of techniques to measure soil structure. X-ray tomography is a non-destructive and non-invasive technique that

Figure 2.4 *Section of a soil sample obtained by X-ray tomography. The cube size (upper right corner) is 10 mm. Courtesy of Gloria Falsone, Department of Agricultural Sciences and Matteo Bettuzzi, Department of Physics and Astronomy, University of Bologna.*

can be successfully used for three-dimensional analysis of soil structure. Quantitative measurements of soil structure elements, especially of soil pores and pore network features, can be obtained. Figure 2.4 shows a three-dimensional section of a soil sample, measured by X-ray tomography. The different structural arrangements between the most superficial layer and the lower layer are clearly visible. Moreover, the heterogeneity of the pore network and of the different aggregate arrangements shows once again the complicated geometry of the soil structure.

It is beyond the scope of this book to describe the basis of X-ray tomographic measurement or the different geometrical approaches employed to exploit the information collected by X-ray tomography. However, the reader can find a number of research papers and reviews on this subject (Vogel, 2002; Taina *et al.*, 2008; Vogel *et al.*, 2010; Dal Ferro *et al.*, 2012).

Often, physical, chemical and biological processes are involved in the formation of structural aggregates. The formation process occurs across various spatial scales, from fractions of micrometres to metres, and it changes over time.

A simpler, macroscopic analysis and classification of the structure (Soil Survey Division Staff, 1993), describes large aggregates as *peds*, while soils without these large aggregates are called structureless. Structureless soils are classified as *single-grain* (when individual particles are easily distinguishable) or *massive* (when individual particles adhere but the mass lacks distinguishable structural units). Three features can then be used to classify structure: *grade*, *size* and *shape*. The *grade* determines the strength of the structural units: for instance, a structure has *weak grade* when the peds are observable in the field but cannot be removed without being destroyed. The *size* refers to the dimension of the peds, and the *shape* refers to the geometry of the peds. For instance, the *shape* can be blocky (angular and subangular), prismatic, columnar, granular or platy. The formation of soil structure is determined by a variety of physico-chemical forces and aggregating mechanisms.

2.3   Fractal Geometry

One of the approaches to describe soil geometry is to employ fractal theory to describe the soil-matrix–pore-space geometry. The term 'fractal' was coined by Mandelbrot (1975) to describe the shape of irregular objects that cannot be described by Euclidean geometry. Fractal objects are geometrical objects described by a specific mathematical formulation that undergoes a recursive iteration generating the fractal object filling the space. Fractal objects are self-similar (which means that the object as a whole is similar to a part of itself) and scale-invariant. The scale invariance is a form of self-similarity where at any scale (or level of magnification) there is a smaller part of the object that is similar to the whole. The fractal dimension D is a quantity that describes how a fractal object fills the space. The fractal dimension is not an integer: it can be, for instance, an intermediate dimension between the Euclidean dimensions 2 (for a surface) and 3 (for a volume).

A large body of literature describes applications of fractal theory to soil science, including fractal aspects of soil texture, structure, porosity and scaling in hydraulic properties (Baveye *et al.*, 1997; Bittelli *et al.*, 1999; Pachepsky *et al.*, 2000). Here we show an example to analyse the geometrical features of a soil thin section using a technique called box counting.

The fractal dimension of an object can be obtained by the box counting technique. First a box of characteristic length L_1 is chosen and then N_1 is defined as the number of boxes that are necessary to cover the object. Then a smaller box size is chosen such that the box has the same shape, but smaller characteristic length L_2. Now the number N_2 of boxes that are necessary to cover the object has to increase. With $L_n \to 0$, the characteristic length of the boxes can be reduced more and more, while the number $N(L)$ required to cover the object increases. The fractal dimension D of the object is then defined as

$$D = \lim_{L \to 0} \frac{\log[N(L)/N(L_0)]}{\log(L_0/L)} \tag{2.2}$$

where L_0 is an arbitrary reference length. This definition means that for small L

$$\left(\frac{L_0}{L}\right)^D \approx \frac{N(L)}{N(L_0)} \tag{2.3}$$

Before applying this definition to soils, it will be demonstrated on three simple objects: a triangle, a cube and the so-called Sierpinski triangle, which is a typical example of a fractal object. Figure 2.5 depicts the three objects before and after the refinement procedure. The three objects at the top of the picture are not yet refined. When the refinement is applied, each object is divided into smaller objects of the same geometry.

This approach, applied to a cube, was presented by Turcotte (1997) and employed by Bittelli *et al.* (1999) to describe the soil particle size distribution, by employing a fragmentation model. The corresponding values are shown, in Table 2.1, where

$$D_i = \frac{\log(N_i / N_0)}{\log(L_0 / L_i)} \tag{2.4}$$

is the evaluation of the dimension based on the ith refinement step. These results illustrates that, for these three objects, the dimension does not depend on the characteristic length of the reference element. It also shows that the fractal dimension is, for simple objects, equal to the Euclidean dimension—indeed a triangle has dimension equal to 2 and a cube a dimension equal to 3. For fractal objects, the fractal dimension does not need to be an integer. Indeed, the word 'fractal' (from the Latin word *fractus*, broken) was introduced by Mandelbrot (1975) to describe an object with fractional (non-integer) dimension. If the values from Table 2.1 are substitured into eqn (2.3), the approximation is perfectly fulfilled.

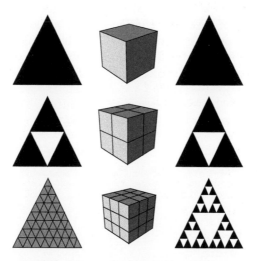

Figure 2.5 *Example of three objects after iterative refinement. This figure was created using the program Visual Python.*

Table 2.1 *Number N_i of reference elements (triangles or cubes) characteristic lengths L_i and fractal dimensions D_i*

	Triangle	Cube	Sierpinski
N_1	1	1	1
N_2	4	8	3
N_3	64	27	27
L_1	1	1	1
L_2	$\dfrac{1}{2}$	$\dfrac{1}{2}$	$\dfrac{1}{2}$
L_3	$\dfrac{1}{8}$	$\dfrac{1}{3}$	$\dfrac{1}{8}$
D_2	2	3	$1.58\ldots$
D_3	2	3	$1.58\ldots$

To determine the fractal dimension of a soil, it is not sufficient to count the number of boxes once of characteristic length L and once of characteristic length $L/2$, but instead it is necessary to go to smaller values of L. This is explained in the following.

The box counting technique is used to analyse the scaling properties of two-dimensional fractal objects (Posadas *et al.*, 2003). The technique covers the object with boxes of size L and counts the number of boxes containing at least one pixel representing the object under study. At the next iteration, the box size is reduced by a given factor (it could be a factor of 2 or a factor of 5) and the number of boxes is again counted. By counting the number of boxes and considering the box size, the fractal dimension is obtained as described above. This technique allows the fractal dimension of the object to be obtained, together with the fraction of surface that is occupied by pores.

The following is the program `PSP_boxCounting` for the box counting technique:

```
#PSP_boxCounting
from __future__ import print_function, division
import math
from PIL import Image
import matplotlib.pyplot as plt

def identifyPores(colorGrid, width, height, myColor):
    print("identifying pores ...")
    nrPores = 0
    rgbThreshold = 128
    for x in range(width):
        for y in range(height):
            color = colorGrid[x, y]
            if ((color[0] > rgbThreshold) and
```

```
                    (color[1] > rgbThreshold) and (color[2] > rgbThreshold)):
                    nrPores += 1
                    colorGrid[x, y] = myColor
        return(nrPores, colorGrid)

    def boxCounting(colorGrid, width, height, size, threshold, myColor):
        nrPixelsThreshold = threshold * size * size
        nrXBoxes = int(width / size)
        nrYBoxes = int(height / size)
        nrOccupiedBoxes = 0
        for i in range(nrXBoxes):
            for j in range(nrYBoxes):
                nrPorePixels = 0
                for dx in range(size):
                    for dy in range(size):
                        x = i * size + dx
                        y = j * size + dy
                        color = colorGrid[x, y]
                        if color == myColor: nrPorePixels += 1
                if (nrPorePixels > nrPixelsThreshold):
                    nrOccupiedBoxes += 1

        #normalized one-dimensional size
        L = 1./math.sqrt(nrXBoxes*nrYBoxes)
        #fractal dimension
        if (nrOccupiedBoxes > 0):
            D = math.log(nrOccupiedBoxes) / math.log(L)
        else:
            D = 0
        print ("box size =", size, " pores =", nrOccupiedBoxes,
               " L =", format(L, '.3f')," D =", format(D, '.3f'))
        return(nrOccupiedBoxes)

    def main():
        picture = Image.open("soil_image.jpg")
        colorGrid = picture.load()
        [width, height] = picture.size
        nrPixels = width * height
        print ("width =", width)
        print ("height =", height)

        myColor = (255, 255, 255)      #red, green, blue 0-255
        nrPores, colorGrid = identifyPores(colorGrid, width, height,
                                            myColor)
        poresPercentage = float(nrPores) / float(nrPixels) * 100.
        print ("% of pores =", format(poresPercentage, '.3f'))
        picture.show()
        plt.xlabel('Box size',fontsize=16,labelpad=5)
```

```
plt.ylabel('N',fontsize=16,labelpad=5)
plt.ion()
size = 200              #pixels
threshold = 0.2
while (size >= 1):
    nrOccupiedBoxes = boxCounting(colorGrid, width,
                          height, size, threshold, myColor)
    if (nrOccupiedBoxes > 0):
        plt.loglog(size, nrOccupiedBoxes, 'ko')
        plt.draw()
    size = int(size / 2)

plt.ioff()
plt.show()

main()
```

The first line of the program is used to compile Python programs written for Python 3.x and use them with a Python 2.x compiler. This statement imports the print function and the division function from version 3.x, allowing the user to compile this program on both Python 3.x and 2.x.

Then the module `math`, the module `Image` from the Python Imaging Library (PIL) and the `pyplot` module from the `matplotlib` library are imported. As described in Appendix A, the `matplotlib` library is a powerful library for plotting two- and three-dimensional scientific graphs.

In Python, the **main**() function must be written at the bottom of the program. In the **main** function, the image file is opened and the variable `colorGrid` is defined by loading the image file. The `picture.load()` method allocates storage for the image and loads it from the file. The object **load** returns a pixel access that can be used to read and modify pixels. It is returned as a cubic matrix, where x is the number of pixels on the x axis, y is the number of pixels on the y axis, and the z are the three colour components (red, green and blue). The instruction [width, height] = picture.size is used to read the width and the height of the image.

An important variable is the `threshold`, which represent the fraction of the pixel that is occupied by pores. This variable is used to evaluate whether a specific box is considered pore or matrix space. This task is performed in the function **boxCounting**() and is described below.

The function **identifyPores**() is used to identify which pixels are considered pores and which are considered matrix. The choice is based on the shades of colours of the pores, which are light grey. A variable `rgbThreshold` is set equal to 128, which is a light grey. Then the image is scanned within two `for` loops, over the x and y coordinates. The instruction `colorGrid[x, y]` = myColor sets the three-component colour value at coordinate x and y to myColor. In the RGB electronic representation of colours, the three primary components (red, green and blue) occupy one byte of memory (2^8)—thus from 0 to 255. If the colours of the pores in Fig. 2.1 were pure white, the numerical representation would be (255, 255, 255). As a matter of fact, the pores in the figure are composed of shades of light grey. Without sophisticated image preprocessing, a simple solution for this problem is to consider pores all the pixels where all three colour components are greater than 128. Indeed, (128,128,128) is light grey. The variable colorGrid is equalled to the variable myColor that is set in the **main**() function equal to a white colour (255, 255, 255). This is done to visualize the pore spaces in white and visually test the identification algorithm. The function returns the variable nrPores occupied by a light grey colour and the colorGrid variables, i.e. the number of pores.

The function **boxCounting**() is written to apply the box counting technique and compute the fractal dimension. The variable nrPixelsThreshold defines the threshold that determines if a box is considered as pore space or matrix space: if the ratio of pore space pixels within one box to the total number of pixels is larger than the threshold, then the box is considered to be pore space. If the ratio value is smaller, then it is matrix space. The numbers of boxes, nrXBoxes and nrYBoxes, are obtained by dividing the width and height of the image by the box size that is initialized; size =200 pixels in the **main** function. Note that only the integer value of the division is used, in case the size of the image is not divisible by the number of boxes. The initial value was chosen based on the size of the image imported, but it can be adjusted depending on the image size.

The size is divided by two at every iteration: size = int(size / 2) in the while loop in the **main** function. Within the for loops over the number of boxes in the x and y ranges, there are two nested for loops of dx and dy increments within each single box size. This loop is used to determine which pixels in each box are grey, and therefore evaluate if that box is considered pore or matrix space. This condition is checked by the **if** statement, if color == myColor: nrPorePixels += 1. At the end of the **boxCounting**() function, the box size L and the fractal dimension D are computed.

It is interesting to note that different values for the threshold variable have an effect on the evolution of the fractal dimension during the refinement. If the threshold is low (e.g. 0.05), the convergence towards the final fractal dimension (1.83) comes from a fractal dimension of 2. A small value of threshold means that even the presence of very few pixels in the image occupied by pores will evaluate the box as a pore, therefore filling the space. As we have seen before, a value of 2 for a two-dimensional image means that all the space is filled. On the other hand, if the threshold is high (e.g. 0.9), the convergence toward the final fractal dimension comes from a fractal dimension of 0. Indeed, a large value of threshold means that the box must have a large fraction of pixels in the image classified as occupied pores to be considered a pore space. Therefore, at the beginning, the fractal dimension will be zero, corresponding to an empty space. If the threshold is 0.2 (and therefore at least one-fifth of the pixel is occupied by pores), the algorithm converges quickly towards the final value of fractal dimension. These results confirm the fractal features of this image, where, no matter whether the computation begins from filled or empty spaces, the final value of the fractal dimension is the same. However, the initial value of the fractal dimension differs depending on the condition of the analysed space.

(a)
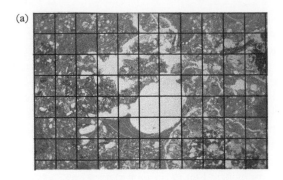

(b)
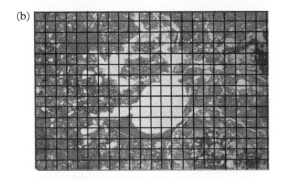

Figure 2.6 *Illustration of the box counting technique applied to the thin-section image of soil. The sizes of the boxes of in (a) and (b) are $L = L_0$, and $L = L_0/2$, respectively.*

The lines at the end of **main()** are used to plot the log–log graph of the fractal dimension as a function of number of boxes, using the program matplotlib. Figure 2.6 shows an example of how the algorithm analyses an image with the box counting technique.

The printed output of this program is as follows:

```
width = 800
height = 531
identifying pores...
\% of pores = 33.734
box size = 200   pores = 7    L = 0.354   D = -1.872
box size = 100   pores = 27   L = 0.158   D = -1.787
box size = 50    pores = 114  L = 0.079   D = -1.866
box size = 25    pores = 403  L = 0.039   D = -1.843
box size = 12    pores = 1554 L = 0.019   D = -1.843
box size = 6     pores = 5517 L = 0.009   D = -1.839
box size = 3     pores = 21522  L = 0.005   D = -1.854
box size = 1     pores = 143302 L = 0.002   D = -1.832
```

(a)

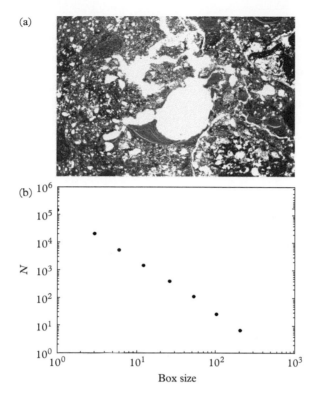

(b)

Figure 2.7 *Results of the box counting technique applied to the thin-section image of soil. In the (a), the areas considered by the program as pore space are coloured white. (b) shows the number of boxes as a function of box size.*

Figure 2.7 shows the visual output. The image with white pixels for the pores is shown in Fig. 2.7(a), while the number of boxes N is plotted as a function of box size in Fig. 2.7(b). For this image, the total porosity is $\phi_f \simeq 33.7\%$, with a fractal dimension $D \simeq 1.83$. There is a fractal relationship between number of boxes and box size, with a well-defined fractal dimension.

While for the image in Fig. 2.7(a), the fractal approach was correctly applied and the geometry could be successfully described, fractal theory may not apply at every scale in soils, and therefore soils need not be fractal. Indeed, the idea of scale invariance may not describe structural units that are characteristic of soils, such as horizons.

2.4 Geometry of the Pore Space

In Section 2.3, we addressed different theories describing the soil geometry of the solid phase. The relation between the pore and solid phases in soil is very important since it provides an opportunity to model hydraulic functions such as the soil water retention curve and the hydraulic conductivity. Certainly the application of fractal geometry to describe the solid–pore space is a very interesting technique for characterizing such complex systems.

The classical methods used for solving soil physical problems are based on using assumptions about the geometry of the pore space to predict macroscopically observable properties. The most common simplified model used to represent the soil pore space is a bundle of parallel capillary tubes (Purcell, 1949). The pore size distribution is then expressed as a density function or a cumulative function of pore sizes (Childs and Collis-George, 1950).

2.4.1   Bundle of Capillaries

One of the most successful models used to describe the geometry of the pore space is the 'bundle of capillaries' model (Childs and Collis-George, 1950). Assuming that the porous medium has interconnected pores, the pore size distribution can be described using a model of capillary bundles, where the distribution of pores is described by a statistical model. If we have a distribution of pores of different radii, we must define a distribution function $F(r)\,dr$, defining the fraction of pores between r and $r + dr$, such that the total porosity is given by

$$\phi_f = \int_0^{r_{max}} F(r)\,dr \qquad (2.5)$$

where r_{max} is the largest capillary tube (or pore).

This model will be discussed in detail in Section 6.3 on water flow, since it is the basic model used to derive hydraulic properties in soils, and is therefore the basic concept applied in the solution of transport equations. However, it is important to keep in mind that it is a simplified model. Since the statistical model implies that the pores are randomly distributed, as an example, the following code shows the generation of capillaries with randomly distributed radii. The first `for` loop is written to guarantee that no midpoint is in another cylinder. The second `for` loop is written to guarantee that the radius is small enough to not cross the other cylinders.

```
#PSP_capillaries
from visual import *
from math import sqrt
from numpy import *
display(background=color.white)
n = 250
x = zeros(n)
z = zeros(n)
r = zeros(n)
for i in range(n):
 check = 0
 while(check == 0):
```

```
x[i] = random.random_sample()-0.5
z[i] = random.random_sample()-0.5
check = 1
for j in range(i):
  if ((x[i]-x[j])**2.+(z[i]-z[j])**2. < r[j]**2.):
    check = 0
if not(x[i]**2.+z[i]**2.<0.2**2): check = 0
r[i] = ((random.random_sample()*4-2)**2.+0.5)/4*0.1
for j in range(i):
  if ((x[i]-x[j])**2.+(z[i]-z[j])**2. < (r[j]+r[i])**2.):
    r[i] = sqrt((x[i]-x[j])**2.+(z[i]-z[j])**2.)-r[j]
for i in range(n):
  cylinder(pos=(x[i],-1.5,z[i]), axis=(0,2,0), radius=r[i])
```

Figure 2.8 shows a distribution of generated capillaries with randomly generated radii, obtained from this program. Clearly, in this figure, all the pores are oriented in the vertical direction, which is fine for one-dimensional problems. When the capillary model is applied to two or three dimensions, the capillaries are also spatially distributed in two or three dimensions. Although it is a simple one, this model has been the theoretical basis for most of the transport models in porous media in recent decades and has often provided satisfactory results when measured and computed flow data were compared.

2.4.2 Pore Size Distribution

On the basis of the capillary model, the size and distribution of soil pores can be obtained. This information is fundamental because it determines water retention, the

Figure 2.8 *Randomly generated capillaries.*

conduction of fluids and the adsorption of solutes on the soil surfaces. It also affects microbial activity and a variety of physical, chemical and biological phenomena.

Similarly to eqn (2.5), the pore size distribution can be represented as a cumulative pore volume V_t:

$$V_t(r) = \sum_i \Delta V_i \tag{2.6}$$

where ΔV_i are the volumes of pores with a pore size smaller than r. The sum of all pore volumes ΔV_i determines the total pore volume. Pore size distribution can also be represented as an incremental pore volume distribution or the volumes of pores in a given size range. Similarly to soil particles, pores can be classified according to their size range. There is a variety of classification systems (Johnson *et al.*, 1960; IUPAC, 1972). For the Soil Survey Division (Soil Survey, 1990), pores range from very coarse (≥ 10 mm), to coarse (5–10 mm), to medium (2–5 mm), to fine (1–2 mm) and to very fine (≤ 0.5 mm). Another way of classifying pores is based on their function, with pores thereby being classified as wormholes, root hairs or bacteria. Clearly, each classification is qualitative and somewhat arbitrary. Only measurement of pore size distributions can provide in-depth information about the soil pore space.

2.5   Specific Surface Area

The surfaces of solid particles play an important role in the adsorption of water, the exchange of cations with the liquid phase (cation-exchange capacity), the adsorption and transport of chemicals, and mechanical properties such as plasticity and cohesion. The specific surface area can be expressed as a function of the solid mass and is defined as surface area per unit mass [m^2 g^{-1}]:

$$A_m = \frac{a_s}{m_s} \tag{2.7}$$

where a_s is the total surface area [m^2] of a soil mass m_s [g]. The surface area per unit volume [m^2 m^{-3}] is

$$A_v = \frac{a_s}{V_s} \tag{2.8}$$

where V_s is the volume of the solid phase [m^3]. The specific surface area depends on the shape and size of soil particles. For instance, the specific surface area of a spherical particle can be computed as the ratio of the sphere surface and the product of the density by the volume:

$$A_{\text{sphere}} = \frac{4\pi r^2}{\rho_s V_s} = \frac{4\pi r^2}{\rho_s (4/3)\pi r^3} = \frac{3}{\rho_s r} \tag{2.9}$$

where r is the sphere radius. Soil particles are often of non-spherical shape, especially in the clay range. Often clay particles are plate-shaped and can be represented by rectangular objects. For instance, a rectangular object with equal length and width $(L = w)$ and with thickness d will have a specific surface area

$$A_{\text{plate}} = \frac{2L^2 + 4Ld}{\rho_s V_s} = \frac{2L^2 + 4Ld}{\rho_s L^2 d} = \frac{2(L + 2d)}{\rho_s L d} \tag{2.10}$$

The specific surface area of a rectangular object is larger than that of a spherical object. Let us assume that we have a sphere and a rectangular object of the same volume $V_s = 1$ cm^3. The sphere would have a radius $\simeq 0.62035$ cm, while we assume that the parallelepiped has dimensions $L = 10$ cm and $d = 0.01$ cm. Assuming a particle density of 2.65 g cm^{-3} and applying eqns (2.9) and (2.10) gives specific surface areas $\simeq 1.82$ and 75.6 cm^2 g^{-1} for the sphere and the rectangular object, respectively. If d is made very small, 0.0001 cm, then L has to become much larger, 100 cm, to have the same volume of 1 cm^3, and the specific surface area becomes 7547.18 cm^2 g^{-1}. From this example, it is clear that if the plates are thin, then plate-like particles can have much higher specific surface areas than spherical particles.

Clearly, in a gram of soil, there are many particles of much smaller radius and therefore the specific surface area of a gram of soil is much larger than the numbers presented in the example above for just one particle. We will see in later chapters the typical sizes of soil particles and their distribution with respect to mass fractions.

In general, sandy soils are characterized by particles that may resemble spheres, while clay soils have particles that may resemble plate shapes. For this reason, clay soils have larger specific surface areas than sandy soils. Tuller and Or (2005) reported measured values ranging from 23 m^2 g^{-1} for a sandy soil to 186 m^2 g^{-1} for a silt loam. For clay minerals (montmorillonite), the specific surface area can reach values of 700–800 m^2 g^{-1} (Cavazza, 1981; Ong and Lion, 1991).

2.6 Averaging

Many physical properties cannot be measured at just one point, but often have to be measured in a control volume. For example, if the density of water has to be measured, a certain amount of water is used to measure its mass and volume in order to determine its density. As water is homogeneous, the size of the control volume is not important. Indeed, if the volume is doubled, so is its mass. Therefore the density does not change. Soil, however, is a highly heterogeneous material at small scales (pore scale) as well as at larger scales (soil layers or soil horizons). Therefore the measurement of soil density can be a complicated task. Let us assume that we want to measure the density around two points (p_1 and p_2) with domains of increasing volume (Ω_1, Ω_2, Ω_3). A schematic diagram is shown in Fig. 2.9. Since the density is given by a mass divided by a volume, we assume the domains Ω_1, Ω_2, Ω_3 to be cubes of incremental volumes. If the pore space is filled only with air, the density of Ω_1 at p_1 is equal to the density of the gas phase, and

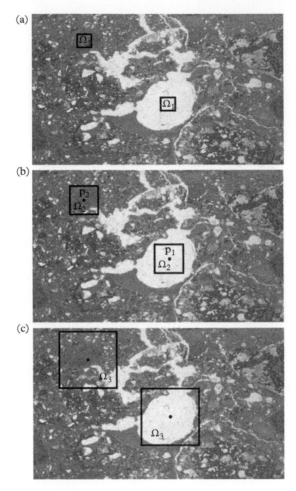

Figure 2.9 *Schematic for measurement of averaged properties. The domain dimension increases from (a) to (b) to (c).*

the density at p_2 is equal to the density of the solid phase (Fig. 2.9(a)). By increasing the volume of investigation using a volume Ω_2, the density at p_1 will still be equal to that of the gas phase, while at p_2 there will be a slight decrease in density since a small part of the volume is now occupied by the gas phase (Fig. 2.9(b)). For Ω_3, both volumes around the points p_1 and p_2 are intermediate values between the density of the gas phase and that of the solid phase (Fig. 2.9(c)).

It can be noted that the density value depends on the position of the point and the volume of the cube. By increasing the characteristic length of the averaging method (i.e. the volume of the cube), the differences in densities at different points p_n rapidly decrease (Fig. 2.10). The density approaches an average value, such that the differences

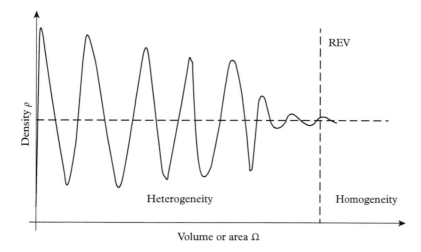

Figure 2.10 *Variation of density as function of characteristic volume.*

among densities sampled at different points are smaller than a given value ϵ. This volume is called the representative elementary volume (REV) and is an important concept for applications in soil physics (Bear, 1972).

This process is necessary to address problems at increasing scales of observation; however, it implies that each bit of information obtained at a smaller scale is lost during the averaging process and cannot be recovered. In fact, when we reach a volume Ω_n such that the density does not change significantly for different points p_n, we also lose all information for volumes smaller than Ω_n. The value of the average density is called the bulk density and will be described in the following section.

2.7   Bulk Density, Water Content and Porosity

After describing the different geometrical soil features and the various approaches employed to describe soil geometry, it is useful to define several variables that describe the physical condition of the three-phase soil system. As seen before, in a given amount of soil, the total mass m_t is divided between mass of gases m_g (usually negligible with respect to the other masses), mass of liquid m_l and mass of solids m_s. The total volume V_t is divided between volume of gases V_g, volume of liquid V_l and volume of solids V_s. The volume of fluid V_f is the sum of V_g and V_l. Using these variables, the following definitions can be developed:

- particle density [kg m^{-3}]

$$\rho_s = \frac{m_s}{V_s} \tag{2.11}$$

- dry bulk density [kg m^{-3}]

$$\rho_b = \frac{m_s}{V_t} \tag{2.12}$$

- porosity [m^3 m^{-3}]

$$\phi_f = \frac{V_f}{V_t} = \frac{V_g + V_l}{V_t} \tag{2.13}$$

- gas-filled porosity [m^3 m^{-3}]

$$\phi_g = \frac{V_g}{V_t} \tag{2.14}$$

- void ratio [m^3 m^{-3}]

$$e = \frac{V_f}{V_s} = \frac{V_g + V_l}{V_s} = \frac{V_f}{V_t - V_f} \tag{2.15}$$

- mass wetness or gravimetric water content [kg kg^{-1}]

$$w = \frac{m_l}{m_s} \tag{2.16}$$

- volume wetness or volumetric water content [m^3 m^{-3}]

$$\theta = \frac{V_l}{V_t} \tag{2.17}$$

- degree of saturation [m^3 m^{-3}]

$$S_e = \frac{V_l}{V_f} = \frac{V_l}{V_g + V_l} \tag{2.18}$$

2.8   Relationships between Variables

It is useful to derive relationships among the variables. For example, it can be shown that

$$\phi_f = 1 - \frac{\rho_b}{\rho_s} \tag{2.19}$$

$$\phi_g = \phi_f - \theta = 1 - \frac{\rho_b}{\rho_s} - \theta \tag{2.20}$$

and

$$e = \frac{\phi_f}{1 - \phi_f} \qquad (2.21)$$

The mass wetness and volume wetness are related by

$$\theta = w \frac{\rho_b}{\rho_l} \qquad (2.22)$$

and

$$w = \theta \frac{\rho_l}{\rho_b} \qquad (2.23)$$

where ρ_l is the density of the liquid phase (usually taken as the density of liquid water). A useful relationship is another definition of the degree of saturation:

$$S_e = \frac{\theta}{\theta_s} \qquad (2.24)$$

where θ_s is the volumetric water content at saturation.

The volumetric fraction of the solid phase (ϕ_s) can be obtained from the porosity by

$$\phi_s = 1 - \phi_f \qquad (2.25)$$

If the soil does not swell, so that ρ_b is not a function of water content, the degree of saturation can also be expressed as the ratio of mass water content to saturation mass water content.

2.9   Typical Values of Physical Properties

The ranges of the variables defined in Sections 2.7 and 2.8 can now be explored using a few simple examples. Table 2.2 gives particle densities of typical soil constituents. The mineral soil particle density used in most calculations is $\rho_s = 2650$ kg m^{-3}. Typical bulk densities (ρ_b) for mineral soils range from around 1000 kg m^{-3} for recently tilled soils to around 1600 kg m^{-3} or higher for compacted soils. A mid-range value of $\rho_b = 1330$ kg m^{-3} would give a total porosity of 0.5 and a void ratio of 1.0.

Water content is usually measured on a mass basis by drying samples at 105 °C and then computing the ratio of mass of water lost to dry mass of the sample (eqn (2.16)). Oven-dry soil is assumed to be at zero water content, although structural water may still be present. Air-dry soil is in equilibrium with atmospheric moisture and therefore has a water content greater than zero. Typical water contents for air-dry soil range from 0.01 to 0.04 kg kg^{-1}, depending on texture and atmospheric humidity.

Table 2.2 *Densities of soil constituents*

Component	Density [kg m^{-3}]
Quartz	2660
Orthoclase	2500
Clay minerals	2650
Mica	2800–3200
Organic matter	1300
Limonite	3400–4000
Fe(OH)$_3$	3730
Water (at 4 °C)	1000
Air (at 20 °C)	1.2
Ice	900

At the other extreme, the saturation water content for a soil with 0.5 m^3 m^{-3} pore space is 0.5 m^3 m^{-3}. Using eqn (2.23), the mass wetness is $w \simeq 0.38$ kg kg^{-1} for a bulk density of 1330 kg m^{-3}. The highest water contents attained under field conditions are somewhat below the saturation value. Some pores always remain air-filled. A typical maximum water content in the field would be 0.45 m^3 m^{-3}.

Near-saturation conditions are usually of short duration in well-drained soils. Following rain or irrigation, the soil quickly approaches a water content termed field capacity. Field capacity water contents vary greatly, but often are roughly half of saturation. A typical field capacity water content for a uniform loam soil might therefore be 0.25 m^3 m^{-3} or 0.19 kg kg^{-1}.

The following is a program to compute basic physical properties as described above:

```
# PSP_basicProperties
from __future__ import print_function, division

def computePorosity(bulkDensity, gravWaterContent):
        waterDensity = 1000
        particleDensity = 2650
        porosity = 1 - (bulkDensity / particleDensity)
        voidRatio = porosity / (1 - porosity)
        waterContent = gravWaterContent * (bulkDensity / waterDensity)
        gasPorosity = porosity - waterContent
        degreeSaturation = waterContent / porosity
        print ("\nTotal porosity [m^3/m^3] = ", format
                (porosity, '.3f'))
        print ("Void ratio [m^3/m^3] = ", format(voidRatio, '.3f'))
        print ("Volumetric water content [m^3/m^3] = ", format
```

```
                (waterContent, '.3f'))
        print ("Gas filled porosity [m^3/m^3] = ", format
                (gasPorosity, '.3f'))
        print ("Degree of saturation [-] =", format
                (degreeSaturation, '.3f'))
        return

def computeSaturationWetness(bulkDensity):
    waterDensity = 1000
    particleDensity = 2650
    porosity = 1 - (bulkDensity / particleDensity)
    return (porosity / (bulkDensity / waterDensity))

def main():
    bulkDensity = float(input("Bulk density [kg/m^3] = "))
    gravWaterContent = float(input("Gravimetric Water Content
                            [kg/kg] = "))

    satMassWetness = computeSaturationWetness(bulkDensity)
    if (gravWaterContent >= 0) and
            (gravWaterContent < satMassWetness):
        computePorosity(bulkDensity, gravWaterContent)
    else:
        print ("Wrong Water Content! value at saturation = ",
                satMassWetness)
main()
```

This program `PSP_basicProperties.py` is based on the implementation of two functions, **computePorosity** and **computeSaturationWetness**, which compute the sample porosity and soil water content. The two instructions at the beginning of the **main()** function are prompt messages asking the user to input a value for the bulk density and a value for gravimetric water content by using the Python instruction `input`.

2.10 Volumes and Volumetric Fractions for a Soil Prism

It is useful to think of soils as unions of modules known as pedons. A pedon is the smallest element of landscape that can be called soil. Its depth limit is the somewhat arbitrary boundary between soil and bedrock. Its lateral dimensions must be sufficiently large to permit a study of any horizons present. An area from 1 to 10 m^2 is commonly used. A horizon may be variable in thickness or even discontinuous. Consider a right prism of soil of 2 m depth with a base of 1 m^2. The total volume of the soil prism is 2 m^3. If, for instance, the liquid content is 0.6 m^3 and the gas content is 0.6 m^3, the solid volume will be 0.8 m^3. The volumetric water content will be $\theta = 0.6/2 = 0.3$ m^3 m^{-3} and the gas-filled porosity $\phi_g = 0.6/2 = 0.3$ m^3 m^{-3}. For practical applications, such as in irrigation and water balance computations, it is convenient to express the volumetric fractions as equivalent water height h_w by multiplying the water volumetric fraction by the prism height h. For instance, the soil water content will be $h_w = \theta \times h = 0.3 \times 2 \text{ m} = 0.6 \text{ m} = 600 \text{ mm}$.

2.11   Soil Solid Phase

Soil is a mixture of solid, liquid and gaseous material. The solid phase is characterized by inorganic and organic materials, the former being composed of the mineral fraction and the latter the soil organic fraction. The liquid phase is composed of liquid water, which fills a fraction or the whole pore space. The solid mineral phase comprises two types of particles: *primary* and *secondary minerals*. The principal *primary mineral* groups are the silica minerals including quartz, feldspars, olivines, pyroxenes, amphiboles and micas. All of these minerals are characterized by different arrangements of connected silica (SiO_2) tetrahedra. *Secondary minerals* are characterized by four main groups of clays: kaolinite, montmorillonite–smectite, illite and chlorite. From the pedological point of view, *primary minerals* are minerals originated mostly by physical weathering, while *secondary minerals* originated from both physical and chemical weathering.

2.12   Soil Texture

Soil texture refers to the size of the particles that make up the soil. After a particle size analysis, it is possible to determine classes of sand, silt and clay based on limits determined by various classification methods; therefore soil texture is used to determine soil classes. Specifically, the textural classification is based on the definition of textural classes that have specific numerical limits. Table 2.3 shows the limits for different particle ranges

Table 2.3 *Systems of classification for Particle Size Distribution*

ISSS[a]		USDA[b]	
Soil fraction	Range [μm]	Soil fraction	Range [μm]
Coarse sand	2000–200	Very coarse sand	2000–1000
Fine sand	200–20	Coarse sand	1000–500
Silt	20–2	Medium sand	500–250
Clay	<2	Fine sand	250–100
		Very fine sand	100–50
		Silt	50–2
		Clay	<2

[a] International Soil Science Society (ISSS, 1966).
[b] United States Department of Agriculture (USDA, 1975).

for the two common classification systems. However, there are many other systems that are used, often depending on the field of applications (engineering, geology, road engineering, geomechanics). Usually the size classes are distinguished into sand, silt and clay. It is conventional to define *soil material* as particles smaller than 2 mm in diameter, while particles larger than 2 mm in diameter are called *skeleton*. The skeleton is then classified as *gravel*, larger rock fragments as *stones* and *cobbles* and, when very large, *boulders*. Soils with large fractions of gravel, stones and cobbles are called *skeletal soils*. The following is a brief description of the three main classes of soil material:

- *Sand* is a granular material that constitutes the coarser fraction of the bulk soil. Depending on the classification system, sand can be subdivided in further subclasses such as very coarse, coarse and fine. Sand can have a variety of shapes, although these all resemble a spherical shape. From the mineralogical standpoint, the most common constituent of sand is silica (SiO_2), usually in the form of quartz. However, many other minerals can be found in these fractions, such as feldspar, mica, zircon and metals. The mineralogical composition depends largely on the bedrock from which the soil originated.

- *Silt* is an intermediate size fraction between sand and clay. It is also a fundamental constituent of soils. Mineralogically, silt usually has a similar composition to sand, although the minerals are of smaller size and with higher specific surface. Silt is generated by a variety of physical weathering processes, including abrasion through transport, as well as glacial grinding.

- *Clay* is the fine fraction, with particles ranging from 2 μm downwards. Clay particles are platelike or needlelike in shape and generally they are mostly composed of minerals grouped as aluminosilicates. However, the clay fraction can have minerals that do not belong to the aluminosilicates, such as iron oxides or carbonates. The clay fraction, because of its very small size and its very high specific surface, is the most reactive soil fraction and has strong physico-chemical activity.

2.12.1 Textural Classification

The textural classification of a soil, called its *textural class*, is determined by the mass ratios of the three main fractions: sand, silt and clay. Soils having different proportions of sand, silt and clay (within certain ranges) are assigned to different classes, as depicted by the textural scheme in Fig. 2.11. This classification scheme allows for soils having different proportions of sand, silt and clay to be in the same textural class, depending on the lower and upper limits of the different classes.

It is sufficient to know two of the three fractions (e.g. the percentages of sand and clay) to classify the soil, since the total fractions add up to 100 %. However, soils that fall into the same textural class may have different properties. For instance, the textural class of 'clay' in the USDA classification scheme contains soil samples that vary in clay content from 40 to 100 %.

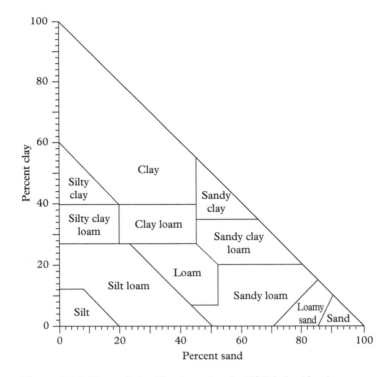

Figure 2.11 *Textural classification using the USDA classification scheme.*

The size definitions of the three main particle fractions of clay, silt, and sand used as diagnostic classification are arbitrary. They do not provide complete information on the soil particle size distribution; however, they are useful for classification properties and to identify classes of soils with 'similar' features.

2.12.2   Particle Size Distribution

A more accurate description of texture is obtained by plotting a particle size distribution. The particle size distribution of a soil sample defines the relative amounts of particles present, sorted according to size. Particle size distribution is also known as grain size distribution and is determined experimentally, through particle size analysis. It is represented using a density or a cumulative curve. The density representation shows particle size plotted against the fraction of soil that falls within a certain range. In the summation or cumulative curve, particle size is plotted against the fraction (or percentage) that is smaller than a given size. Often the particle size is plotted on a logarithmic scale. Figure 2.12 shows the density and the cumulative distribution fractions for four soils having different textural composition (Pieri *et al.*, 2006). For the sandy soil (L-soil), the density distribution depicts that the largest fraction is between ~50 and 1000 μm, while for

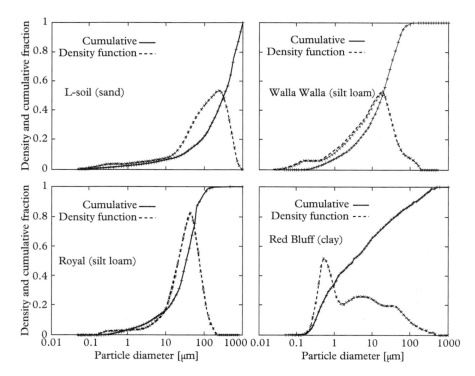

Figure 2.12 *Density and cumulative distribution fractions for four soils with different textural composition, obtained by laser diffraction technique. Taken from Pieri et al. (2006). Reprinted with permission from Elsevier.*

the clay soil (Red Bluff), a large fraction of the total mass is between ∼0.1 and 1 μm. Moreover, the clay soil shows a multimodal distribution.

2.12.3   Particle Size Distribution Functions

After having experimentally obtained a particle size distribution, it is useful to character-ize it by employing a mathematical model allowing specific parameters of the distribution to be obtained. These parameters can then be used to compute other soil physical prop-erties through functions called pedotransfer functions. A *pedotransfer function* uses basic pedological information such as textural fractions, particle size distribution parameters, bulk density and others to compute other soil physical properties. There is therefore a *transfer* of information. Commonly, the obtained properties are hydrological properties such as the soil water retention curve or hydraulic conductivity. Since the latter are more difficult and expensive to obtain experimentally, it is convenient to derive them from other basic properties that are readily easier to measure and more commonly available in soil databases.

Particle size distribution has then been represented by a variety of mathematical functions (Buchan *et al.*, 1993), also called particle size distribution models, to obtain specific parameters to be used as input for the pedotransfer functions. Software is available for derivation of hydrological properties by using pedotransfer functions (Schaap *et al.*, 2001; Acutis and Donatelli, 2003).

Gaussian distribution

Shiozawa and Campbell (1991) and Pieri *et* al. (2006) found that particle size distribution functions are typically bimodal and that each mode can be represented by a Gaussian function. The Gaussian function used to describe the particle size distribution is

$$f(x) = \left[\frac{1}{\sigma(2\pi)^{1/2}}\right] \exp\left[-\frac{(x-\mu)^2}{2\sigma^2}\right] \tag{2.26}$$

where μ is the mean of x and σ is the standard deviation. For a log-normal distribution, the particle diameter x is replaced by its logarithm. Usually, the particle size distribution is represented by a cumulative curve, which, for a Gaussian function, is obtained by integration of eqn (2.26) with respect to x:

$$F(x) = \frac{1}{2}\left\{1 + \text{erf}\left[\frac{(x-\mu)}{\sigma\sqrt{2}}\right]\right\} \quad \text{for } x > \mu \tag{2.27}$$

$$F(x) = \frac{1}{2}\left\{1 - \text{erf}\left[\frac{(x-\mu)}{\sigma\sqrt{2}}\right]\right\} \quad \text{for } x \leq \mu \tag{2.28}$$

where erf[ ] is the error function. For the computation of the error function, a numerical approximation from Abramowitz and Stegun (1970) can be used:

$$\text{erf}(x) = 1 - (0.3480242T - 0.0958T^2 + 0.7478556T^3)\,\exp(-x^2) \tag{2.29}$$

where $T = 1/(1 + 0.47047x)$.

In a bimodal distribution, the sample is divided into two fractions: the primary fraction (sand and silt) and the secondary fraction (clay). The distribution is then a weighted sum of the two fractions:

$$F(x) = \epsilon F_1(x) + (1 - \epsilon)F_2(x) \tag{2.30}$$

where $F_1(x)$ represents the cumulative Gaussian function for the secondary minerals (clay), $F_2(x)$ represents the Gaussian function for the primary minerals (sand and silt), and ϵ is the weighting factor between the secondary minerals (clay fraction) and the primary minerals (sand and silt).

The mean and the standard deviation of the bimodal distribution are also weighted sums of the mean and standard deviation of the two fractions, called bimodal parameters. The model can be fitted to experimental particle size distribution data using a nonlinear least squares fitting procedure (Marquardt, 1963), where μ_1, μ_2, σ_1, and σ_2 are fitting parameters, while ϵ is set equal to the clay fraction at 2 μm. The subscripts 1 and 2 for the mean and the standard deviation are the Gaussian parameters for the clay (F_1) and the silt and sand fraction (F_2) distributions, respectively. A nonlinear least squares fitting algorithm will be presented in Chapter 5. The size distribution function for primary minerals, with sizes between 2 and 2000 μm, is approximately log-normal. Secondary minerals, with sizes less than 2 μm, form a second distribution (Pieri *et al.*, 2006).

If a complete particle size distribution is not available, means and standard deviations may be obtained from size fractions. The appropriate means and standard deviations are log means and log standard deviations, and they can be computed from

$$d_g = \exp\left(\sum m_i \ln d_i\right) \tag{2.31}$$

and

$$\sigma_g = \exp\left[\sqrt{\sum m_i (\ln d_i)^2 - (\ln d_g)^2}\right] \tag{2.32}$$

where m_i and d_i are respectively the mass fraction geometric mean diameter of separate i. Equations (2.31) and (2.32) are appropriate for any number of size classes.

If only three textural classes are available, it is possible to derive the geometric means and standard deviations. The geometric means (in μm), d_y, d_t and d_d, are the geometric means for clay, silt and sand (the subscripts y, t and d refer to the last letters of the words clay, silt and sand). They are calculated from set values:

$$d_y = \sqrt{0.01 \times 2} = 0.14 \tag{2.33}$$

$$d_t = \sqrt{2 \times 50} = 10 \tag{2.34}$$

and

$$d_d = \sqrt{50 \times 2000} = 316.2 \tag{2.35}$$

The lower and the upper limits for the clay fraction were set at 0.01 and 2, while 2 and 50 represent the lower and upper limits for the silt fraction, and 50 and 2000 represent those for sand, based on the USDA classification (Shiozawa and Campbell, 1991). The geometric mean of the distribution (μ_m) is given by

$$d_g = \exp[m_y(-1.96) + m_t(2.3) + m_d(5.76)] \tag{2.36}$$

The geometric standard deviation (σ_m) of the distribution is given by

$$\sigma_g = \exp\left[\sqrt{m_y(-1.96)^2 + m_t(2.3)^2 + m_d(5.76)^2 - (\ln d_g)^2}\right] \tag{2.37}$$

Table 2.4 *Typical sand, silt and clay fractions and geometric mean particle diameter and geometric standard deviation for the 12 textural classes*

Texture	m_d	m_t	m_y	d_g [μm]	σ_g [μm]
Sand	0.92	0.05	0.03	211.75	4.43
Loamy sand	0.81	0.12	0.07	122.05	8.71
Sandy loam	0.65	0.25	0.1	61.74	12.21
Loam	0.42	0.4	0.18	19.81	16.34
Silt loam	0.2	0.65	0.15	10.52	9.57
Silt	0.06	0.87	0.07	9.11	4.08
Sandy clay loam	0.6	0.13	0.27	25.17	28.40
Clay loam	0.32	0.34	0.34	7.09	23.10
Silty clay loam	0.08	0.58	0.34	3.09	10.96
Sandy clay	0.53	0.07	0.4	11.36	39.70
Silty clay	0.1	0.45	0.45	2.07	13.75
Clay	0.2	0.2	0.6	1.55	22.81

Referring back to eqns (2.31) and (2.36), it should be clear that the logarithm of each size class is just multiplied by the mass fraction for that size class, and the result is summed to get the logarithm of the mean size (Shiozawa and Campbell, 1991). Table 2.4 shows typical silt and clay fractions (USDA) for each of the textural classes, along with their geometric mean diameter and standard deviation.

Fractal distribution

Particle size distribution can be described by other mathematical formulations. For instance, Bittelli *et al.* (1999) proposed to analyse the particle size distribution with a power-law distribution, where the exponent is interpreted as a fractal dimension. A description of this method was presented earlier. The authors utilized a mass-based approach, since this is compatible with the data obtained from experiments. The mass ratio is expressed as

$$\frac{M(r < R)}{M_T} = \left(\frac{R}{R_{L,\text{upper}}}\right)^{\nu} \tag{2.38}$$

where $M(r < R)$ is the mass of soil particles with a radius r smaller than R, M_T is the total mass of particles with radius less than $R_{L,\text{upper}}$, $R_{L,\text{upper}}$ is the upper size limit for fractal behaviour and ν is a constant exponent. The authors showed that $D = 3 - \nu$, where D is the fractal dimension of the distribution. Therefore, cumulative particle size distributions in soils can be represented by a power-law distribution, consistent with a fractal fragmentation model, but with scale invariance valid only over a limited domain range. Three domains—clay, silt and sand—were identified in which power-law scaling was applicable (Bittelli *et al.*, 1999). The boundaries between the domains were relatively constant for different soil types, but did not coincide with the traditional boundaries between clay, silt and sand.

2.13   Sedimentation Law

An important aspect of particle size analysis is sedimentation theory. This theory is widely employed for particle size analysis, as described by Gee and Or (2002). In this book, we are not describing experimental methods for soil physical analysis, but it is worth describing sedimentation theory, since it is employed in many studies related to sedimentation processes and erosion, among other things.

The sedimentation process is described by Stokes' law, which establishes a relationship between settling terminal velocity v and particle size (radius or diameter). A particle of density ρ_s and spherical geometry, settling into a liquid of density ρ_l and viscosity η, is subject to three forces: the force of gravity (F_g) acting downwards, the force of buoyancy (F_b) acting upwards, and the viscous drag force (F_d) against velocity, acting upwards. The force of gravity is given by

$$F_g = m_s g = \rho_s \left(\frac{4\pi r^3}{3} \right) g \tag{2.39}$$

where m_s is the particle mass (calculated from knowledge of particle density ρ_s and particle volume), g is the gravitational acceleration and r is the particle radius. The buoyancy force is given by

$$F_b = m_l g = \rho_l \left(\frac{4\pi r^3}{3} \right) g \tag{2.40}$$

where m_l the liquid mass and ρ_l is the liquid density. When the particle reaches a terminal velocity v, the drag force is equal to the motion force, which is the difference between the gravitational attraction and the buoyancy:

$$F_d = F_g - F_b \tag{2.41}$$

This causes the net force on the object to be zero, resulting in acceleration equal to zero ($dv/dt = 0$). It is therefore possible to calculate the terminal velocity by equating the net forces to zero:

$$\sum F_i = F_g - F_b - F_d = 0 \tag{2.42}$$

When the particle is settling into a laminar flow region, the drag force is

$$F_d = \pi r^2 \rho_l \frac{v^2}{2} C_d \tag{2.43}$$

where πr^2 is the orthographic projection of the sphere onto a plane perpendicular to the direction of movement and C_d is the drag coefficient.

Dimensional analysis of the problem shows that the drag coefficient depends on the particle size and the flow velocity, as described by the Reynolds number

$$Re = \frac{\rho_l v 2 r}{\eta} \tag{2.44}$$

where η is the dynamic viscosity of the fluid and v is the terminal velocity. Stokes found that for small Reynolds numbers ($Re < 0.3$), the relationship between drag coefficient and Reynolds number is (Allen, 1981):

$$C_d = \frac{24}{Re} \tag{2.45}$$

Substituting eqns (2.45) and (2.44) into eqn (2.43) yields

$$F_d = 6\pi r \eta v \tag{2.46}$$

This equation is valid only for low Reynolds numbers ($Re < 0.3$), which also determines a critical radius (or diameter) above which Stokes' equation should not be used (Allen, 1981).

By inserting eqns (2.39)–(2.46) into eqn (2.42), we get

$$\rho_s \left(\frac{4\pi r^3}{3} \right) g - \rho_l \left(\frac{4\pi r^3}{3} \right) g - 6\pi r \eta v = 0 \tag{2.47}$$

Rearranging and simplifying leads to

$$g \left(\rho_s - \rho_l \right) \frac{4 r^2}{3} \frac{1}{6\eta} = v \tag{2.48}$$

which can be further simplified to

$$v = g \left(\rho_s - \rho_l \right) \frac{2}{9} \frac{r^2}{\eta} \tag{2.49}$$

Often the particle size is expressed as equivalent particle diameter ($d = 2r$). Thus eqn (2.49) is written as

$$v = g \left(\rho_s - \rho_l \right) \frac{d^2}{18\eta} \tag{2.50}$$

If the particle's falling distance S is known, then it is possible to determine the falling time t:

$$t = \frac{18\eta S}{g\,(\rho_s - \rho_l)\,d^2} \qquad (2.51)$$

These equations are valid only under certain assumptions:

1. The particles are smooth, spherical and rigid.
2. The particles are small enough not to interact with each other.
3. The particle density is known and constant.
4. The particles are uniformly distributed within the liquid.
5. The liquid flow is laminar, no turbulent flow is present and the fall of the particles is unhindered by the proximity of the container walls.
6. The particles are large in comparison with the liquid molecules, so that Brownian movements do not affect the fall.
7. The liquid viscosity is constant during the fall; therefore temperature is also constant, since the liquid viscosity is affected by the liquid temperature.

The program PSP_sedimentation simulates the falling of a soil particle in a soil–water suspension, for different temperatures, concentrations of sodium hexametaphosphate (HMP) and particle diameters. The **main** function imports data saved in the file waterDensity.dat. HMP is a dispersing agent used for particle size analysis. The data provide different values of the density and viscosity of water at temperatures ranging from 16 to 26 °C. These values are used in the program to compute the changes in density and viscosity as functions of HMP concentration. The HMP density ρ_{hmp} as a function of concentration is given by (Gee and Bauder, 1986)

$$\rho_{hmp} = \rho_l \left(1 + 0.63\frac{c}{1000}\right) \qquad (2.52)$$

where ρ_l is the density of water and c is the HMP concentration [g L^{-1}].

The HMP viscosity η_{hmp} is given as a function of concentration by (Gee and Bauder, 1986)

$$\eta_{hmp} = \eta_l \left(1 + 4.25\frac{c}{1000}\right) \qquad (2.53)$$

where η_l is the viscosity of water [kg m^{-1} s^{-1}].

A visual program is also included where the falling of a particle (of variable radius) in a cylinder is simulated. The program imports the visual module and a file called PSP_readDataFile, which is used to read data from an ASCII file. This file will be used many times throughout the book, and it is described in Appendix A. In the **main()** function of the program, the function **readDataFile**, written in the file PSP_readDataFile, is used to open and read the file. The temperature, concentration and particle size are input by the user after being prompted by the program. The settling particle is then visualized by employing the instructions of the program Visual Python, and the total number of seconds is printed.

The function **computeSedimentationTime** calculates the sedimentation time using Stokes' law, while the function **computeSedimentationDepth** calculates the position (depth) of the settling particle at different times. The computation of solution density and viscosity as functions of HMP concentration is written in the functions **getSolutionDensity** and **getSolutionViscosity**, as described in eqns (2.52) and (2.53).

```
#PSP_sedimentation
from __future__ import print_function, division
import visual
from PSP_readDataFile import readDataFile

particleDensity = 2580
g = 9.80665

def getSedimentationTime(solutionDensity, solutionViscosity,
                                      particleDiameter, z):
    numer = 18 * solutionViscosity * z
    denom = (g * (particleDensity - solutionDensity)
                * (particleDiameter * 0.000001)**2)
    return (numer / denom)

def getSedimentationDepth(solutionDensity, solutionViscosity,
                                      particleDiameter, time):
    numer = (time * g * (particleDensity - solutionDensity)
                * (particleDiameter * 0.000001)**2)
    denom = 18 * solutionViscosity
    return (numer / denom)

def getSolutionDensity(liquidDensity, concentration):
        return (liquidDensity * (1 + 0.63*(concentration / 1000)))

def getSolutionViscosity(liquidViscosity, concentration):
        return (liquidViscosity * (1 + 4.25*(concentration / 1000)))

def initializeScene(heightCylinder, radiusParticle):
    scene = visual.display(width = 600, height = 600, exit=True)
    scene.background = visual.color.white
    scene.center = (0, 0, -heightCylinder/2)
    scene.forward = (0, 1, -0.5)

    cylinder = visual.cylinder(pos = (0,0,0), radius = 0.05)
    cylinder.axis = (0,0,-heightCylinder)
    cylinder.color = visual.color.white
    cylinder.opacity = 0.2
```

```
        particle = visual.sphere(pos = (0,0,0), color = visual.color.black)
        particle.radius = min(radiusParticle * 0.001, 0.02)
        return(particle)

def main():
    A, isFileOk = readDataFile("waterDensity.dat", 2, '\t', False)
    if ((not isFileOk) or (len(A[0]) < 3)):
        print("Wrong file!\nMissing data or wrong delimiter in line:",
                A+1)
        return()

    temperature = A[:,0]
    density = A[:,1]
    viscosity = A[:,2]
    print ("Available temperature:", temperature)

    isGoodChoice = False
    while not isGoodChoice:
        t = float(input("\nInput temperature: "))
        for i in range(len(temperature)):
            if (temperature[i] == t):
                isGoodChoice = True
                liquidDensity = density[i]
                liquidViscosity = viscosity[i]
        if not isGoodChoice:
            print ("Warning: not available value")

    concentration = float(input("Input concentration [g/l]: "))
    diameter = float(input("Input particle diameter [micrometers]: "))

    solutionDensity = getSolutionDensity(liquidDensity, concentration)
    solutionViscosity = getSolutionViscosity(liquidViscosity,
                                                concentration)

    heightCylinder = 0.1
    particle = initializeScene(heightCylinder, diameter/2.0)

    time = 0
    depth = 0
    while (depth < heightCylinder):
        visual.rate(1000)
        depth = getSedimentationDepth(solutionDensity,
                            solutionViscosity, diameter, time)
        particle.pos.z = -depth
        time += 1

    print ("seconds:", time)
main()
```

Figure 2.13 *Visualization of a particle settling in a fluid according to Stokes' law as simulated by Python and visualized with Visual Python.*

Figure 2.13 shows a visualization, using Visual Python, of a 10 μm particle settling in a fluid (particle size is not to scale, but was magnified for visualization purposes).

....

2.14 EXERCISES

2.1. Use the program PSP_boxCounting by running **main**. Set different values for the threshold variable and observe the evolution of the fractal dimension during refinement. Change the colour of the pore space in the output image generated by the program to red.

2.2. A soil sample collected in the field with a brass cylinder (8 cm diameter, 6 cm hight) has mass of 45 g. The wet sample (including the brass cylinder) weighed 445 g. After oven drying for 24 h at 105 °C, the dry sample (including the cylinder) weighed 375 g. Assuming that the soil particle density ρ_s is 2650 kg m^{-3} and the density of water ρ_l is 1000 kg m^{-3}, compute the mass basis water content w, the dry bulk density ρ_b, the volumetric water content θ and the total porosity ϕ_f. For the constant π, use four decimal places.

2.3. Use the program PSP_basicProperties.py . Input the values of bulk density and mass basis water content obtained from Exercise 2.2 to calculate the total porosity, the void ratio, the volumetric water content, the gas porosity and the degree of saturation. Try different values of bulk density and mass basis water content to check the output variable changes. Modify the program to enter the particle density chosen by the user. Pass the variable as an argument to the function and use the new particle density for computing the porosity.

2.4. Assume that the soil pedon is described by a right prism of 1.5 m depth. Calculate the equivalent water height h_w in millimetres, by knowing that the volumetric soil water content θ is $0.25 \, \text{m}^3 \, \text{m}^{-3}$.

2.5. Compute the geometric mean diameter and the geometric standard deviation for a silt loam soil having textural fractions of silt = 60%, sand = 20% and clay = 20%.

2.6. Derive the dimensions of the dynamic viscosity η, by using eqn (2.47).

2.7. Using Stokes' law, compute the settling time from the liquid surface at a depth of 0.2 m, for particles of 15 μm in diameter. Assume that the density of the solid phase is $2650 \, \text{kg m}^{-3}$, the temperature is 20 °C and the liquid is pure water. Repeat the computation for the same particle diameter, but settling in a solution of sodium hexamataposphate. For pure liquid water, use a liquid density of $998.23 \, \text{kg m}^{-3}$ and a viscosity $\eta = 0.001002 \, \text{Pa s}^{-1}$. For the sodium hexamataposphate solution, the liquid density is $998.5 \, \text{kg m}^{-3}$ and the viscosity is $0.001023 \, \text{Pa s}^{-1}$.

2.8. Based on Stokes' law and using `PSP_sedimentation.py`, compute the sedimentation times required for spherical particles of sizes 100, 50 and 2 μm to settle in pure water, at 20 °C, down to a depth of 0.3 m. Repeat the computation for a solution of hexametaphosphate at concentrations of 5 and $10 \, \text{g L}^{-1}$. The required variables, such as liquid and solid density, as well as liquid viscosity, are provided in the program. Check the effect on the settling times by changing the particle diameter and the concentrations of hexametaphosphate.

3

Soil Gas Phase and Gas Diffusion

The gas-phase composition of soil is similar to that of the atmosphere in terms of molecules characterizing the gas phase, but molecules are present in different concentrations to those in the atmosphere. Oxygen is consumed and carbon dioxide is generated by the metabolic activity of microorganisms and plant roots in the soil. Oxygen, carbon dioxide and water vapour all diffuse into and out of the soil. Soil air is typically around 79 % N_2. The remaining 21 % is mostly CO_2 and O_2, which vary reciprocally. In well-aerated soil, the CO_2 content of the air is around 0.25 % and the O_2 content 20.73 %. In poorly aerated soil, O_2 levels can approach zero. The absorption of nutrients by roots and the beneficial activity of microorganisms depend on an adequate supply of oxygen. Soils should therefore be managed to provide adequate aeration.

The gas phase is characterized by a high compressibility; therefore, as a first approximation, we can consider the gas phase to have a constant volume. It has low viscosity (about 1/50 of the water viscosity); therefore, computation of energy losses during convective movement of gas are commonly neglected. The soil gas phase has a low heat capacity; hence it is often ignored in computation of thermal dynamics and heat transport. It also has a very low density; therefore its contribution is usually neglected in the computation of the total soil mass or in relationships between masses of the different soil phases.

The rate of respiration of living organisms in the soil depends primarily on the availability of oxygen and carbon in the soil, and on soil temperature and soil moisture. Oxygen consumption is greatest when organic matter is incorporated into moist, warm soil. Respiration rates in winter are suppressed because of low temperature and sometimes low oxygen concentration. In the summer, low soil water potential may inhibit respiration. Respiration rates are usually highest in spring months, when roots are growing rapidly, root exudates are plentiful for microorganism growth, and moisture and temperature conditions are favourable.

In this chapter, we will first present simple models for gas exchange for individual roots or microorganism colonies, as well as gas exchange between the soil profile and the atmosphere. To introduce the first concepts of numerical solutions for mass exchanges in soils, only steady-state exchange will be considered. Useful results relating to soil aeration can be obtained at steady state, and the simpler steady-state solutions will allow us to start developing numerical tools. The transient flow problems in later chapters should provide clear examples of how the transient gas flow problem can be solved if such solutions are needed.

Soil Physics with Python. First Edition. Marco Bittelli, Gaylon S. Campbell and Fausto Tomei.
© Marco Bittelli, Gaylon S. Campbell and Fausto Tomei 2015. Published in 2015 by Oxford University Press.

The solution of gas flow by diffusion was selected as a first example for presenting a numerical solution of mass flow because of its simplicity in the numerical scheme. The term 'diffusion' refers to the transport of a material within a single phase by random molecular motion. Diffusion can result from gradients in pressure, temperature, external forces (forced diffusion) or concentration. Here we discuss only diffusion due to concentration gradients. We assume an isobaric and isothermal system, which means pressure and temperature are constant in time and space. Detailed discussions of diffusion fluxes and coefficients have been provided by Bird *et al.* (1960), Cussler (1997) and Poling *et al.* (2000).

3.1   Transport Equations

Diffusion of gases in soil follows Fick's law

$$f_g = \frac{Q_g}{A} = -D_g \frac{dc}{dx} \tag{3.1}$$

where Q_g is flux [g s^{-1}], A is area [m^2], f_g is flux density [g m^{-2} s^{-1}], D_g is the diffusion coefficient [m^2 s^{-1}], c is concentration [g m^{-3}] and x is distance [m]. Equation (3.1) can be rearranged as

$$\frac{Q_g}{A} dx = -D_g\, dc \tag{3.2}$$

Equation (3.2) is integrated to give

$$\int_{x_1}^{x_2} \frac{Q_g}{A} dx = -\int_{c_1}^{c_2} D_g\, dc \tag{3.3}$$

Under steady conditions, the flux of gas, Q_g, across any surface is constant, and, assuming D_g to be constant, eqn (3.3) can be rearranged by moving Q_g and D_g out of the integrals,

$$Q_g \int_{x_1}^{x_2} \frac{dx}{A(x)} = -D_g \int_{c_1}^{c_2} dc \tag{3.4}$$

Carrying out the integration on the right-hand side gives

$$Q_g \int_{x_1}^{x_2} \frac{dx}{A(x)} = -D_g(c_2 - c_1) \tag{3.5}$$

Rearranging this leads to

$$Q_g = \frac{-D_g(c_2 - c_1)}{\int_{x_1}^{x_2} \dfrac{dx}{A(x)}} \tag{3.6}$$

This gives a general definition of conductance as

$$K_g = \frac{D_g}{\int_{x_1}^{x_2} \frac{dx}{A(x)}}$$

(3.7)

Substituting this into eqn (3.6) leads to

$$Q_g = -K_g(c_2 - c_1)$$

(3.8)

where the subscripts on the concentrations indicate locations at the different concentrations. Under steady conditions, the flux of gas, Q_g, across any surface is constant, although the flux density may change if the area available for flow changes. For simple geometries, the integration can be done and the conductances determined. When diffusion is purely one-dimensional, the area available for flow remains constant with distance, so $A(x) = 1\,\text{m}^2$. The conductance for planar diffusion is therefore

$$K_g = \frac{D_g}{x_2 - x_1}$$

(3.9)

When diffusion is spherical (an approximation applicable for a microorganism colony in soil), $A(x) = 4\pi x^2$, where x is the radial distance from the centre of the sphere. Integration of eqn (3.7) gives the conductance for spherical diffusion:

$$K_g = \frac{4\pi D_g r_1 r_2}{r_2 - r_1}$$

(3.10)

Diffusion to a root or cylindrical surface is a third interesting case. Here $A(x) = 2\pi x$, per metre of root, where x is the distance from the centre of the root cylinder. Again, using eqn (3.7), we obtain the conductance for cylindrical diffusion:

$$K_g = \frac{2\pi D_g}{\ln(r_2/r_1)}$$

(3.11)

Note that the units of K_g have been allowed to change depending on system geometry. For planar diffusion, where flux per unit area is constant, conductance has units of m s^{-1} and Q_g is per square metre. For cylindrical systems, the flux per unit length is constant, so K_g has units of m^2 s^{-1} and flux is per metre of length. For spherical systems, K_g has units of m^3 s^{-1} and flux is per sphere.

It is interesting to note that for a spherical organism, the conductance becomes independent of r_2 when r_2 is large compared with the radius of the organism. In other words, when $r_2 \gg r_1$, eqn (3.10) becomes $K_g \simeq 4\pi D_g r_1$.

3.2   The Diffusivity of Gases in Soil

Diffusivity in the gas phase of a porous medium is usually expressed as the product of two terms: the binary diffusion coefficient for the gas in air and some function of gas-filled porosity. A function accounting for gas-filled porosity is given by

$$D_g = D_0 \zeta (\phi_g) \tag{3.12}$$

where D_g is gas diffusivity, D_0 is the binary diffusion coefficient for gas in air and ζ is a variable that depends on the gas-filled porosity ϕ_g. The diffusivity of a gas in air depends on the diffusing species, as well as the temperature and pressure of the air (Poling *et al.*, 2000). Some binary diffusion coefficients for gases in air under standard conditions are given in Table 3.1.

Diffusivity increases with temperature at a rate of about 0.7 % K^{-1} and decreases with increasing atmospheric pressure at a rate of about 1 % kPa^{-1}. The following relationship is often used to express the pressure and temperature dependence of diffusivity:

$$D_0(T_K, P) = D_0(273.16\,\mathrm{K}, 101.3\,\mathrm{kPa}) \left(\frac{T_K}{273.16} \right)^{1.75} \left(\frac{101.3}{P} \right) \tag{3.13}$$

where T_K is temperature in kelvin and P is atmospheric pressure in kilopascals. Over some part of the diffusion path, O_2 and CO_2 usually must diffuse through liquid water. Diffusivities in water are much smaller than those in air. Typical values are around 2×10^{-9} m^2 s^{-1} for both CO_2 and O_2 at 20 °C. Diffusivity of gases in water increases with temperature at about the same rate as viscosity decreases with temperature (Bird *et al.*, 1960). The temperature correction can be made using $D_w = D_{wo}(T_K / 293)$, where T_K is the kelvin temperature of the water. Depending on the experimental methods employed for measurements of binary diffusion coefficients, slightly different values among different methods were presented in the literature (Poling *et al.*, 2000).

The fact that diffusivities in the liquid phase are four orders of magnitude smaller than those in the gas phase indicates that gas exchange in a soil profile without a continuous air phase is, for practical purposes, zero. Gas transport in moist soil profiles will therefore be assumed to occur only through the gas phase. Gas transport through the liquid phase will only be considered when dealing with gas exchange of microorganisms and roots.

The diffusion rate of a gas in a porous medium is reduced relative to that in free space because some of the space is occupied by solid particles and liquids and because the path

Table 3.1 *Binary diffusion coefficients for gases in air at 101.3 kPa and 273.16 K*

Gas	D_0(NTP) [m^2 s^{-1}]	Reference
Oxygen	1.77×10^{-5}	Campbell (1977)
Water vapour	2.12×10^{-5}	Campbell (1977)
Carbon dioxide	1.39×10^{-5}	Pritchard and Currie (1982)
Nitrous oxide	1.43×10^{-5}	Pritchard and Currie (1982)
Ethane	1.28×10^{-5}	Pritchard and Currie (1982)
Ethylene	1.37×10^{-5}	Pritchard and Currie (1982)

Table 3.2 *Values for the constants b and m in eqn (3.14) from various sources. Where a material is indicated, the values are from nonlinear least squares fits to data from the indicated source*

Material	b_g	m_g	Data source
	0.66	1	Penman (1940)
	1	1.5	Marshall (1959)
Aggregated Yolo silt loam	0.90	2.36	Sallam *et al.* (1984)
Hamble silt loam	0.81	2.29	Ball (1981)
Sandy loam	1.30	1.70	Lai *et al.* (1976)

the molecules must follow is more tortuous than that of molecules in free space. For dry porous materials, Currie (1965) has shown that an equation of the form

$$\zeta(\phi_g) = b_g \phi_g{}^{m_g} \tag{3.14}$$

fits data reasonably well. The constant m_g has a value that depends on the shape of the soil particles, but generally falls between 1 and 2 for dry materials. The constant b_g ranges from 0.5 to 1.0 and depends on the value chosen for m_g. Penman (1940) found that $b_g = 0.66$ and $m_g = 1$ provided a good fit to data. Marshall (1959) found $b_g = 1$ and $m_g = 1.5$.

When water is added to the porous material, the cross-section for flow is reduced and the tortuosity of the flow path is increased. As water content approaches saturation, dead-end pores may be formed that contribute to ϕ_g, but do not aid gas diffusion. Equation (3.14) can still be used, but the constants b_g and m_g may not be the ones which work best for dry materials. Table 3.2 gives values for b_g and m_g from several data sets. In spite of some variation, it appears that values of b_g around 0.9 and m_g around 2.3 are good approximations for undisturbed samples, especially at high water content.

3.3 Computing Gas Concentrations

The composition of air is generally reported in terms of the fraction of the total volume that would be occupied by each gas in its pure state or in terms of the partial pressure of that gas. The partial pressure can be converted to the concentration, which is the mass of gas per unit volume of air, using the gas law:

$$c_i = \frac{M_i P_i}{R T_K} \tag{3.15}$$

where P_i is the partial pressure of the gas [Pa], M_i is the molecular mass, R is the gas constant (8.3143 J mol^{-1} K^{-1}) and T_K is the kelvin temperature. The partial pressure of

Table 3.3 *Partition coefficients K_H for gases in water*

Temperature [°C]	N_2	O_2	CO_2	NH_3
0	0.024	0.049	1.71	1094
10	0.019	0.038	1.19	852
20	0.016	0.031	0.88	653
30	0.013	0.026	0.67	497
40	0.012	0.023	0.53	383

the gas is the volume fraction (0.21 for O_2, 0.79 for N_2, etc.) multiplied by atmospheric pressure (1.013×10^5 Pa at sea level). Concentrations of O_2, CO_2 and N_2 in air are therefore $c_{O_2} = 279\,g\,m^{-3}$, $c_{CO_2} = 0.76\,g\,m^{-3}$, and $c_{N_2} = 920\,g\,m^{-3}$.

For computing the concentrations of gases in cells, in anaerobic micro sites and in the soil solution, it is necessary to consider the equilibrium between the dissolved gas and the gas in the soil air spaces. Henry's law is used to approximate the partitioning between gas and liquid phases. It says that the concentration of gas in solution is linearly related to the concentration in the gas phase.

$$c_{il} = K_H c_{ig} \tag{3.16}$$

where c_{il} is the concentration [$g\,m^{-3}$ of water] of species i in the liquid phase and c_{ig} is the gas-phase concentration [$g\,m^{-3}$ of air]. The partition coefficient K_H (sometimes called the dimensionless Henry's law constant) depends strongly on the species of gas and is also dependent on temperature.

Table 3.3 shows values for several gases that are present in soil. Note that ammonia and carbon dioxide have much higher partition coefficients than oxygen and nitrogen. However, their concentrations in the gas phase are normally much smaller than the oxygen and nitrogen concentrations. As a consequence of the large partition coefficients of these gases, their concentrations in the liquid phase are more similar to those of oxygen and nitrogen than their concentrations in the gas phase.

3.4   Simulating One-Dimensional Steady-State Oxygen Diffusion in a Soil Profile

As a first example of a numerical solution of the transport equations, we will consider oxygen flux into soil and O_2 profiles within a soil profile. The problem is a steady-state diffusion problem. A system is in a steady state if all variables are constant in time. This implies that for the variable y, its partial derivative with respect to time is zero:

$$\frac{\partial y}{\partial t} = 0 \tag{3.17}$$

To solve the one-dimensional steady-state Fick's equation (3.1), a numerical solution is implemented. To do this, the soil is divided into layers, or elements. Within each element, we assume that the flux is linearly related to the concentration difference across the element. Respiratory production of CO_2 and consumption of O_2 are assumed to occur. Even though production and consumption are continuous throughout a layer or element, we will approximate their effects by concentrating them at the nodes, or junctions between the elements. Figure 3.1 shows two elements with their conductances $k(i)$, nodal concentrations $co(i)$ and sink functions $u(i)$.

The flux (positive downwards) within element i is

$$j(i) = -k(i)[co(i+1) - co(i)] \tag{3.18}$$

where

$$k(i) = \frac{df(i)}{z(i+1) - z(i)} \tag{3.19}$$

$df(i)$ is the gas diffusivity in element i. Note that eqn (3.19) gives the element conductances, defined in eqn (3.9). The mass balance for gas at node i is

$$j(i) - j(i-1) - u(i) = 0 \tag{3.20}$$

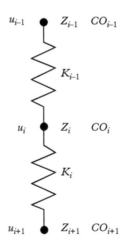

Figure 3.1 *Elements and nodes for computing gas concentration and flux in a soil profile, showing the numbering system, concentrations, and sources or sinks.*

The source term at node i, $u(i)$, is defined as

$$u(i) = \int_{i-1/2}^{i+1/2} \alpha_s(z)\, dz \tag{3.21}$$

where $\alpha_s(z)$ is the source strength [g m^{-3} s^{-1}] (+ for production, – for consumption). If the source strength is assumed constant for short distances, then eqn (3.21) becomes

$$u(i) = \alpha(z)\frac{z(i+1)-z(i-1)}{2} \tag{3.22}$$

By substituting each flux defined in eqn (3.18) for two elements into eqn (3.20), we obtain

$$\{-k(i)\,[co(i+1)-co(i)]\} - \{-k(i-1)\,[co(i)-co(i-1)]\} - u(i) = 0 \tag{3.23}$$

Multiplying gives

$$-k(i)co(i+1) + k(i)co(i) - [-k(i-1)co(i) + k(i-1)co(i-1)] - u(i) = 0 \tag{3.24}$$

which can be rearranged as

$$-k(i)co(i+1) + k(i)co(i) + k(i-1)co(i) - k(i-1)co(i-1) - u(i) = 0 \tag{3.25}$$

Grouping under a common factor (the concentrations) leads to

$$-co(i-1)k(i-1) + co(i)\,[k(i)+k(i-1)] - co(i+1)k(i) - u(i) = 0 \tag{3.26}$$

Now, the equation can be written for each node, giving m equations in $m+2$ unknowns. The following is an example for three nodes, where we have three equations in five unknowns:

$$-co(0)k(0) + co(1)\,[k(0)+k(1)] - co(2)k(1) = u(1)$$
$$-co(1)k(1) + co(2)\,[k(1)+k(2)] - co(3)k(2) = u(2) \tag{3.27}$$
$$-co(2)k(2) + co(3)\,[k(2)+k(3)] - co(4)k(3) = u(3)$$

A set of equations like this is not solvable. Its solution requires that we increase the number of equations by two, defined by the boundary conditions.

3.4.1 Boundary Conditions

Boundary conditions are used to reduce the number of unknowns (in this case by two) so that the system of equations can be solved. The boundary condition at the top of the soil is established by setting $co(0) = 280\,\mathrm{g\,m^{-3}}$ for O_2 or $0.59\,\mathrm{g\,m^{-3}}$ for CO_2. The node at $z(1)$ is at the soil surface, and the element conductance $k(0)$ is the atmospheric boundary-layer conductance. The boundary condition at the bottom of the soil profile is probably best handled by setting $k(m) = 0$ so that there is no flow out of the bottom of the profile and the concentration $co(m + 1)$ has no effect on the other concentrations. To solve this system of equations, the equations are transformed into matrix form, and solved using the Thomas algorithm. In Appendix B, two common direct methods for solving linear systems are presented: the Gauss elimination and the Thomas algorithm.

Before solving the linear system by using the linear algebra algorithm described in Appendix B, eqn (3.27) stills needs some rearrangement to move all the unknowns to one side of the equation to later obtain a matrix of known terms:

$$
\begin{aligned}
co(1)\,[k(0) + k(1)] - co(2)k(1) &= u(1) + co(0)k(0) \\
-co(1)k(1) + co(2)\,[k(1) + k(2)] - co(3)k(2) &= u(2) \\
-co(2)k(2) + co(3)\,[k(2) + k(3)] &= u(3) + co(4)k(3)
\end{aligned}
\tag{3.28}
$$

Equation (3.28) displays the known terms on the right-hand side. We now define the following variables to formulate the matrix:

$$
\begin{aligned}
a(i + 1) &= -k(i) \\
b(i) &= k(i - 1) + k(i) \\
c(i) &= -k(i) \\
d(i) &= u(i)
\end{aligned}
\tag{3.29}
$$

for $i \neq 1$

3.4.2 Matrix Formulation

Now, eqns (3.28) and (3.29) are combined to be written in matrix form:

$$
\begin{aligned}
b(1)co(1) + c(1)co(2) &= d(1) \\
a(2)co(1) + b(2)co(2) + c(2)co(3) &= d(2) \\
a(3)co(2) + b(3)co(3) &= d(3) + co(4)k(3)
\end{aligned}
\tag{3.30}
$$

Writing eqn (3.30) for each node and then arranging the equations in matrix form for $m = 4$ gives

$$
\begin{vmatrix}
b(1) & c(1) & 0 & 0 \\
a(2) & b(2) & c(2) & 0 \\
0 & a(3) & b(3) & c(3) \\
0 & 0 & a(4) & b(4)
\end{vmatrix}
\begin{vmatrix}
co(1) \\
co(2) \\
co(3) \\
co(4)
\end{vmatrix}
=
\begin{vmatrix}
d(1) \\
d(2) \\
d(3) \\
d(4)
\end{vmatrix}
\tag{3.31}
$$

For node 1, $d(1) = u(1) + k(0)co(0)$. If $k(m)$ were not zero, then $k(m)co(m + 1)$ would have to be added to $d(m)$. Now we utilize the Thomas algorithm to solve the matrix and obtain the unknown concentrations.

3.5 Numerical Implementation

The program `PSP_gasDiffusion` is a Python code for determining the oxygen concentration using eqns (3.18)–(3.29). This program utilizes the Thomas algorithm described in Appendix B to solve the linear system of equations. The reader should read Appendix B.3.4, where the Thomas algorithm is described and a numerical example is provided.

The **main**() function is defined at the bottom of the program, where constants are defined and where the solver is called. The first lines of the program are used to import the functions for the Thomas algorithm and the setting of the computational grid. The following lines are written to define the numpy arrays used for building the tridiagonal matrix. The function **linear**() is written in the file `PSP_grid.py` and it builds the grid used for the numerical solution. The Thomas algorithm written with the implementation of the boundary conditions is called by the function **ThomasBoundaryCondition**(). The program is as follows:

```
#PSP_gasDiffusion
from __future__ import print_function, division
from math import exp
from PSP_ThomasAlgorithm import ThomasBoundaryCondition
import PSP_grid as grid
import matplotlib.pyplot as plt
import numpy as np

def gasSolver(boundaryLayerCond, boundaryOxygenConc, dg,
                            respRate, totalDepth, n):
    a  = np.zeros(n+2, float)
    b  = np.zeros(n+2, float)
    c  = np.zeros(n+2, float)
    d  = np.zeros(n+2, float)
    g  = np.zeros(n+2, float)
    u  = np.zeros(n+2, float)
    co = np.zeros(n+2, float)

    g[0]  = boundaryLayerCond
    co[0] = boundaryOxygenConc
    # vector depth [m]
    z = grid.linear(n, totalDepth)

    for i in range(1, n+1):
        u[i] = respRate * exp(-z[i] / 0.3) * (z[i + 1] - z[i - 1])
                / 2.0
```

```
            if i < n:
                g[i] = dg / (z[i + 1] - z[i])
            else:
                g[i] = 0
            a[i + 1] = -g[i]
            b[i] = g[i - 1] + g[i]
            c[i] = -g[i]
            d[i] = u[i]

    d[1] = d[1] + g[0] * co[0]

    ThomasBoundaryCondition(a, b, c, d, co, 1, n)

    return(z, co)

def main():
    R = 8.3143
    n = 20
    totalDepth = 1.0
    bulkDensity = 1300.
    particleDensity = 2650.
    waterContent = 0.2
    respRate = -0.001
    oxygenDiff = 1.77e-5
    temperature = 25.
    atmPressure = 101.3
    boundaryLayerCond = 0.01

    # O2 concentration in air [g/m^3]
    boundaryOxygenConc = (0.21 * atmPressure * 1000. * 32. /
                          (R * (temperature + 273.15)))
    porosity = 1. - bulkDensity / particleDensity
    gasPorosity = porosity - waterContent

    #   binary diffusion coefficient [m2/s]
    binaryDiffCoeff = (oxygenDiff * (101.3 / atmPressure)
                       * ((temperature + 273.15) / 273.15)**1.75)

    bg = 0.9
    mg = 2.3
    dg = binaryDiffCoeff * bg * gasPorosity**mg

    z, co = gasSolver(boundaryLayerCond, boundaryOxygenConc,
                      dg, respRate, totalDepth, n)

    print ("node   depth [m]   Co [g\m^3]")
    for i in range(n + 2):
        print ("%3d    %6.2f     %.2f" %(i, z[i], co[i]))
```

```
    # plot results
    for i in range(n+1):
        plt.plot(co[i], -z[i], 'ko')

    plt.xlabel('Concentration [g m$^{-3}$]',fontsize=14,labelpad=2)
    plt.ylabel('Depth [m]',fontsize=14,labelpad=2)

    plt.show()
main()
```

The file `PSP_ThomasAlgorithm.py` contains the Thomas algorithm (as also described in *Appendix B*). The first `for` loop contains the forward elimination scheme, while the second contains the backward substitution. Finally, the function **Thomas** is called and the results are printed. The function **ThomasBoundaryCondition** implements the Thomas algorithm with the option of defining the boundary conditions and therefore knowing the first and the last element of the matrix. This is the algorithm used in the numerical solutions presented here:

```
#PSP_ThomasAlgorithm.py
from __future__ import division
import numpy as np

def Thomas(a, b, c, d):
    n=len(d)
    x =np.zeros(n, float)

    for i in range(0, n-1):
        c[i] /= b[i]
        d[i] /= b[i]
        b[i + 1] -= a[i + 1] * c[i]
        d[i + 1] -= a[i + 1] * d[i]

    # back substitution
    x[n-1] = d[n-1] / b[n-1]
    for i in range(n-2, -1, -1):
        x[i] = d[i] - c[i] * x[i + 1]

    return(x)


def ThomasBoundaryCondition(a, b, c, d, x, first, last):

    for i in range(first, last):
        c[i] /= b[i]
        d[i] /= b[i]
        b[i + 1] -= a[i + 1] * c[i]
        d[i + 1] -= a[i + 1] * d[i]
```

```
# back substitution
x[last] = d[last] / b[last]
for i in range(last-1, first-1, -1):
    x[i] = d[i] - c[i] * x[i + 1]
```

The grid is generated in a separate file called PSP_grid.py shown below. The program allows one to generate a linear grid (where the elements are of equal dimension) or a geometric grid (where the elements are thinner closer to the soil surface and geometrically increasing with depth). The returned variable z is the incremental depth, which is called z in the function.

```
from __future__ import division
import numpy as np

def linear(n, depth):
    z = np.zeros(n+2, float)
    dz = depth / n
    z[0] = 0
    z[1] = 0
    for i in range(1, n+1):
        z[i + 1] = z[i] + dz
    return z

def geometric(n, depth):
    z = np.zeros(n+2, float)
    mySum = 0.0
    for i in range(1, n+1):
        mySum = mySum + i * i
    dz = depth / mySum
    z[0] = 0.0
    z[1] = 0.0
    for i in range(1, n+1):
        z[i + 1] = z[i] + dz * i * i
    return z
```

Figure 3.2 shows results of the simulation for a soil column at uniform water content. Typical literature values of O_2 uptake rate for surface soil samples are $10^{-4} - 10^{-3}$ g m^{-3} s^{-1}. In this simulation, a rate of 5×10^{-4} g m^{-3} s^{-1} was used at the surface. Using these values, the flux at the surface was 145 mg m^{-2} s^{-1}. Typical values are 36 mg m^{-2} s^{-1} for an undisturbed sandy clay loam soil, 170 mg m^{-2} s^{-1} for a peat soil, and 36, 93 and 50 mg m^{-2} s^{-1} for potato, kale and tobacco crops. The simulation therefore seems reasonable. Oxygen concentration remains near atmospheric levels when the gas-filled porosity is 0.2 or above, but decreases quickly with depth as the gas filled porosity drops to 0.1 or below. Negative O_2 concentrations would be obtained for values deeper in the profile. These are an artefact of the model caused by the incorrect assumption that the oxygen consumption rate is independent of O_2.

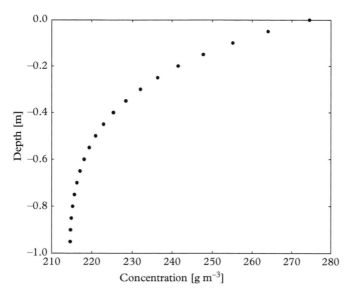

Figure 3.2 *Simulated oxygen concentration as a function of depth in a soil profile.*

3.6 EXERCISES

3.1. Derive the integration steps to obtain eqns (3.10) and (3.11) from eqn (3.7). Show all the steps.

3.2. Use the program in Python to determine the effects of volumetric water content, bulk density and boundary-layer conductance on oxygen and carbon dioxide concentrations in the soil profile.

3.3. Alter the program `main.py` so that it computes and prints percent oxygen and percent CO_2 at each level in the soil profile. Then use the program to determine what gas-filled porosity is required to keep the CO_2 concentration below 5 % and the O_2 concentration above 15 % in the root zone.

4

Soil Temperature and Heat Flow

Temperature affects all physical, chemical, and biological processes in soils. Biological processes such as the uptake of nutrients and water by roots, the decomposition of organic matter by microbes and the germination of seeds are strongly affected by soil temperature. Rates of some of these processes more than double for each 10 °C increase in temperature. In some cases, growth of above-ground plant parts is more closely correlated with soil temperature than with air temperature. Physical processes, such as water movement and soil drying, are also strongly influenced by temperature. At larger scales, soil thermal dynamics are important in the conversion of solar radiation into long-wave radiation and latent heat of evaporation. Indeed, the diurnal temperature cycle regulates heat exchange between earth and atmosphere. The modulation of different forms of heat exchange affects the evolution of the atmospheric boundary layer, determining weather patterns. For these reasons, most simulations of soil processes require submodels that predict soil temperature.

The simplest model for soil temperature would be to assume that it is constant or that it is equal to a daily, weekly or monthly mean air temperature. Such models may be adequate for some purposes, but they are limited in providing a realistic description of natural conditions. More accurate models require analysis of heat flow within the soil and heat exchange at the soil surface to determine soil temperature. The transport or loss of latent heat of either fusion or vaporization and the movement of soil moisture are important in determining soil temperature. Analysis of these aspects of the problem must follow the study of moisture movement, and will be considered in Chapters 6–10.

From a thermodynamic standpoint, heat is a transfer of energy as a result of a temperature difference. The transfer of energy can occur by conduction and radiation. The transfer by conduction is by means of energy carriers within the material. For instance, in a fluid, the energy carriers are usually individual molecules, while in a solid, they can be electrons or photons. When water evaporates or condenses and carries with it the latent heat of vaporization, the transport is referred as latent heat transport, but it still is a form of heat conduction, where the carriers are the water molecules in the vapour phase.

In this chapter, the differential equations for heat transport and the parameters needed to solve them will be given, followed by methods for solving the equations using numerical procedures. For simplicity, in this chapter, we assume that water content is constant and there is no water flux in the soil. In the following chapters, the two fluxes (heat and water) will be coupled, to obtain a realistic description of soil fluxes.

4.1   Differential Equations for Heat Conduction

Heat conduction is determined by random transfer of molecular kinetic energy. A variety of textbooks about heat transfer theory present analytical and numerical solutions to heat flow problems (Carslaw and Jaeger, 1959; Nellis and Klein, 2009).

The relationship between heat flux density [W m^{-2}] and temperature gradient [K m^{-1}] is Fourier's law:

$$f_h = -\lambda \frac{dT}{dz} \tag{4.1}$$

where λ is thermal conductivity [W m^{-1} K^{-1}], T is temperature [K] and z is vertical distance [m]. The minus sign on the right-hand side of the equation indicates that heat flows from points having higher temperature to points having lower temperature.

Equation (4.1) applies to heat flow in the vertical direction only, but it is easily generalized to three dimensions by adding conductivities and temperature gradients in the x and y directions. Equation (4.1) can be combined with the continuity equation to obtain the time-dependent differential equation

$$C_h \frac{\partial T}{\partial t} = \frac{\partial}{\partial z}\left(\lambda \frac{\partial T}{\partial z}\right) \tag{4.2}$$

where C_h is the volumetric specific heat of the soil [J m^{-3} K^{-1}]. Solutions to eqn (4.2) describe soil temperature as a function of depth and time. A closed-form solution to eqn (4.2) can only be obtained for simple sets of soil properties and boundary conditions that do not realistically represent soils under field conditions. It is useful, however, to examine at least one of these solutions in detail.

If thermal conductivity and heat capacity do not vary with depth, then λ may be taken out of the derivative and combined with C_h to give

$$\frac{\partial T}{\partial t} = D_h \frac{\partial^2 T}{\partial z^2} \tag{4.3}$$

where D_h is the thermal diffusivity, which is given by the thermal conductivity divided by the volumetric specific heat, $D_h = \lambda/C_h$. Equation (4.3) is a partial differential equation, which can be solved, for specific initial and boundary conditions, by decomposition into a Dirac δ function or into a periodic sine function.

For a soil column that has a surface temperature given by

$$T(0, t) = \overline{T} + A(0) \sin \omega t \tag{4.4}$$

and is infinitely deep, the temperature at any depth and time is given by

$$T(z, t) = \overline{T} + A(0) \exp\left(\frac{z}{z_d}\right) \sin\left(\omega t - \frac{z}{z_d}\right) \tag{4.5}$$

Here $\overline{T}$ is the mean soil temperature, $A(0)$ is the amplitude of the temperature wave at the soil surface (the difference between $\overline{T}$ and T_{max} or T_{min}), ω [s^{-1}] is the angular frequency of the oscillation (2π divided by the period in seconds) and z_d is the damping depth, given by

$$z_d = \sqrt{\frac{2D_h}{\omega}} \tag{4.6}$$

Equation (4.5) shows how temperature would vary with depth and time in a uniform soil subject to the assumed boundary conditions. Note that the amplitude is attenuated exponentially with depth and that the phase of the wave is shifted with depth. At $z = z_d$, the amplitude is $e^{-1} = 0.37$ times its value at the surface. The temperature at 2–3 damping depths would therefore be expected to be about the mean temperature for the period of oscillation, because temperature fluctuations would be only 5–10 % of the temperature fluctuation at the surface. Typical diurnal damping depths are 10–15 cm for mineral soils, and typical annual damping depths are 3–4 m.

4.2   Soil Temperature Data

The behaviour described above can be detected in measured values of soil temperatures as a function of time and soil depth. As described above, the main features are that the diurnal and annual variations are sinusoidal and that the amplitude decreases with depth and the times for the maximum and minimum temperatures are shifted with depth. Clearly, the periods of the annual and daily temperature cycles are 365 days and 24 hours, respectively.

Figures 4.1 and 4.2 depict soil temperature fluctuations measured over a two-day and a four-year period in the experimental station of S. Pietro Capofiume (Bologna, Italy), at latitude 44.65°N and longitude 11.62°E. Temperatures were measured at different depths from −0.1 to −1.8 m below the soil surface. Here we are showing temperatures at two depths. Note the attenuation of the temperature and the phase shift with time for both diurnal and annual cycles. For the daily cycle, the damping depth is at about −0.4 m. The amplitudes at both −1.35 and −1.8 m in Fig. 4.2 are still large so the damping depth for the annual cycle must be more than 2 m.

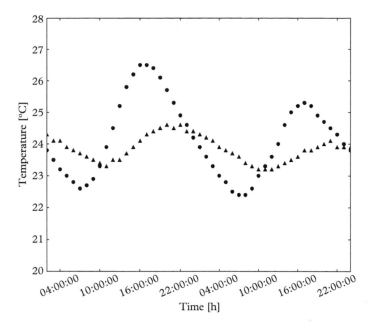

Figure 4.1 *Field-measured soil temperature variations over a two-day period at −0.1 m (dots) and −0.25 m (triangles) depths.*

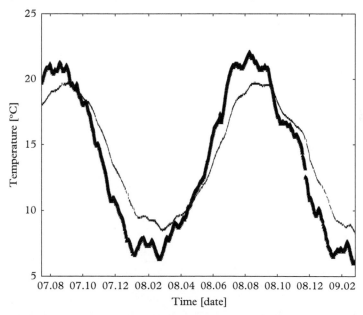

Figure 4.2 *Field-measured soil temperature variations between 15 July 2007 and 25 February 2009 at −0.1 m (triangles) and −1.35 m (line) depths. Data are from the experimental station of S. Pietro Capofiume, Bologna, Italy. Regional Agency for Environmental Protection, Emilia–Romagna region, Bologna, Italy.*

4.3   Numerical Solution of the Heat Flow Equation

A numerical solution of eqn (4.2) can be applied more generally than the analytical solution that has just been presented. No assumptions need to be made about constant thermal properties and the boundary conditions can be made realistic. Setting up the equations for a numerical solution of the heat flow problem is similar to setting up the gas diffusion problem. Now, however, we will consider the time-dependent problem, so solutions need to involve both depth and time. Also, heat will be stored or taken from storage within the soil. Figure 4.3 shows how the soil is divided into elements.

Depth is indicated by z, temperature by T, heat storage by C_h and conductances by K. Note the numbering system for nodes and elements. The element number is that of the node just above it. Heat flow within any element is assumed to be steady, so that eqn (4.1) describes the heat flux density. Storage of heat is assumed to occur only at the nodes.

Here we are presenting two numerical methods to solve the second-order partial differential equation: the finite difference method and the cell-centred finite volume method. Others will also be presented when we discuss solutions of the liquid water, water vapour and solute transport equations later.

4.3.1   Finite Difference Method

The energy balance equation for node i is

$$\frac{C_h(T_i^{j+1} - T_i^j)(z_{i+1} - z_{i-1})}{2\Delta t} = K_i(\overline{T}_{i+1} - \overline{T}_i) - K_{i-1}(\overline{T}_i - \overline{T}_{i-1}) \tag{4.7}$$

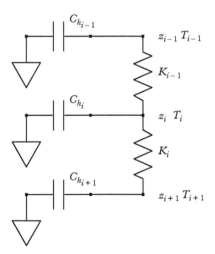

Figure 4.3 *Elements and nodes for computing temperature and heat flux in a soil profile, showing numbering system, conductances and storage.*

where Δt is the time increment, C_h is the soil volumetric specific heat and K is the conductance:

$$K_i = \frac{\lambda_i}{z_{i+1} - z_i} \tag{4.8}$$

where λ_i is the thermal conductivity of the ith element. The superscript j indicates the time at which the temperature is determined. The first term on the left-hand side of eqn (4.7) is the specific heat per unit volume of soil multiplied by the change in temperature of the soil and divided by the time over which the temperature change occurs; it is therefore the rate of heat storage at node i. The terms on the right-hand side represent difference between heat flow in and heat flow out of node i.

In eqn (4.7), the heat flux between nodes is proportional to the temperature difference. When the time increment is infinitesimally small or when the temperature does not change with time, it is obvious which temperature to use in computing the heat flux through an element. However, eqn (4.7) indicates that temperature does change with time, so the choice of nodal temperature to be used for computing the heat flux through an element must be specified. The overbar on the T's on the right-hand side of eqn (4.7) is to indicate that the flux is computed from an appropriate mean temperature. If superscript j indicates present time and $j + 1$ is one time step in the future, an appropriate mean temperature must be somewhere between T^j and T^{j+1}. Therefore,

$$\overline{T} = \eta\, T^{j+1} + (1 - \eta)\, T^j \tag{4.9}$$

Here, η is a weighting factor that may range from 0 to 1:

- If $\eta = 0$, the flux is determined by the temperature difference at the beginning of the time step. The numerical procedure that results from this choice is called a forward difference or explicit Euler method, since an explicit expression for T^{j+1} can be written using eqn (4.9).

- If $\eta = 0.5$, the average of the old and new temperatures is used to compute the heat flux. This is called a time-centered, or Crank–Nicholson, scheme.

- If $\eta = 1$, the equation for computing T^{j+1} is implicit Euler, since each T^{j+1} depends on the values of the new temperatures.

Most heat flow models use either $\eta = 0$ or $\eta = 0.5$. The best value to use is determined by considerations of numerical stability and error analysis. The reader is encouraged to investigate the effect of η on accuracy in the exercises at the end of this chapter. A few general comments, however, are given here. The explicit scheme, with $\eta = 0$, predicts more heat transfer between nodes than would actually occur, and it can therefore become unstable if the time steps are too large. Stable numerical solutions are only obtained when

$$\Delta t < \frac{C_h \Delta z^2}{2\lambda} \tag{4.10}$$

When $\eta > 0.5$, stable solutions to the heat flow problem will always be obtained, but if η is too small, the solutions may oscillate. The reason for this is that the simulated heat transfer between nodes in one time step is too large and the new temperature overshoots. On the next time step, the excess heat must be transferred back, so the predicted temperature at that node is too small. On the other hand, if η is too large, the temperature difference will be too small and not enough heat will be transferred. Simulated temperatures will never oscillate under these conditions, but the simulation will underestimate the heat flux. The best accuracy is obtained with η around 0.4, while the best stability is at $\eta = 1$. A good compromise is $\eta = 0.6$. In the program described below, we implemented an adaptive time step scheme, to allow the reader to utilize every value of η, including a fully explicit solution.

When using a fully explicit solution, it is important to consider the choice of the grid. Indeed, it is advisable to use a linear grid. If the geometric grid is used, the top cell is very small and the heat flux into the top cell then becomes very large, since dz in eqn (4.1) is small. In the explicit solution for the first time steps, there is no flux into the cells below the top one, since the temperature is equal for the whole soil profile. The excessive flux into the first nodes determines numerical oscillations and the algorithm does not converge.

Equations (4.7) and (4.9) can be combined to give a single equation for T^{j+1} in terms of the known values of K and C_h and unknown values of T^{j+1}_{i+1}. If similar equations are written for each node in the soil, the set of equations describing the system is similar to gas diffusion. For four nodes,

$$
\begin{vmatrix}
B(1) & C(1) & 0 & 0 \\
A(2) & B(2) & C(2) & 0 \\
0 & A(3) & B(3) & C(3) \\
0 & 0 & A(4) & B(4)
\end{vmatrix}
\begin{vmatrix}
TN(1) \\
TN(2) \\
TN(3) \\
TN(4)
\end{vmatrix}
=
\begin{vmatrix}
D(1) \\
D(2) \\
D(3) \\
D(4)
\end{vmatrix}
\tag{4.11}
$$

where

$$
B(i) = \eta \left[K(i) + K(i-1) \right] + \frac{C_h(i)\,[z(i) - z(i-1)]}{2\Delta t} \tag{4.12}
$$

$$
A(i+1) = C(i) = -\eta K(i) \tag{4.13}
$$

and

$$
D(i) = (1 - \eta)K(i-1)T(i-1)
$$
$$
+ \left\{ \frac{C_h(i)[z(i+1) - z(i-1)]}{2\Delta t} - (1-\eta)\left[K(i) + T(i) + K(i-1)\right] \right\} \tag{4.14}
$$
$$
+ (1 - \eta)K(i)T(i+1)
$$

The new $(j+1)$ temperatures are $TN(i)$; the old (j) temperatures are $T(i)$.

4.3.2 Boundary conditions

The temperature at the bottom of the soil column may be set constant or to some measured deep soil temperature. This then becomes the value of $TN(m + 1)$. The last value for D at the bottom boundary node is therefore

$$D(m) = D'(m) + \eta K(m) TN(m + 1) \tag{4.15}$$

where $D'(m)$ is the value obtained from eqn (4.14). For a no-flux condition, $K(m) = 0$ and nothing is added. The boundary condition at the soil surface is more complex since convection, evaporation and radiation may be important. Considering only the convective heat transport at the surface,

$$D(1) = D'(1) + \eta K(0) TN(0) \tag{4.16}$$

where $D'(1)$ is the value computed from eqn (4.14), $K(0)$ is the boundary-layer conductance and $TN(0)$ is the air temperature at the end of the time step.

When significant radiative and/or latent heat transfer occur, they are added as heat sources at node 1 to give

$$D(1) = D'(1) + \eta K(0) TN(0) - R_n + L_v E \tag{4.17}$$

where R_n is the net radiation at the soil surface and $L_v E$ is the latent heat flux, where L_v is the latent heat of vaporization and E is the evaporation rate. The net radiation will be discussed in detail in Chapter 15. The evaporation rate depends on the water content and temperature of the soil surface, on the concentration of water vapour in the air and on atmospheric boundary-layer conductance. Computation of the evaporation rate will be covered in Chapter 11.

4.3.3 Cell-Centred Finite Volume

The cell-centred finite volume (also called integrated finite difference) method is a method used for multidimensional solutions. For a regular two-dimensional geometry, a rectangular grid is convenient, as shown in Fig. 4.4; however, the methods can be applied to grids of different geometries. The method is based on integration of the partial differential equation over a finite volume V_t. Then the volume integral is transformed into an integral over the boundaries of each of the cells of the grid by using Gauss' theorem.

The heat balance equation can be written as

$$\frac{\partial (C_h T)}{\partial t} = \nabla (f_h) + u \tag{4.18}$$

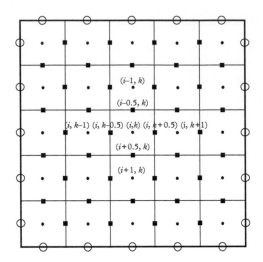

Figure 4.4 *Grid representing rows, columns, nodes and faces for the cell-centred finite volume scheme.*

where f_h is the heat flux, C_h is the volumetric specific heat, u is the sink–source term and ∇ is the vector differential operator

$$\nabla = \hat{x}\frac{\partial}{\partial x} + \hat{y}\frac{\partial}{\partial y} + \hat{z}\frac{\partial}{\partial z}$$

The solution of this equation is based on its integration over the total volume V_t of the grid cell at each node:

$$\int_{V_t} \frac{\partial(C_h T)}{\partial t} = \int_{V_t} \nabla(f_h) + \int_{V_t} u \tag{4.19}$$

Gauss' theorem states that the integral of $\nabla(f_h)$ over the total volume of the grid cell can be replaced by the fluxes over the boundary of the grid cell, leading to

$$\int_{V_t} \frac{\partial(C_h T)}{\partial t} = \int_{\partial V_t} f_h + \int_{V_t} u \tag{4.20}$$

This equation simply means that the change in total thermal energy within one grid cell is equal to the net heat fluxes across its boundaries into the cell and the thermal sources and sinks within the grid cell.

In the following, it will be demonstrated how this equation can be solved for a one-dimensional case. The concept, however, can also be used for higher-dimensional problems. Equations (4.1) and (4.20) lead to

$$\int_{V_t} \frac{\partial (C_h T)}{\partial t} = \int_{\partial V_t} -\lambda \frac{dT}{dz} + \int_{V_t} u \qquad (4.21)$$

In one dimension, the volume of grid cell i is $V_t = A(z_{i+1} - z_{i-1})/2$, where A is the cross-section of the domain and the area of the cell boundary is $\partial V_t = 2A$. For each cell, there are two boundaries where dT/dz needs to be evaluated. The gradient of the temperature can be approximated by finite differences:

$$\frac{T_i - T_{i-1}}{z_i - z_{i-1}} \text{ and } \frac{T_{i+1} - T_i}{z_{i+1} - z_i}$$

We assume in the following that there are no sources or sinks: $u = 0$. Inserting into eqn (4.21) leads to

$$\frac{z_{i+1} - z_{i-1}}{2} A C_h \frac{\partial T_i}{\partial t} = A\lambda_i \frac{T_{i+1} - T_i}{z_{i+1} - z_i} - A\lambda_{i-1} \frac{T_i - T_{i-1}}{z_i - z_{i-1}} \qquad (4.22)$$

The change in temperature at node i can be expressed by the temperature at time step j and $j + 1$:

$$\frac{\partial T}{\partial t} = \frac{T_i^{j+1} - T_i^j}{\Delta t}$$

As discussed in Section 4.3.1, it is not obvious at what time the temperature used to calculate thermal fluxes should be evaluated. In the program presented here, they are evaluated at the end of the time step; hence, the time scheme is an implicit Euler method:

$$\frac{(z_{i+1} - z_{i-1})C_h}{2\Delta t}(T_i^{j+1} - T_i^j) = \frac{\lambda_i}{z_{i+1} - z_i}(T_{i+1}^{j+1} - T_i^{j+1}) - \frac{\lambda_{i-1}}{z_i - z_{i-1}}(T_i^{j+1} - T_{i-1}^{j+1}) \qquad (4.23)$$

For a one-dimensional case, the finite volume and finite difference methods lead to the same linear equation system. However, in multidimensional unstructured grids, the finite difference method cannot be applied in this form, but the finite volume method can still be used. This method will be used for multidimensional solutions, although, for simplicity, it was presented here in its one-dimensional form.

When using the cell-centered finite volume method, the node is positioned at the centre of the element (Fig. 4.4) and not on the boundary as is done in the finite difference solution (Fig. 4.3). Therefore, for the cell-centered finite volume method, an important aspect of the solution is the choice of the mean used to compute an average thermal conductivity between elements. This choice is particularly important when the simulation involves materials having different soil thermal properties between elements. We tested different means, and the ones that provided numerically stable results were the geometric and the logarithmic means. In particular, for heat transport, the logarithmic mean provided the more stable results. The logarithmic mean is given by (Carlson, 1972)

$$\bar{\lambda} = \frac{\lambda_1 - \lambda_2}{\ln \lambda_1 - \ln \lambda_2} \quad \text{for} \quad \lambda_1 \neq \lambda_2 \tag{4.24}$$

the geometric mean is given by

$$\bar{\lambda} = \sqrt{\lambda_1 \lambda_2} \tag{4.25}$$

and the harmonic mean is given by

$$\bar{\lambda} = \frac{2}{1/\lambda_1 + 1/\lambda_2} \tag{4.26}$$

4.4   Soil Thermal Properties

4.4.1   Heat Capacity

Before soil heat fluxes and temperature profiles can be computed, the specific heat and thermal conductivity of the soil must be known. It is possible to obtain these values experimentally, but for modelling purposes, it is better to predict them from basic physical properties of the soil.

The volumetric specific heat of the soil is given by the contributions of the specific heat of each soil constituent:

$$C_h = C_s \phi_s + C_l \theta + C_g \phi_g + C_o \phi_o \tag{4.27}$$

where ϕ_s is the volumetric solid fraction, θ the volumetric liquid fraction, ϕ_g the volumetric gas fraction and ϕ_o the organic matter volumetric fraction, while C_s, C_l, C_g and C_o are the volume specific heats of the solid matrix, liquid, gas and organic constituents, respectively.

Table 4.1 gives typical values of C for each of these substances. The contribution of soil air to heat capacity is usually ignored. In mineral soils, the contribution from organic matter is also often negligible because the volumetric specific heats of mineral and organic materials are similar. The volumetric specific heat of mineral soil can therefore be approximated as

$$C_h = C_s(1 - \phi_f) + C_l \theta \tag{4.28}$$

where ϕ_f is the total porosity. Therefore, to obtain the total soil heat capacity, it is sufficient to know the heat capacities of the solid matrix and the liquid phase and their volume fractions. We have seen in Chapter 2 that these volume fractions are easily obtained from basic soil properties (in non-swelling soils). Heat capacity can therefore be obtained from

Table 4.1 *Thermal properties of soil materials (T is Celsius temperature)*

Material	Density [Mg m^{-3}]	Specific heat [J g^{-1} K^{-1}]	Thermal conductivity [W m^{-1} K^{-1}]	Volumetric specific heat [MJ m^{-3} K^{-1}]
Soil minerals	2.65	0.87	2.5	2.31
Granite	2.64	0.82	3.0	2.16
Quartz	2.66	0.80	8.8	2.13
Glass	2.71	0.84	1.0	2.28
Organic matter	1.30	1.92	0.25	2.50
Water	1.00	4.18	$0.56 + 0.0018T$	4.18
Ice	0.92	$2.1 + 0.0073T$	$2.22 - 0.011T$	$1.93 + 0.0067T$
Air (101 kPa)	$(1.29 - 0.0041T) \times 10^{-3}$	1.01	$0.024 + 0.00007T$	$(1.3 - 0.0041T) \times 10^{-3}$

$$C_h = \frac{2.4 \times 10^6 \rho_b}{2650} + 4.18 \times 10^6 \theta \qquad (4.29)$$

where the bulk density ρ_b has units of kg m^{-3}. In swelling soils, the problem is more difficult, since the volume fractions are functions of space, time and the volumetric water content fraction itself.

4.4.2   Thermal Conductivity

The conduction of heat within a material is due to the interactions of energy carriers. The energy carriers can be different depending on the material. For instance, the energy carriers in metals are electrons; in gases, they are gas molecules; in a liquid fluid like water they are the molecules of the liquid. Thermal conductivity is therefore proportional to the number and volume density of energy carriers. The thermal conductivity of pure metals is very high because the electron density is high. On the other hand, gas thermal conductivity is small because the number density of gas molecules in a gas is small (Nellis and Klein, 2009).

Soil is a three-phase system composed of solids, liquids and gases. Its thermal conductivity is therefore determined by the contribution of the different phases, which have different material properties and therefore energy carriers. Since the proportion of these components varies with space and time, the thermal conductivity of soil depends on properties that vary with space and time, such as its bulk density, water content, quartz content and organic matter content. At low water content, the air space controls the thermal conductivity. At high water content, the thermal conductivity of the solid phase becomes more important. The large differences in thermal conductivity among different materials are the result of differences in bulk density and composition. The transition from low to high conductivity occurs at low water content in sands and at high water content in soils high in clay.

Thermal conductivity model

Here we present a model to compute thermal conductivity from bulk density, water content, clay content and temperature. The model was originally published by Campbell *et al.* (1994) and it is a modification of the original de Vries (1963) model. The model is based on the assumption that the thermal conductivity of soil is given by a weighted sum of the thermal conductivities of its components:

$$\lambda = \frac{k_w \theta \lambda_w + k_g \phi_g \lambda_g + k_s \phi_s \lambda_s}{k_w \theta + k_g \phi_g + k_s \phi_s} \qquad (4.30)$$

where k_g, k_s and k_w are weighting factors for gas, solid and water. λ is the thermal conductivity and λ_g, λ_s and λ_w are thermal conductivities of gas, solid and water. The volumetric fractions of gas, solid and water are ϕ_g, ϕ_s and θ. The thermal conductivity of the gas phase accounts for both the conductivity of air λ_a and the latent heat component given by evaporation and condensation. In most soil models, thermal conductivity is considered to be independent of temperature. However, Campbell *et al.* (1994) showed that λ is temperature-dependent, with the dependence being mostly due to the variation of the latent heat transport with temperature. With respect to the de Vries (1963) model, the model of Campbell *et al.* (1994) provides a more rational treatment of the transition from high to low vapour contributions as the soil dries. Indeed, the de Vries (1963) model accounted for the thermal conductivity of continuous phases: when the soil is saturated, the continuous phase is the liquid water, whereas when the soil is dry, the continuous phase is gas. However, the de Vries (1963) model did not account for the intermediate conditions of partial saturation occurring in soils. This was introduced by Campbell *et al.* (1994) by defining a continuous function through a 'fluid' thermal conductivity variable that can be used over the whole range of water contents. The authors defined a variable called 'fluid' thermal conductivity:

$$\lambda_f = \lambda_g + f_w(\lambda_w - \lambda_g) \tag{4.31}$$

where f_w is an empirical parameter computed by

$$f_w = \frac{1}{1 + \left(\dfrac{\theta}{\theta_0}\right)^{-q}} \tag{4.32}$$

ranging from 0 in dry soil to 1 in saturated soil. The parameters θ_0 and q are specific soil properties that determine when water content begins to affect thermal conductivity and the rate of the transition from air- to water-dominated conductivity. The parameters q and θ_0 were found to be highly correlated to clay content and the following regression equations were presented:

$$q = 7.25m_y + 2.52 \tag{4.33}$$

$$\theta_0 = 0.33m_y + 0.078 \tag{4.34}$$

where m_y is the fractional clay content [0–1]. Using eqn (4.31), the weighting factors are then computed:

$$k_g = \frac{1}{3}\left[\frac{2}{1 + \left(\dfrac{\lambda_g}{\lambda_f} - 1\right)g_a} + \frac{1}{1 + \left(\dfrac{\lambda_g}{\lambda_f} - 1\right)g_c}\right] \tag{4.35}$$

$$k_s = \frac{1}{3}\left[\frac{2}{1 + \left(\dfrac{\lambda_s}{\lambda_f} - 1\right)g_a} + \frac{1}{1 + \left(\dfrac{\lambda_s}{\lambda_f} - 1\right)g_c}\right] \tag{4.36}$$

$$k_w = \frac{1}{3} \left[\frac{2}{1 + \left(\frac{\lambda_w}{\lambda_f} - 1 \right) g_a} + \frac{1}{1 + \left(\frac{\lambda_w}{\lambda_f} - 1 \right) g_c} \right] \tag{4.37}$$

where g_a and g_c are shape factors. The value of g_a for mineral soils is 0.088 and $g_c = 1 - 2g_a$. Therefore it is sufficient to obtain only one shape factor.

The thermal conductivity of the gas phase, λ_g, is given by

$$\lambda_g = \lambda_a + \frac{L_v \Delta f_w \hat{\rho}_a D_v}{P - e_a} \tag{4.38}$$

where λ_a is the thermal conductivity of air (reported in Table 4.1), L_v is the latent heat of vaporization, Δ is the slope of the saturation vapour pressure function, D_v is the vapour diffusivity for soil, $\hat{\rho}_a$ is the molar density of air, P is the atmospheric pressure and e_a is the actual vapour pressure. The second term in eqn (4.38) is the latent heat term and is responsible for almost all the temperature dependence of soil thermal conductivity.

Below we present the program to estimate thermal conductivity for the theory described above. The project is called PSP_thermalConductivity and it is made of three files:

1. main.py
2. PSP_thermalCond.py
3. PSP_readDataFile.py

The file PSP_readDataFile.py was already described in Appendix A and is used to read a row of soil temperatures. The file main.py contains instructions to read temperature data from the file soilTemperature.txt; the user is then asked to input a bulk density and a clay content value. Then incremental values of water content are created in the first **for** loop, while in the second **for** loop, the function **thermalConductivity** is called. The last instructions are written to plot the results.

```
#main.py
from __future__ import print_function, division
import matplotlib.pyplot as plt
import PSP_thermalCond as soil
from PSP_readDataFile import *

def main():
    A, isFileOk = readDataFile("soilTemperature.txt", 0, ',', False)
    if (isFileOk == False):
        print ("Incorrect format")
        return()
```

```
        soilTemperature = A[0]
        print ("Temperatures = ", soilTemperature)
        nrTemperatures = len(soilTemperature)

        myStr = "bulk density [kg/m^3]: "
        bulkDensity = float(input(myStr))
        myStr = "clay [0 - 1]: "
        clay = float(input(myStr))

        particleDensity = 2650
        porosity = 1 - (bulkDensity / particleDensity)

        step = 0.02
        nrValues = int(porosity / step) + 1
        waterContent = np.zeros(nrValues)
        thermalConductivity = np.zeros(nrValues)
        for i in range(nrValues):
            waterContent[i] = step*i

        fig = plt.figure(figsize=(10,8))
        plt.xlabel('Water Content [m$^{3}$ m$^{-3}$]',
                   fontsize=20,labelpad=8)
        plt.ylabel('Thermal Conductivity [W m$^{-1}$C$^{-1}$]',
                   fontsize=20,labelpad=8)
        plt.tick_params(axis='both', which='major', labelsize=20,pad=8)
        plt.tick_params(axis='both', which='minor', labelsize=20,pad=8)
        #plt.xlim(0, 0.7)
        for t in range(nrTemperatures):
            for i in range(nrValues):
                thermalConductivity[i] = soil.thermalConductivity
                        (bulkDensity, waterContent[i], clay,
                        soilTemperature[t])
            if (t == 0): plt.plot(waterContent, thermalConductivity,'k')
            if (t == 1): plt.plot(waterContent, thermalConductivity,'--k')
            if (t == 2): plt.plot(waterContent, thermalConductivity,'-.k')
            if (t == 3): plt.plot(waterContent, thermalConductivity,':k')
        plt.show()
main()
```

The file PSP_thermalCond.py contains the program with the **thermalConductivity** function. It takes as arguments bulk density, water content, clay content and temperature. Therefore, from knowledge of four basic soil properties, it is possible to obtain an estimate of thermal conductivity. The function **heatCapacity**() computes the soil heat capacity.

```python
#PSP_thermalCond.py
from __future__ import division
from math import exp
import numpy as np

GEOMETRIC = 0
LOGARITHMIC = 1
bulkDensity = 1300

def kMean(meanType, k1, k2):
    if (meanType == GEOMETRIC):
        k = np.sqrt(k1 * k2)
    elif (meanType == LOGARITHMIC):
        if (k1 == k2):
            k = k1
        else:
            k = (k1-k2) / np.log(k1/k2)
    return k

def thermalConductivity(bulkDensity, waterContent, clay, temperature):
    ga = 0.088
    thermalConductivitysolid = 2.5
    atmPressure = 100

    q = 7.25 * clay + 2.52
    xwo = 0.33 * clay + 0.078
    solidContent = bulkDensity / 2650
    porosity = 1 - solidContent
    gasPorosity = max(porosity - waterContent, 0.0)

    temperatureK = temperature + 273.16
    Lv = 45144 - 48 * temperature
    svp = 0.611 * exp(17.502 * temperature / (temperature + 240.97))
    slope = 17.502 * 240.97 * svp / (240.97 + temperature)**2.0
    Dv = 0.0000212 * (101.3 / atmPressure) * (temperatureK / 273.16)
                **1.75
    rhoair = 44.65 * (atmPressure / 101.3) * (273.16 / temperatureK)
    stcor = max(1 - svp / atmPressure, 0.3)

    thermalConductivitywater = 0.56 + 0.0018 * temperature
    if waterContent < 0.01 * xwo :
        wf = 0
    else:
        #empirical weighting function D[0,1]
        wf = 1 / (1 + (waterContent / xwo)**(-q))

    thermalConductivitygas = (0.0242 + 0.00007 * temperature +
                            wf * Lv * rhoair * Dv * slope /
                            (atmPressure * stcor))
```

```
    gc = 1 - 2 * ga
    thermalConductivityfluid = (thermalConductivitygas +
    (thermalConductivitywater - thermalConductivitygas) *
    (waterContent / porosity)**2.0)
    ka = (2 / (1 + (thermalConductivitygas / thermalConductivityfluid
                - 1) * ga) + 1 /
        (1 + (thermalConductivitygas / thermalConductivityfluid
              - 1) * gc)) / 3
    kw = (2 / (1 + (thermalConductivitywater /
                    thermalConductivityfluid - 1) * ga) +
          1 / (1 + (thermalConductivitywater /
                    thermalConductivityfluid - 1) * gc)) / 3
    ks = (2 / (1 + (thermalConductivitysolid /
                    thermalConductivityfluid - 1) * ga) +
          1 / (1 + (thermalConductivitysolid /
                    thermalConductivityfluid - 1) * gc)) / 3

    thermalConductivity = ((kw * thermalConductivitywater *
                    waterContent + ka *  thermalConductivitygas *
                    gasPorosity + ks * thermalConductivitysolid *
                    solidContent) / (kw * waterContent + ka *
                    gasPorosity + ks * solidContent))
    return(thermalConductivity)

def heatCapacity(bulkDensity, waterContent):
    return (2.4e6 * bulkDensity / 2650 + 4.18e6 * waterContent)
```

Figure 4.5 shows the dependence of thermal conductivity on water content for four different temperatures, generated using a bulk density of 1300 kg m^{-3} and a clay content of 0.3. Note that the temperature dependence is mainly from the latent heat transport, which decreases as the soil dries. At about 65 °C, the apparent thermal conductivity from latent heat transport is equal to the thermal conductivity of water, so increasing water content does not affect the soil conductivity, at water content above 0.3 m^3 m^{-3}.

4.5 Numerical Implementation

An example program for simulating heat flow in soil is given. The program implements the two numerical solutions described above, to introduce the main numerical schemes used in the following chapters.

The project is called PSP_heat and is made of five files:

1. main.py
2. PSP_heat.py
3. PSP_heatSoil.py
4. PSP_ThomasAlgorithm.py
5. PSP_grid.py

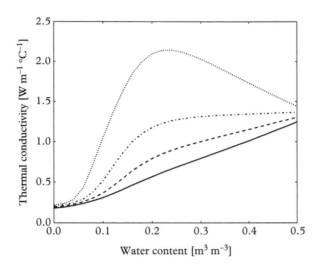

Figure 4.5 *Soil thermal conductivity as function of water content and temperature. Specifically, the different lines represents incremental temperature values at 0, 30, 60 and 90 °C. The lower solid line is at 0 °C and the higher dotted line is at 90 °C.*

The modules PSP_ThomasAlgorithm.py and PSP_grid.py were described in Chapter 3. The module PSP_heatSoil.py contains the computation of thermal conductivity as described already. Depth is input in metres, water content in $m^3\ m^{-3}$ and temperature in °C. To run the program, the user should run the file main.py.

When the user runs the program, a prompt command is displayed to select the numerical scheme:

```
1 Finite Difference
2 Cell-Centered Finite Volume
Select solver:
```

Then the prompt asks the user to select the input values for the soil water content, the average air temperature and the amplitude of the oscillations around the average temperature. This choice is needed to set up the upper boundary conditions. Indeed, in this example, the upper boundary condition is a periodic boundary condition given by the oscillations in air temperature, as described in Section 4.3.2. For simplicity, the boundary-layer conductance is assumed to be known.

```
water content (m^3/m^3): 0.2
mean temperature [C]: 20
amplitude of change in temperature [C]: 4
```

If the user selects the `Finite Difference` solution, the prompt asks him or her to select the weighting factor for the discretization scheme. This option allows the user to investigate the effects of using explicit, implicit, Crank–Nicholson or other intermediate values.

```
weighting factor for time discretization:(0:explicit,1:implicit
                                       Euler)=0.5
```

Then the user is asked to input the total number of simulation hours:

```
nr of simulation hours: 48
```

Below is the file `main.py`. The first lines are written to import the modules employed in the program and the files used for the numerical solution `PSP_heat.py` and to generate the grid. Then the program prompts the selections described above. Variables are defined and the plot instructions are listed. The variable omega is the angular frequency used for simulation of air temperature, which is simulated by a sine function within the time loop, `while (time < endTime)`.

```python
#main.py
from __future__ import print_function, division
import matplotlib.pyplot as plt
from PSP_heat import *
from PSP_grid import *

def main():
    global z
    print (FIN_DIFF, 'Finite Difference')
    print (CELL_CENT_FIN_VOL, 'Cell-Centered Finite Volume')

    solver = int(input("Select solver: "))
    myStr = "water content (m^3/m^3): "
    thetaIni = float(input(myStr))
    myStr = "mean temperature [C]: "
    meanT = float(input(myStr))
    myStr = "amplitude of change in temperature [C]: "
    ampT = float(input(myStr))
    omega = 2.0 * np.pi/(24 * 3600.0)
    airT0 = meanT
    timeShift = 8

    if (solver == FIN_DIFF):
        myStr = "weighting factor for time discretization:"
        myStr += " (0: explicit, 1: implicit Euler) = "
        factor = float(input(myStr))
```

```
z = initialize(airT0, thetaIni, solver)
simulationLenght = int(input("nr of simulation hours: "))

endTime = simulationLenght * 3600.0
timeStepMax = 3600.0
dt = timeStepMax / 8.0
time = 0.0
sumHeatFlux = 0
totalIterationNr = 0

f, plot = plt.subplots(3, figsize=(8,8), dpi=80)
plt.subplots_adjust(hspace = 0.3)
plot[1].set_xlabel("Time [h]",fontsize=14,labelpad=2)
plot[1].set_ylabel("Temperature [C]",fontsize=14,labelpad=2)
plot[2].set_xlabel("Time [h]",fontsize=14,labelpad=2)
plot[2].set_ylabel("Heat flux [W m$^{-2}$]",fontsize=14,labelpad=2)
plot[1].set_xlim(timeShift, simulationLenght+timeShift)
plot[1].set_ylim(meanT-ampT, meanT+ampT)
plot[2].set_xlim(timeShift, simulationLenght+timeShift)

while (time < endTime):
    dt = min(dt, endTime - time)
    airT = airT0 + ampT * np.sin((time+dt)*omega)
    if (solver == FIN_DIFF):
        success, nrIterations, heatFlux = (
            finiteDifference(airT, meanT, dt, factor))
    elif (solver == CELL_CENT_FIN_VOL):
        success, nrIterations, heatFlux = (
            cellCentFiniteVol(airT, meanT, dt))
    totalIterationNr += nrIterations

    if success:
        #Convergence achieved
        for i in range(n+1):
            oldT[i] = T[i]
        sumHeatFlux += heatFlux * dt
        time += dt

        t = time/3600. + timeShift

        plot[0].clear()
        plot[0].set_xlabel("Temperature [C]",fontsize=14,labelpad=2)
        plot[0].set_ylabel("Depth [m]",fontsize=14,labelpad=2)
        plot[0].set_xlim(meanT-ampT, meanT+ampT)
        plot[0].plot(T[1:len(T)], -z[1:len(T)], 'k')
        plot[0].plot(T[1:len(T)], -z[1:len(T)], 'ko')
```

```
            plot[1].plot(t, T[getLayerIndex(z, 0.0)], 'ko')
            plot[1].plot(t, T[getLayerIndex(z, 0.15)], 'ks')
            plot[1].plot(t, T[getLayerIndex(z, 0.4)], 'k^')
            plot[2].plot(t, heatFlux, 'ko')
            plt.pause(0.0001)
            #increment time step when system is converging
            if (float(nrIterations/maxNrIterations) < 0.25):
                    dt = min(dt*2, timeStepMax)
            #print(heat.T[getLayerIndex(heat.z, 0.15)])
        else:
            #No convergence
            dt = max(dt / 2, 1)
            for i in range(n+1): T[i] = oldT[i]
            print ("dt =", dt, "No convergence")

    print("nr of iterations per hour:", totalIterationNr /
            simulationLenght)
    plt.ioff()
    plt.show()
main()
```

The two numerical schemes are written in the file PSP_heat.py shown below and they are implemented in the functions **finiteDifference()** and **cellCentFiniteVol()**. The following list of variables are the elements of the matrix initialized as numpy arrays, temperatures and thermal properties.

```
#PSP_heat
from __future__ import print_function, division
import PSP_grid as grid
from PSP_ThomasAlgorithm import *
import numpy as np
from PSP_heatSoil import *

area = 1
maxNrIterations = 100
tolerance = 1.e-2

n = 20
z = np.zeros(n+2, float)
zCenter = np.zeros(n+2, float)
dz = np.zeros(n+2, float)
vol = np.zeros(n+2, float)
wc = np.zeros(n+2, float)
a = np.zeros(n+2, float)
b = np.zeros(n+2, float)
c = np.zeros(n+2, float)
d = np.zeros(n+2, float)
```

```python
T = np.zeros(n+2, float)
dT = np.zeros(n+2, float)
oldT = np.zeros(n+2, float)
C_T = np.zeros(n+2, float)
lambda_ = np.zeros(n+2, float)
k_mean = np.zeros(n+2, float)
f = np.zeros(n+2, float)

def initialize(T_0, thetaIni, solver):
    global z, dz, zCenter, vol, wc, T, oldT
    # vector depth [m]
    z = grid.geometric(n, 1.0)

    vol[0] = 0
    for i in range(n+1):
        dz[i] = z[i+1]-z[i]
        if (i > 0): vol[i] = area * dz[i]
    for i in range(n+2):
        zCenter[i] = z[i] + dz[i]*0.5

    if (solver == CELL_CENT_FIN_VOL):
        for i in range(n+1):
            dz[i] = zCenter[i+1]-zCenter[i]

    for i in range(1, n+2):
        T[i] = T_0
        oldT[i] = T_0
        wc[i] = thetaIni
    return z

def finiteDifference(airT, boundaryT, dt, factor):
    g = 1.0 - factor
    energyBalance = 1.
    for i in range(1, n+2):
        T[i] = oldT[i]
    nrIterations = 0
    while ((energyBalance > tolerance) and (nrIterations <
                                    maxNrIterations)):
        for i in range(1, n+2):
            lambda_[i] = thermalConductivity(bulkDensity, wc[i], clay,
                                    T[i])
            C_T[i] = heatCapacity(bulkDensity, wc[i])*vol[i]
        f[0] = 0.
        for i in range(1, n+1):
            f[i]=area* lambda_[i] / dz[i]
        for i in range(1, n+1):
            if (i == 1):
                a[i] = 0.
```

```
                b[i] = 1.
                c[i] = 0.
                d[i] = airT
            elif (i < n):
                a[i] = -f[i-1]*factor
                b[i] = C_T[i]/dt + f[i-1]*factor + f[i]*factor
                c[i] = -f[i]*factor
                d[i] = C_T[i]/dt * oldT[i] +(1.-factor)*(f[i-1]
                                      *oldT[i-1]+f[i]
                                      *oldT[i+1]-(f[i-1]+f[i])
                                      *oldT[i])
            elif (i == n):
                a[n] = 0.
                b[n] = 1.
                c[n] = 0.
                d[n] = boundaryT
        ThomasBoundaryCondition(a, b, c, d, T, 1, n)
        dSum = 0
        for i in range(2, n):
            dSum += C_T[i]*(T[i]-oldT[i])
        energyBalance = (abs(dSum - factor*dt*(f[1]*(T[1]-T[2])
                    - f[n-1]*(T[n-1]-boundaryT))
                    - g*dt*(f[1]*(oldT[1]-oldT[2])
                    - f[n-1]*(oldT[n-1]-boundaryT))))
        nrIterations += 1

    if (energyBalance < tolerance):
        flux = f[1]*(T[1]-T[2])
        return True, nrIterations, flux
    else:
        return False, nrIterations, 0

def cellCentFiniteVol(airT, boundaryT, dt):
    energyBalance = 1.
    for i in range(1, n+2):
        T[i] = oldT[i]
    nrIterations = 0
    while ((energyBalance > tolerance) and
           (nrIterations < maxNrIterations)):
        for i in range(1, n+2):
            lambda_[i] = thermalConductivity(bulkDensity, wc[i],
                                    clay, T[i])
            C_T[i] = heatCapacity(bulkDensity, wc[i])*vol[i]
        f[0] = 0.
        for i in range(1, n+1):
            f[i] = area * kMean(LOGARITHMIC, lambda_[i],
                            lambda_[i+1]) / dz[i]
        for i in range(1, n+1):
```

```
        if (i == 1):
            a[i] = 0.
            b[i] = 1.
            c[i] = 0.
            d[i] = airT
        elif (i < n):
            a[i] = -f[i-1]
            b[i] = C_T[i]/dt + f[i-1] + f[i]
            c[i] = -f[i]
            d[i] = C_T[i]/dt * oldT[i]
        elif (i == n):
            a[n] = 0.
            b[n] = 1.
            c[n] = 0.
            d[n] = boundaryT
    ThomasBoundaryCondition(a, b, c, d, T, 1, n)
    dSum = 0
    for i in range(2, n):
        dSum += C_T[i]*(T[i]-oldT[i])
    energyBalance = (abs(dSum - f[1]*(T[1]-T[2])*dt
                    + f[n-1]*(T[n-1]-boundaryT)*dt))
    nrIterations += 1

if (energyBalance < tolerance):
    flux = f[1]*(T[1]-T[2])
    return True, nrIterations, flux
else:
    return False, nrIterations, 0
```

The heat storage terms are computed in the instruction `C_T[i] = soil.heatCapacity()` and the element conductances in `lambda_[i] = soil.thermalConductivity()`. The function **ThomasBoundaryCondition()** is the Thomas algorithm for setting up and solving the system of equations for the new temperatures. The variable `factor` passed to the function **finiteDifference()** is used to choose the numerical procedure that results from this choice (explicit Euler, implicit, Crank–Nicolson) as described above. The first node ($z(1)$) is at the soil surface ($z = 0$). Since temperature changes rapidly near the surface and very little at depth, the best simulation will be obtained with short elements near the soil surface and longer ones deeper in the soil. The element lengths should go in a geometric progression. The user can select between a linear and a geometric grid. Ten to twelve nodes are probably sufficient for short-term simulations (daily or weekly). Fifteen nodes would probably be sufficient for annual cycle simulation, where a deeper grid is needed.

Figure 4.6 shows the output of the simulation. The user can observe the evolution of the simulation at run time. This feature is performed by the interactive property of the pyplot interface implemented in `matplotlib`, called `plt.ion()`. Figure 4.6(a) shows the temperature variation with depth, 4.6(b) the variation of temperature at three depths with time and 4.6(c) the heat flux at the upper boundary. This figure was created by

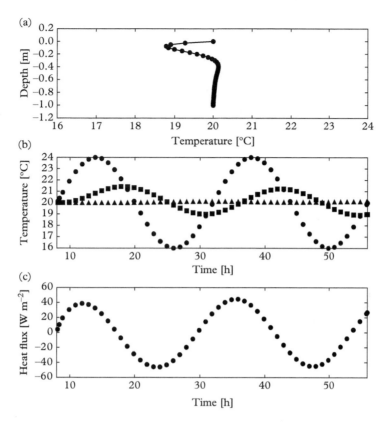

Figure 4.6 *Simulated soil temperatures. (a) Temperature as function of depth at 1.00 a.m. (b) Temperature as function of time for three depths, 0, 0.15 and 0.5 m, corresponding to the circles, squares and triangles. (c) Heat flux at the upper boundary.*

simulating heat flow for an average air temperature of 20 °C, with an amplitude of 4°C, for a period of 48 hours. Note that at 0.5 m, the oscillations have almost disappeared, indicating that the damping depth for this soil is $A(z)/A(0) \approx 0.37$. The model reproduces the amplitude decrease with depth and the time shift for the maximum and minimum temperatures with depth. The program outputs the nr of iterations per hour as an additional parameter to evaluate the algorithm's performance.

....

4.6 EXERCISES

4.1. Use the function **thermalConductivity** to investigate the temperature dependence of thermal conductivity. At what temperature is the thermal conductivity of wet soil independent of water content? Explain why the temperature dependence of conductivity disappears at this temperature.

4.2. What is the possible range of volumetric specific heat in mineral soils?

4.3. Use the program to predict soil temperature as a function of time and depth. What effect would a surface residue mulch have on soil temperature? At what depth are temperature changes 180 degrees out of phase with the surface temperature?

4.4. Modify the program `PSP_heat.py` so that it allows water content to change with depth. Then compare simulations for uniform, dry soil with simulations of a moist soil with a dry dust mulch on the surface.

4.5. Modify the program `main.py` to print on screen soil temperatures at 0.15 and 0.3 m depths.

4.6. Modify the values of clay content and bulk density in the program `PSP_soil.py` and investigate the effects of these variables on soil temperature. Compare simulations for a sand, a silt loam and clay soil and for a compacted soil ($\rho_b = 1500 \, \text{kg m}^{-3}$) and a recently tilled soil ($\rho_b = 900 \, \text{kg m}^{-3}$).

4.7. Modify the program `PSP_heat.py` and investigate the effects of explicit Euler, implicit Euler and Crank–Nicolson numerical solutions.

5

Soil Liquid Phase and Soil–Water Interactions

5.1 Properties of Water

5.1.1 Thermal Properties

The specific heat capacity of water (often shortened to specific heat) is a measure of the energy required to increase the temperature of a unit quantity of water by one degree. The specific heat capacity of water is $4183\,\mathrm{J\,kg^{-1}\,K^{-1}}$ at 20 °C. Water has a higher specific heat capacity than most liquids (second only to ammonia), owing to the extensive hydrogen bonding between its molecules. These properties, along with a high latent heat of vaporization, make water a key molecule for thermal regulation and heat transfer.

The latent heat of vaporization L_v is the energy required to evaporate a unit of water and is $2453\,\mathrm{kJ\,kg^{-1}}$ at 20 °C. The latent heat of vaporization is temperature-dependent. Harrison (1963) gives the following equation:

$$L_v = 2.501 - 2.361 \times 10^{-3}\,T \qquad (5.1)$$

where L_v is in MJ kg^{-1} and T is in °C. This equation shows that at 0 °C, $L_v = 2.501\,\mathrm{MJ\,kg^{-1}}$, with decreasing values at increasing temperatures.

The latent heat of fusion is the energy required to melt a unit of ice. At the normal freezing point for pure water (0 °C), the heat of fusion is $333.68\,\mathrm{kJ\,kg^{-1}}$. In both cases, the change is endothermic, with absorption of energy going from solid to liquid to gas. The energy is required to overcome the molecular forces of attraction between water particles; therefore the process of transition from liquid water to vapour requires input of energy, causing a decrease in temperature in its surroundings. This property of water is the basis for cooling down the bodies of living organisms and it is also called *evaporative cooling*. In plants, when liquid water is converted into water vapour, heat is removed from the plant surface in the process of transpiration. Opening and closing of stomates in plant leaves changes evaporation rates and thus affects leaf temperature.

Soil Physics with Python. First Edition. Marco Bittelli, Gaylon S. Campbell and Fausto Tomei.
© Marco Bittelli, Gaylon S. Campbell and Fausto Tomei 2015. Published in 2015 by Oxford University Press.

5.1.2 Surface Tension

Molecules of a fluid are exposed to different molecular attractions depending on their spatial positions with respect to each other. For instance, if the molecules are within bulk water and far away from any interface, they are subject to the hydrogen bonding of the adjacent molecules, which acts in all directions, resulting in a net force of zero. On the other hand, the molecules at or close to an interface (e.g. water–air) are subject to an inward attraction from the liquid because the density of water molecules is higher in the liquid phase than in the vapour phase. The inward attraction is balanced by the water's limit to compressibility, and this unequal energy distribution causes the liquid to behave as an elastic sheet, with the liquid squeezing itself to reach the locally lowest surface area possible. This energy is called surface tension γ and it is measured in force per unit length [N m^{-1}] or energy per unit area [J m^{-2}]. Surface tension depends on temperature. Increasing the temperature decreases the surface tension until it reaches a value of 0 at the critical temperature, a temperature at which distinct liquid and gas phases do not exist. A general formula for surface tension of liquids as function of temperature is (Guggenheim, 1945)

$$\gamma = \gamma_0 \left(1 - \frac{T}{T_C}\right)^n \tag{5.2}$$

where γ_0 is an empirical constant, T is temperature in °C, T_C is the critical temperature and n is an empirical parameter.

5.1.3 Contact Angle

If the liquid phase is present in a system characterized by three phases (solid, liquid and gas), the angle between the liquid–solid interface and the liquid–gas interface is called the contact angle. Depending on the wetting properties of the fluid and the wettability properties of the surface, the contact angle varies. Figure 5.1 shows an example where a water droplet wets a surface (a), or is repelled by the surface (b). The contact angle is measured from the solid surface through the liquid as shown in the figure. Commonly, for hydrophilic surfaces, $\beta < 90°$, while for hydrophobic surfaces, $\beta > 90°$. Surface tension and wetting angles will be seen later when we discuss capillary forces and water in soil.

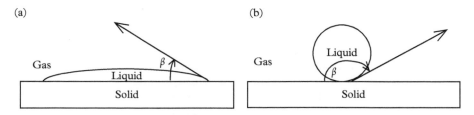

Figure 5.1 *Example of water wetting the surface (a) and repelling the surface (b).*

5.1.4 Electromagnetic Properties

In the last few decades, the electromagnetic properties of water and soils have been studied extensively. The growing application of electromagnetic measurements was fostered by the fact that soil water content can be detected by measuring bulk soil electromagnetic properties.

Before discussing the interactions between an electromagnetic wave and a material, we describe some basic concepts of dielectric theory. A material is called a dielectric if it stores energy when it is subject to an external electric field. The relative dielectric permittivity (here shortened as dielectric permittivity) ϵ_r is described as the ratio of the permittivity of the material ϵ [F m^{-1}] and the permittivity of free space ϵ_0 (8.85×10^{-12} F m^{-1}):

$$\epsilon_r = \frac{\epsilon}{\epsilon_0} \tag{5.3}$$

The dielectric permittivity is mathematically described by a complex number, with real part ϵ_r' and imaginary part ϵ_r''. The real part describes the energy stored in the dielectric (at a given frequency and temperature), while the imaginary part describes the dielectric losses or the energy dissipation.

The relative magnetic permeability (here shortened as magnetic permeability) is the measure of the ability of a material to allow the formation of a magnetic field within the material itself. Therefore it expresses the degree of magnetization that a material can assume in response to a magnetic field. The magnetic permeability is the ratio of the magnetic permeability of a specific medium μ [H m^{-1}] to the magnetic permeability of free space μ_0 (1.26×10^{-6} H m^{-1}):

$$\mu_r = \frac{\mu}{\mu_0} \tag{5.4}$$

The magnetic permeability is also mathematically described by a complex number, with real part μ_r' and imaginary part μ_r''.

The measurement of dielectric properties of materials is a very powerful technique to infer information about the physical state of the material under investigation. A number of books are available on this subject (Ramo *et al.*, 1994; Raju, 2003). Among these methods, electromagnetic properties of wet porous media are measured to obtain the state of water and its content. A common approach is to employ travel time analysis of a generated electromagnetic wave, and a large body of literature is available on the subject (Hasted, 1973; Santamarina *et al.*, 2001).

5.1.5 Measuring Soil Water Content with Time-Domain Reflectometry

One of the most common techniques for measuring soil water content is based on measurement of soil dielectric permittivity and is called time-domain reflectometry (TDR).

The velocity v [m s^{-1}] of an electromagnetic wave is related to the dielectric permittivity ϵ_r and the magnetic permeability μ_r by

$$v = \frac{c}{\sqrt{\mu_r \epsilon_r}} \tag{5.5}$$

where c is the speed of light, 2.997×10^8 m s^{-1}. From a mechanical standpoint, the velocity v of an electromagnetic wave travelling through a probe of length L [m], is given by

$$v = \frac{2L}{t} \tag{5.6}$$

where t is time [s]. For the TDR-measurement, the number 2 in front of the probe length is included because the wave is reflected at the end of the probe. For most soils $\mu_r = 1$ (Roth *et al.*, 1990); therefore eqn (5.5) can be written as

$$v = \frac{c}{\sqrt{\epsilon_r}} \tag{5.7}$$

By equating the definitions of velocity,

$$\frac{c}{\sqrt{\epsilon_r}} = \frac{2L}{t} \tag{5.8}$$

and solving for ϵ_r,

$$\epsilon_r = \left(\frac{ct}{2L}\right)^2 \tag{5.9}$$

Equation (5.9) allows the relative dielectric permittivity to be obtained from measurements of the travel time t, since the length of the probe, L and the speed of light c are known.

Among the different techniques available for measuring travel time, TDR has become established for soil water content measurement. TDR exploits the difference in dielectric permittivity values between the solid, gas and liquid phases. At TDR frequencies, pure liquid water has a dielectric permittivity of about 80 (depending on temperature and electrolyte concentration), air has a dielectric permittivity of about 1 and the solid phase a dielectric permittivity of about 4 (Hallikainen *et al.*, 1985; Bittelli *et al.*, 2004). This contrast makes the dielectric permittivity of soil very sensitive to variations in soil water content. Table 5.1 shows the dielectric permittivity of different materials at a frequency of 1.5 GHz. The dielectric permittivity of water is temperature-dependent. Table 5.1 reports a linear equation, with variations ranging from 87.17 at 1 °C to 76.74 at 30 °C.

When a TDR probe is inserted into the soil, the travel time is measured to obtain the dielectric permittivity of the bulk soil (hereinafter called ϵ_b) by employing eqn (5.9). After measuring the travel time and deriving the dielectric permittivity, the latter is used to obtain the volumetric soil water content.

Table 5.1 *Dielectric permittivity of materials at 1.5 GHz*

Material	Dielectric permittivity
Vacuum	1
Air	1.0005
Fresh water	$78.54 \times [1 - 4.579 \times 10^{-3}(T - 25)]$
Fresh water ice	3.2
Quartz	4–6
Granite	5

A variety of equations have been proposed to compute the water content from know-ledge of the soil bulk dielectric permittivity (Topp *et al.*, 1980; Ledieu *et al.*, 1986; Roth *et al.*, 1990; Malicki *et al.*, 1996). Ledieu *et al.* (1986) proposed

$$\theta = 0.1138\sqrt{\epsilon_b} - 0.1758 \tag{5.10}$$

where ϵ_b is the measured bulk dielectric permittivity. Malicki *et al.* (1996) included bulk density in the following equation for water content:

$$\theta = \frac{\sqrt{\epsilon_b} - 0.819 - 0.168\rho_b - 0.159\rho_b^2}{7.17 + 1.18\rho_b} \tag{5.11}$$

where ρ_b is the bulk density [g cm^{-3}].

A different approach was proposed by Roth *et al.* (1990), by using a dielectric mixing model. This model is based on the same idea of the model used for thermal conductivity presented in Chapter 4, where the soil thermal conductivity is given by a weighted sum of the thermal conductivities of its components. Indeed, the de Vries (1963) model for thermal conductivity was developed by considering a dielectric model for a mixture of granules. The dielectric mixing model therefore computes the bulk dielectric permittivity as a weighted sum of the dielectric permittivity of each soil constituent:

$$\epsilon_b = (\phi_s\epsilon_s^\alpha + \theta\epsilon_l^\alpha + \phi_g\epsilon_g^\alpha)^{1/\alpha} \tag{5.12}$$

where ϕ_s, θ and ϕ_g are the solid-, liquid- and gas-phase volumetric fractions. The corresponding dielectric permittivities are ϵ_s, ϵ_l and ϵ_g, while α is a geometrical parameter related to the geometrical orientation of soil particles with respect to the electromagnetic field. A default value of 0.5 was used as suggested by the authors. The volumetric solid fraction can be also written as $\phi_s = 1 - \phi_f$, where ϕ_f is the porosity and the volumetric fraction of the gas phase is $\phi_g = \phi_f - \theta$. Since the model is used to quantify water content, eqn (5.12) is solved for water content:

$$\theta = \frac{\epsilon_b^\alpha - \left[(1 - \phi_f)\epsilon_s^\alpha + \phi_f\epsilon_g^\alpha\right]}{\epsilon_l^\alpha - \epsilon_g^\alpha} \tag{5.13}$$

The dielectric mixing model of eqn (5.13) can be written as a function of bulk density and the square root of the bulk dielectric permittivity:

$$\theta = \frac{\sqrt{\epsilon_b} - \left(\frac{\rho_b}{\rho_s}\sqrt{\epsilon_s}\right) - \left(1 - \frac{\rho_b}{\rho_s}\right)\sqrt{\epsilon_g}}{\sqrt{\epsilon_l} - \sqrt{\epsilon_g}} \tag{5.14}$$

To employ the dielectric mixing model, knowledge of the bulk density and dielectric permittivity of the solid phase is needed, as well as the dielectric permittivity of the liquid phase. Table 5.1 provides the dielectric permittivity for different materials.

Changes in dielectric properties of wet porous materials are complex, involving a variety of relaxation phenomena that depend on mineralogical properties, surface properties, geometrical arrangement and shape of soil particles, water content, solute concentration, frequency, and temperature. Therefore a dielectric measurement is influenced by all soil constituents and not just water. Empirical equations based only on bulk dielectric measurement (like the Topp *et al.* (1980) or Malicki *et al.* (1996) models) can therefore never be accurate for all soils. For a more detailed and extensive discussion of these phenomena, the reader should see Santamarina *et al.* (2001) and Bittelli *et al.* (2008*b*).

5.1.6 Travel Time Analysis and Soil Water Content Measurement

The computation of the travel time from a TDR waveform is usually obtained using computer programs. The following is a description of a Python code written to analyse a TDR waveform and obtain soil water content.

The reflected distance of the electromagnetic wave travelling from the TDR device, through cables to the head of the probe and back again, is fairly constant, and this is the first reflection of the waveform. This first reflection (peak) can be determined by calibration as will be described below. The second reflection is used to determine the travel time, back and forth, along the length of the probe from the first reflection. With knowledge of the travel time, it is possible to determine the relative dielectric permittivity. The graph in Fig. 5.2 shows an example of the first peak and of the end reflection.

The project for travel time analysis is called TDRPy and comprises five files:

1. `main.py`
2. `PSP_readDataFile.py`
3. `PSP_travelTime.py`

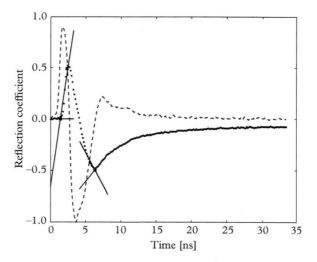

Figure 5.2 *Time domain reflectometry waveform. The peak and the reflection point are indicated by squares. The dots are the experimental waveform, the dashed line is the first derivative and the solid lines are the tangent lines.*

4. `PSP_TTwaterContent.py`

5. `PSP_TTplot.py`

The `main.py` file contains the user interface written using the module `tkinter`, and the calls to the functions contained in the files `PSP_travelTime.py` and `PSP_TTplot.py`. The functions used for travel time analysis are written in the file `PSP_travelTime.py` and will be described here. For brevity, the code written in the files `main.py` and `PSP_TTplot.py` is not shown, since it is made of a series of instructions used to generate the interface. The file `PSP_readDataFile.py`, used to read the experimental data, is described in Appendix A.

Figure 5.3 depicts the user interface, obtained by running the program written in the file `main.py`. The waveform is shown in Fig. 5.2 with time on the *x*-axis and reflection coefficient on the *y*-axis. The output from the tangent lines procedure to obtain the travel time is also shown. The dashed line is the first derivative, while the solid straight lines arc the tangent lines to the curve, used to identify the peak and the reflection point. The three points indicated by squares are the inflection points. Specifically, they are as follows, from left to right: the first point is the transition from the cable into the probe handle, the second point corresponds to the end of the probe handle (beginning of the metal rods) and the third point is the reflection at the end of the probe. For travel time calculation, the second and third points are used. The correction for the travel time in the epoxy handle is obtained by knowing the relative dielectric permittivity of the epoxy. Note that in the windows interface, there is a box for the permittivity of the handle

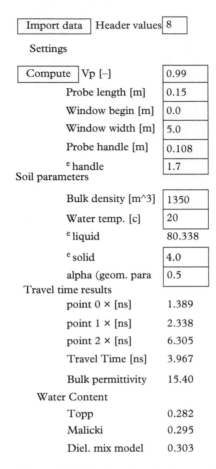

Import data	Header values	8

Settings

Compute	Vp [–]	0.99
	Probe length [m]	0.15
	Window begin [m]	0.0
	Window width [m]	5.0
	Probe handle [m]	0.108
	^{e} handle	1.7

Soil parameters

	Bulk density [m^3]	1350
	Water temp. [c]	20
	^{e} liquid	80.338
	^{e} solid	4.0
	alpha (geom. para	0.5

Travel time results

	point 0 × [ns]	1.389
	point 1 × [ns]	2.338
	point 2 × [ns]	6.305
	Travel Time [ns]	3.967
	Bulk permittivity	15.40

Water Content

	Topp	0.282
	Malicki	0.295
	Diel. mix model	0.303

Figure 5.3 *User interface for the travel time analysis program.*

(e_{handle}). If this information is lacking, the correction for the handle must be performed through calibrations with dielectrics of known permittivity.

The following is an explanation of the user interface. The first button (Import data), must be clicked to import the experimental data. A window pops-up and the file with the experimental data can be imported. In this example, a file dataTDRSoil.dat is provided. Depending on the commercial TDR, the output data can be organized either in columns or in rows. The file PSP_readDataFile.py was written to be able to import both formats. Usually, output files obtained from TDR analysis contain a number of header lines, specifying the settings of the data acquisition device (i.e. the datalogger). For instance, for the Campbell Scientific TDR100, the output file contains eight header lines with the following parameters: wave average, wave velocity, number of output points, cable length, window length, probe length, probe offset and multiplier. In this program, this information (needed for travel time analysis) is input by the user in the interface. Therefore these lines need not to be read by the program. In the window on the right side of the button `Import data`, the number of header lines must be specified, such that the program reads only the experimental data and skips the header lines. This number must be entered before clicking `Import data`.

The section `Settings` is used to specify the experimental settings such as the wave velocity, the probe length, the window beginning, the window width, the handle dimensions and the dielectric permittivity of the material used to build the probe handle (often specified by the manufacturer). This information is needed to account for the wave travel time in the probe handle, to be subtracted from the travel time calculation. Therefore, before clicking the `Compute` button, this information must be written into the boxes. The section `Soil Parameters` is written to input soil parameters needed to compute soil water content from knowledge of dielectric permittivity. These parameters will be described in more detail below.

The sections `Travel Time Results` and `Water Content` print the results of the computation. The output from travel time analysis comprises the points described above and, specifically, `Point 0 x` is the point where the wave enters the probe, `Point 1 x` is where the handle ends and the metal rods begin, and `Point 2 x` is the reflection point corresponding to the end of the metal rods. The names `Topp`, `Malicky` and the `dielectric mix model` are three different models used to compute volumetric water content from dielectric permittivity.

The computation of the relative dielectric permittivity is based on the measurement of the travel time t as described in eqn (5.9). The travel time is given by the difference between `Point 1 x` and `Point 2 x`. Therefore the core of a travel time analysis algorithm is the correct measurement of the *peak* and the *reflection point* to identify these points.

The first reflection is usually called a *spike* or a *peak* and is determined by the change in impedance from the cable to the probe. This point can be determined manually or by knowing the dielectric properties of the handle material. By looking at Fig. 5.2, the spike is identified as the abrupt change of slope of the signal at the beginning of the waveform. The algorithm identifies that point by the tangent lines method. However, this point corresponds to the change in impedance going from the cable into the TDR handle, and therefore it is not the point that should be used for travel time analysis. This point must be shifted to the right, by a given amount of time corresponding to the travel time over the probe handle, as indicated by the second dot from the left in the plot. To know how far to shift this point to the right, it is necessary to know the dielectric permittivity of the material used to build the probe handle. Commonly, this is an epoxy or plastic material of known dielectric permittivity. The second reflection is determined by a tangent fitting procedure as shown in Fig. 5.2, by computing the first derivative with a Savitzky–Golay algorithm (Press *et al.*, 1992). After obtaining the travel time and the dielectric permittivity, different equations are used to obtain the volumetric water content as described above.

The code of the file `PSP_travelTime.py` is shown below, since it contains the main variables and functions for travel time analysis. For readability, the code will be described in sections.

```
#PSP_travelTime
from __future__ import division
import math
import numpy as np

c = 299792458
NODATA = -9999
MAXDELTAINDEX = 6
SX = 0
DX = 1

class CLine:
    a = NODATA
    b = NODATA

class CPoint:
    x = NODATA
    y = NODATA

flatLine = line1 = line2 = line3 = CLine()
p0 = p1 = p2 = CPoint()
indexP0 = indexP2 = NODATA

timeVector = []
reflecCoeff = []
dy =[]

deltaSpace = 0
deltaTime = 0
```

The variable MAXDELTAINDEX is used to define the maximum number of points to calculate the tangent lines, while the variables SX and DX are used to define the direction. This is used to choose the weight to be assigned to different points when using a weighted linear regression as detailed in the following, depending if the line comes from the right (DX) side or from the left (SX) side of the inflection point. The class cLine is defined for the slope a and the intercept b. The straight lines used for travel time analysis will be obtained and plotted by using the instances of the class cLine. The class cPoint is written to define the points that will be computed with the travel time analysis. The variables deltaSpace and deltaTime are initialized. These variables will be used to define the incremental values of space and time read from the experimental data.

A variety of functions are written to compute the tangent lines. The function **indexofMaxVector**() finds the index of the vector x with maximum value, within the imported experimental values. On the other hand, the function **indexofMinVector**() finds the index of the vector x with minimum value. The average vector (**avg**()) is then computed, while the function **normalizeVector**() is used to compute a normalized vector, to express the reflection coefficient values between −1 and 1.

```
def indexOfMaxVector(y, first, last):
    myMax = max(y[first:last])
    for i in range(first, last):
        if (y[i] == myMax):
            return(i)

def indexOfMinVector(y, first, last):
    myMin = min(y[first:last])
    for i in range(first, last):
        if (y[i] == myMin):
            return(i)

def avg(y, index1, index2):
    if (index2 < index1):return(NODATA)
    first = max(index1, 0)
    last = min(index2+1, len(y))
    nrValues = last - first
    return sum(y[first:last]) / nrValues

def normalizeVector(y):
    y = (y-min(y))/(max(y) - min(y))
    avgFirstValues = avg(y, 1, 6)
    return (y - avgFirstValues)
```

The following functions are used to compute the waveform parameters. The variable Vp is the wave velocity and `probeHandle` is the physical length of the probe handle. The variables `windowBegin` and `windowWidth` define the beginning and the width of the TDR window used for the measurement; then `nrPoints` defines the number of points acquired during the experiment. To obtain the space increment between each point (`deltaSpace`), `windowWidth` is divided by `nrPoints-1`. Then, the time increment (`deltaTime`) is obtained by multiplying `deltaSpace` by two and then dividing it by the speed of light times the velocity parameter. The multiplication by two is performed because the wave travels back and forth (reflection measurement). The speed of light is multiplied by the velocity parameter (ranging from 0.6 to 0.99) that can be set in the TDR device. This parameter defines the wave velocity across the cable. Then a time vector is defined (`timeVector`), which is an incremental value of time given by the multiplication of the incremental time by the number of points.

A running average function **runningAverage**() is written to smooth the experimental data and thus allow for identification of the dominant reflection point, rather than minor oscillations due to scattered experimental data points. The function has two arguments y and `nrPoints`. The first is the variable containing the experimental data, while the second defines the number of points used to compute the running average.

An important part of this program relies on the computation of the derivative of the travel time data. A five-point derivative function **firstDerivative5Points**() is employed. Discussion of numerical derivatives and an example code are provided in Appendix B.1. Computation of the derivative allows for identification of slope changes and inflection points. Figure 5.2 shows the change in the derivative values (dashed line) and their correspondence to the reflection points of the experimental waveform.

```
def WF_parameters(Vp, probeHandle, windowBegin, windowWidth,
                                                    nrPoints):
    global deltaTime, deltaSpace, timeVector
    #abs. time [s] corresponding to the 1st point
    firstPointTime = 2. * windowBegin / (c*Vp)
    deltaSpace = windowWidth /(nrPoints - 1)
    deltaTime = 2. * deltaSpace /(c*Vp)
    timeVector = np.zeros(nrPoints, float)
    for i in range(nrPoints):
        timeVector[i] = firstPointTime + deltaTime * i

def runningAverage(y, nrPoints):
    smooth = np.zeros(len(y), float)
    for i in range(len(y)):
        smooth[i] = avg(y, i-nrPoints, i+nrPoints)
    return smooth / max(abs(smooth))

def firstDerivative5Points(y):
    dy = np.zeros(len(y), float)
    for i in range(2):
        dy[i] = 0.
    for i in range(2, len(y)-2):
        dy[i] = (1./(12.)) * (y[i-2] - 8.*y[i-1] + 8.*y[i+1] - y[i+1])
    for i in range(len(y)-2, len(y)):
        dy[i] = 0.
    return dy / max(abs(dy))
```

After obtaining the raw data, transforming the data points into travel time increments and computing the running average and a numerical derivative, the next step is the identification of the reflection points. This task is based on computation of regression lines that are fitted to the curve. To identify an accurate value corresponding to the reflection point, two weighted linear regression lines are fitted to the experimental data, corresponding to the change of slope indicated by the derivative. After the parameter of the line (slope and intercept) has been found, the intersection point is obtained.

The function used to compute the weighted linear regression is called **weightedLinearRegression()**. The equation of the slope for a weighted linear regressions is described by Kleinbaum *et al.* (2008). In this algorithm, the weight assigned to the points, differed depending on the direction. If the points were weighted towards the left (SX), and therefore from right to left, the weight assigned to the point was different with respect to the points that went towards the right (DX), and therefore from left to right. The flexibility of a weighted regression line is necessary because the slopes of the two branches may be substantially different (see the example in Fig. 5.4); therefore, depending on how many points are used, the intercept between the two regression lines may differ.

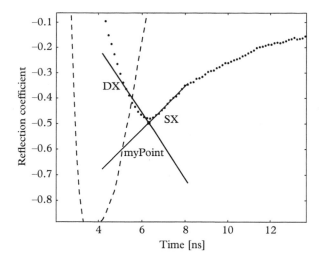

Figure 5.4 *Detail of the waveform and weighted linear regression for travel time analysis. The symbols DX and SX indicates the direction for weighting the regression points. Specifically DX indicates towards the right and SX towards the left.*

A function was written to identify the index (position) of the flat point, called **checkIndexFlatPoint**(), while another function was written to check the index closest to the intersection of the *x*-axis. This function is called **checkIndexZeroValue**() and is used to find the index when the derivative is zero or closest to zero.

Finally, the intersection point of the two regression lines was computed by the function **lineIntersection**(). This function returns the variable myPoint, which is the point used to compute the travel time. Figure 5.4 shows a detail of the waveform around the reflection point, indicating the two regression lines obtained from the function **weightedLinearRegression**() and the identification of the intersection point.

The functions that implement the equation described above are written in the following code:

```
# return a line structure with intercept (b) and slope (a)
def weightedLinearRegression (x, y, index1, index2, versus):
    sumX = sumY = 0.
    sumX2 = sumXY = 0.

    if(index1 == index2):
        index1 -= 1
        index2 += 1

    #check index range
    if (index1 < 0):
        index1 = 0
```

```
    if (index2 >= len(y)):
        index2 = len(y)-1

    nrPoints = index2-index1+1
    if (versus == SX):
        for i in range(nrPoints-1, -1, -1):
            for j in range (i+1):
                sumX += x[index1+i]
                sumY += y[index1+i]
                sumX2 += (x[index1+i] * x[index1+i])
                sumXY += x[index1+i] * y[index1+i]
    else:
        for i in range(nrPoints):
            for j in range (i+1):
                sumX += x[index1+i]
                sumY += y[index1+i]
                sumX2 += (x[index1+i] * x[index1+i])
                sumXY += x[index1+i] * y[index1+i]

    n = (nrPoints*(nrPoints+1))/2
    line = CLine()
    line.a = (sumXY - sumX * sumY/n) / (sumX2 - sumX * sumX/n)
    line.b = (sumY - line.a * sumX)/n
    return(line)

#backward function
def checkFlatPoint(y, indexMaxDy):
    index = indexMaxDy
    dy = abs(y[index] - y[index-1])
    threshold = dy / 1000.
    while ((dy > threshold) and (index > 0)):
        index -= 1
        dy = abs(y[index]-y[index-1])
    return (index)

#backward function
def checkZeroValue(y, indexMaxY):
    index = indexMaxY
    while ((y[index] > 0) and (index > 0)):
        index -= 1
    if ((index == 0) and (y[index] > 0)):
        return(NODATA)
    else:
        if (abs(y[index]) < abs(y[index+1])):
            return (index)
        else:
            return (index+1)
```

```
def lineIntersection(line1, line2):
    myPoint = CPoint()
    if (line1.a != line2.a):
        myPoint.x = (line2.b - line1.b) / (line1.a - line2.a)
        myPoint.y = myPoint.x * line1.a + line1.b
    else:
        myPoint.x = NODATA
        myPoint.y = NODATA

    return(myPoint)
```

After the point (myPoint) where the wave is reflected back from the end of the probe has been obtained, the travel time can finally be computed. This task is performed in the function **computeTravelTime**(), shown below. This function identifies the point where the derivative has maximum value. The first reflection is used to identify the first peak. The first peak is found by searching for the first flat section of the waveform, while the reflection point is obtained as detailed above.

```
def computeTravelTime(probeHandle, permittivity, Vp):
    global dy, flatLine, line1, line2, line3
    global indexFlatLine, indexRegr1, indexRegr2, indexRegr3
    global p0, p1, p2

    dy = firstDerivative5Points(reflecCoeff)
    dy = runningAverage(dy, 5)
    indexMaxDerivative = indexOfMaxVector(dy, 0, len(dy))
    indexMinDerivative = indexOfMinVector(dy, 0, len(dy))
    #check first maximum
    if (indexMaxDerivative > indexMinDerivative):
        indexMaxDerivative = indexOfMaxVector(dy, 0,
                                          indexMinDerivative)

    #search first reflection
    indexFlatLine = checkFlatPoint(reflecCoeff, indexMaxDerivative)
    nrPoints = len(reflecCoeff)
    step = int(8.0 * (nrPoints / 256.0))
    average = avg(reflecCoeff, indexFlatLine - step, indexFlatLine)
    flatLine.a = 0
    flatLine.b = average

    delta = min((indexMaxDerivative - indexFlatLine), MAXDELTAINDEX)
    indexRegr1 = indexFlatLine + delta
    line1 = weightedLinearRegression(timeVector, reflecCoeff,
                    indexRegr1 - delta, indexRegr1 + delta, SX)

    p0 = lineIntersection(flatLine, line1)
    dt0 = (2. * probeHandle * math.sqrt(permittivity)) / (c*Vp)
```

```
p1.x = p0.x + dt0
index = int(p1.x / deltaTime)
p1.y = reflecCoeff[index]

#search second reflection
indexSecondMaxDerivative = indexOfMaxVector(dy,
                          indexMinDerivative, len(dy))
indexZeroDerivative = checkZeroValue(dy, indexSecondMaxDerivative)
delta = min((indexSecondMaxDerivative - indexZeroDerivative),
            MAXDELTAINDEX)
indexRegr2 = indexZeroDerivative - delta
indexRegr3 = indexZeroDerivative + delta

line2 = weightedLinearRegression(timeVector, reflecCoeff,
                    indexRegr2 - delta, indexRegr2 + delta, DX)

line3 = weightedLinearRegression(timeVector, reflecCoeff,
                    indexRegr3 - delta, indexRegr3 + delta, SX)

p2 = lineIntersection(line2, line3)
```

The next step is to calculate the dielectric permittivity, which is used to obtain the soil water content. The functions used to compute the bulk dielectric permittivity and then calculate the water content are written into a file called `PSP_TTwaterContent.py`.

The first function, **getLiquidPermittivity**(), is written to account for the liquid water dielectric permittivity as a function of temperature. This equation is used in the dielectric mixing model to account for the variation of the liquid-phase dielectric permittivity with temperature. If this information is available from experiments, it can be input by the user in the section `Soil Parameters` of the window user interface. Otherwise, the program utilizes a default value of 20 °C.

The following function **getBulkPermittivity**() computes the dielectric permittivity from knowledge of the travel time as described in eqn (5.9):

```
#PSP_TTwaterContent.py
from __future__ import division
from math import sqrt

c = 299792458
airPermittivity = 1.00058986

#Lide, Handbook of Chemistry and Physics, CRC 1992
def getLiquidPermittivity(temperature):
    deltaT = temperature - 25.
    return(78.54 * (1-4.579E-03 * deltaT))

def getBulkPermittivity(probleLenght, travelTime, Vp):
    return(((c * Vp * travelTime) / (2. * probleLenght))**2)
```

```
def getWaterContentTopp(bulkPermittivity):
    return(-5.3E-02 + 2.92E-02 * bulkPermittivity - 5.5E-04 *
           bulkPermittivity**2 + 4.3E-06 * bulkPermittivity**3)

def getWaterContentMalicki(bulkPermittivity, bulkDensity):
    bulkDensity /= 1000.
    return((sqrt(bulkPermittivity) - 0.819 - 0.168*bulkDensity -
            0.159*bulkDensity**2) / (7.17 + 1.18*bulkDensity))

def getWaterContentMixModel(bulkPermittivity, bulkDensity,
                    solidPermittivity, liquidPermittivity, alpha):
    porosity = 1. - bulkDensity/2650.
    numerator = bulkPermittivity**alpha - ((1. - porosity) *
        solidPermittivity**alpha + porosity * airPermittivity**alpha)
    denominator =  liquidPermittivity**alpha - airPermittivity**alpha
    return(numerator/denominator)
```

The function **getWaterContentTopp**() computes the volumetric water content using the Topp *et al.* (1980) equation, while the Malicki *et al.* (1996) equation is implemented in the function **getWaterContentMalicki**(). Since the Malicki *et al.* (1996) equation utilizes the bulk density, the user should include a value of bulk density in the section Soil Parameters of the window user interface. A value of 1350 kg m^{-3} is used as a default.

A third available dielectric model is the dielectric mixing model (Roth *et al.*, 1990). As already described, to employ the dielectric mixing model, knowledge of the dielectric permittivity of the solid phase is needed, as well as the dielectric permittivity of the liquid phase. The former is included as an input parameter in the user interface, and it varies depending on soil mineralogy. Note (Fig. 5.3) that for the same travel time data, the three models provide different values, with higher water content values for the Malicki *et al.* (1996) and Roth *et al.* (1990) models. These differences depend on the values of the model parameters, but, in general, when bulk density data are available, the Malicki *et al.* (1996) and the dielectric mixing model should be used.

5.2   Soil Water Potential

Water potential plays a key role in water flow theory similar to the role played by temperature in heat flow problems, or voltage in electrical circuit theory. Water flows in response to gradients in water potential. Darcy's law (Darcy, 1856) states that

$$f_w = -K\frac{d\psi}{dx} \tag{5.15}$$

where f_w is the water flux density [kg m^{-2}s^{-1}], K is the hydraulic conductivity [kg s m^{-3}], ψ is the water potential [J kg^{-1}], x is the space dimension [m] and $d\psi/dx$ is the water potential gradient that drives the flow.

The soil water potential is the potential energy of water in soil. Since we cannot define an absolute scale for potential energy, the soil water potential is quantified relative to a standard state where water has no solutes, is free from external forces except gravity, and is at a reference pressure, a reference temperature and a reference elevation. The soil water potential is then defined as the energy state of water in soil with respect to the energy of water at the standard state. The driving force for water flow is the uneven distribution of water potential.

When pure free water at a standard state is brought into contact with soil water, across a semipermeable membrane, a pressure difference develops across the membrane. The pressure difference corresponds to the water potential. The standard-state values, usually specified, are temperature T_0, pressure P_0 and vertical position z_0. The value of the water potential in these conditions is equal to zero.

Soil water potential represents energy per unit quantity of water. The quantity can be energy per unit mass [J kg^{-1}] or per unit volume of water [J m^{-3}]. The former is preferable because there is then no need to include in the computation the changes of water volume with temperature. Soil water potential is also expressed as energy per unit weight, which is equivalent to a head of water. The energy is equivalent to the pressure exerted by a water column of a given height. For instance, a column of water of about 10 m correspond to a pressure of 100 kPa. This unit is common because it appears simpler and allows visualization of the gravitational and pressure potentials, which are often expressed in metres. The base 10 logarithm of the head expressed in centimetres is called pF, which is another common unit for water potential.

The unit J m^{-3} is the same as N m^{-2}, which is the SI unit of pressure, the pascal [Pa]. Energy per unit volume is therefore a pressure. Pressure units have long been used in soil physics to measure the soil water potential in tensiometers and pressure plate apparatus (Richards, 1948). Table 5.2 shows water potential in various units for a range of water potentials.

The soil water potential is usually expressed as a negative number, because it represents the energy required to transfer the soil water to the reference state of pure, free water described above. The terms 'suction' and 'tension' are definitions developed to avoid using the negative sign and to represent the soil water potential as a positive number. They are common terms used in geotechnical engineering and soil mechanics. Soil water potential can range over several orders of magnitude, from a few joules per kilogram when the soil is close to saturation to minus thousands of joules per kilogram when the soil is very dry.

The total soil water potential ψ_t is determined by a variety of forces acting on the soil water, including gravitational (ψ_g), matric (capillary and adsorptive, ψ_m), osmotic (ψ_o), hydrostatic (ψ_h) and overburden pressure (ψ_Ω) components:

$$\psi_t = \psi_g + \psi_m + \psi_o + \psi_h + \psi_\Omega \tag{5.16}$$

Usually, only one or two of the component potentials needs to be considered in any given flow problem, but gradients in any of these potentials can result in water flow when conditions are right.

Table 5.2 *Water potential in various units for a range of water potentials*

Water potential			Head [cmH$_2$O]	pF	Pore diameter [μm]	h_r[a]	FPD.[b] [°C]
[J kg^{-1}]	[MPa]	[bar]					
−1	−0.001	−0.01	10	1.0	290.80000	0.99999	−0.001
−10	−0.01	−0.1	102	2.0	29.08000	0.99993	−0.008
−30	−0.03	−0.3	306	2.5	9.69333	0.99978	−0.025
−100	−0.1	−1	1 020	3.0	2.90800	0.99926	−0.082
−1000	−1	−10	10 204	4.0	0.29080	0.99262	−0.820
−1500	−1.5	−15	15 306	4.2	0.19387	0.98895	−1.230
−10 000	−10	−100	102 041	5.0	0.02908	0.92860	−8.197
−100 000	−100	−1 000	10 20 408	6.0	0.00291	0.47676	(na)
−1 000 000	−1000	−10 000	10 204 082	7.0	0.00029	0.00061	(na)

[a] h_r = relative humidity.
[b] FPD = freezing-point depression.

5.2.1 Gravitational Potential

The gravitational component of the water potential is fundamentally different to any of the other components, since it is the result of 'body forces' applied to the water as a consequence of the water being in a gravitational field. The gravitational potential is calculated from

$$\psi_g = g(z - z_0) \tag{5.17}$$

where g is the gravitational acceleration (9.8 m s^{-2}) and z is height. The reference level is z_0, at which ψ_g is taken as zero. The reference level is usually taken as the soil surface or the surface of a water table. In flow problems, we are interested in the gradient of the gravitational potential, which is $dg/dz = g$, a constant.

5.2.2 Matric Potential

The matric potential is one of the most important components of the water potential in soil and plant systems. It is defined as the amount of work, per unit mass of water, required to transport an infinitesimal quantity of soil water from the soil matrix to a reference pool of the same soil water at the same elevation, pressure and temperature. The reduction in potential energy of water in porous materials is primarily the result of physical forces that bind the water to the porous matrix. The water potential under a curved air–water interface, such as might exist in a capillary tube or a soil pore, is given by the capillary rise equation

$$\psi_m = \frac{2\gamma \, \cos\beta}{\rho_l \, r_c} \tag{5.18}$$

where β is the contact angle between the water and the wetted surface, γ is the surface tension of water [N m^{-1}], ρ_l is the density of water [kg m^{-3}] and r_c is the radius of curvature [m]. Both the contact angle and the surface tension of the liquid phase were described earlier.

Equation (5.18) can be used to find the equivalent diameter of pores in a soil that corresponds to a given matric potential. These are shown in Table 5.2 for 20 °C and zero contact angle. For example, a soil at $\psi_m = -100$ J kg^{-1}, according to the capillary equation, will have pores larger than 2.9 μm filled with air and pores smaller than this value filled with water. At some water potential, in the range -3000 J kg^{-1} $< \psi_m < -100\,000$ J kg^{-1}, the capillary analogy breaks down because most of the water is absorbed in layers on particle surfaces rather than being held in pores between particles. The potential where the capillary equation is no longer valid is not a fixed value but depends on various soil properties such as structure and texture.

Although the total amount of adsorbed water is typically small when compared with the volumetric contribution of capillary water, its contribution is important for processes such as microbial activity, plant water uptake and evaporation in dry environments. The adsorption of water on soil particles is mainly due to van der Waals forces that promote the formation of liquid films around soil particles. Clearly, adsorbed water is strictly linked to the soil specific surface area and is important in determining processes related to contaminant adsorption, ion exchange reactions, microbial attachment to solid particles and heat transfer. Based on the contribution of van der Waals forces controlling adsorbed water films in soils, it is possible to postulate a relationship between the amount of soil water in the 'dry end' and the soil specific surface area (Tuller and Or, 2005):

$$\theta_d = h^* A_m \rho_l \tag{5.19}$$

where θ_d is volumetric water content in the 'dry end', A_m [m^2 kg^{-1}] is the specific surface area and h^* [m] is the thickness of the water film (Grismer, 1987). The thickness of the water film can be obtained from knowledge of the measured water potential (Iwamatsu and Horii, 1996):

$$h^* = \sqrt[3]{\frac{A_{svl}}{6\pi g \rho_l \psi_m}} \tag{5.20}$$

where A_{svl} [J] is the Hamaker constant for solid–vapour interactions and ψ_m is the absolute value of the matric potential [m]. This equation is valid for planar surfaces and neglects contributions of capillary condensation. Tuller and Or (2005) utilized the relationship to estimate the soil specific surface area from measurements of soil water content and soil water potential at low matric potential. Substituting eqn (5.20) into eqn (5.19) leads to

$$\theta_d = \sqrt[3]{\frac{A_{svl}}{6\pi g \rho_l \psi_m} A_m \rho_l}$$ (5.21)

The methodology is based on measurements of soil water content θ_d and matric potential ψ_m at low water contents, and derivation of the soil specific surface area A_m by fitting eqn (5.21) to experimental data, using only A_m as a fitting parameter. For the Hamaker constant, the authors recommend using a value of -6×10^{-20} J. Limitations of these techniques could be due to additional effects such as formation of water molecule clusters around cationic charge sites determining other adsorption mechanisms in addition to the van der Waals effects or different values of the Hamaker constant for different materials (Tuller and Or, 2005). Moreover measurement of the matric potential in dry soil is performed with dew-point techniques where the measurement is affected by both the matric and the osmotic potential. In soils with high solute concentration, error may be introduced by the osmotic contribution.

5.2.3   Osmotic Potential

The osmotic potential is equivalent to the work required to transport water reversibly and isothermally from a solution to a reference pool of pure water at the same elevation. In practical terms, it is the energy one must add to a solution to equilibrate the solution with pure water across a perfect semipermeable membrane. If the concentration of solute in a solution is known, the osmotic potential can be calculated from

$$\psi_o = -cv\alpha R T_K$$ (5.22)

where c is the solute concentration [mol kg^{-1}], v is the number of particles in solution per molecule of solute ($v = 1$ for non-ionizing solutes; v = number of ions per molecule for ionizing solutes), α is the osmotic coefficient, R is the gas constant (8.31 J mol^{-1} K^{-1}) and T_K is temperature in kelvins. The osmotic coefficient is a function of solution concentration and solute species. Osmotic coefficients for common solutes are given by Robinson and Stokes (1965).

   If mixtures of solutes are present, the total osmotic potential is the sum of the contributions from the components. In other words, the interaction between species is apparently small. When detailed data on the chemical composition of the soil solution are not available, the osmotic potential can still be estimated if the electrical conductivity of the saturation extract is known. A rule of thumb for a soil solution of typical composition is

$$\psi_{os} = -36\sigma$$ (5.23)

where ψ_{os} is the osmotic potential of the saturation extract [J kg^{-1}] and σ is the electrical conductivity [dS m^{-1}] of the saturation extract. If the soil dries without a change in the amount of solutes present, then (ignoring changes in c, anion exclusion effects and precipitation of sparingly soluble salts) we can write

$$\psi_o = \psi_{os}\frac{\theta_s}{\theta} \tag{5.24}$$

where θ_s is the saturation water content. These are rough approximations, but are adequate for many purposes if used with care.

The osmotic potential is an important component of water potential in plant cells and affects plant water uptake. Osmotic potential gradients in soil are usually unimportant as driving forces for flow, because the salts move with the water. Osmotic potential is always negative or zero.

5.2.4   Hydrostatic Potential

The hydrostatic potential describes the effects on water of changing the hydrostatic or pneumatic pressure applied to the water. This pressure changes the energy of soil water relative to the reference level. The relationship between potential and pressure is

$$\psi_h = \frac{P}{\rho_l} \tag{5.25}$$

where P is pressure [Pa]. Pressure can be either higher or lower than the reference pressure, so ψ_h can be either positive or negative. The pressure potential is an important component of the water potential below a water table, in plant cells and in tensiometers, which are used for measuring matric potential. The pressure component is used to describe the status of water under various laboratory conditions when pressure or suction is applied to equilibrate an external phase with soil or plant water. The suction plate, pressure plate (Richards, 1948) and pressure bomb (Scholander *et al.*, 1965) are examples of equipment that utilize this principle.

5.2.5   Overburden Potential

The overburden potential is similar to the pressure potential, in that it is the increase in potential of water in a porous system resulting from the application of pressure to the water in the system. The difference is that with the pressure potential, the pressure is applied directly to the water. This can only occur if the system is saturated. With the overburden potential, the pressure is applied to the water by the matrix. When mechanical pressure is applied to the matrix, some of the pressure is borne by the solid structure of the matrix itself, but part of it may be transferred to the water in the matrix. The overburden potential is a function of the pressure applied to the matrix and the fraction of the load that is transmitted to the water in the matrix. If the matrix has no resistance to deformation, then

$$\psi_\Omega = \frac{P}{\rho_l} \tag{5.26}$$

where P is the applied pressure. The overburden potential can be present in an unsaturated porous medium. It is mainly important in soil at depth, where the pressure of the overburden can be substantial. It is relatively unimportant in sand, but can be an appreciable fraction of the overburden pressure in wet clay.

5.3   Water Potential–Water Content Relations

Two components of the soil water potential depend on water content: the matric and the osmotic. In plant cells, a third potential, the turgor pressure, depends on water content. We will discuss the turgor potential in the chapter on plant transpiration. We have already shown how osmotic potential changes with water content when the solute content of the system stays constant.

5.3.1   Soil Water Retention Curve

The relationship between matric potential and water content is called the soil water retention or soil moisture characteristic curve. Figure 5.5 shows experimental soil water retention curves for three soils having different textural composition (Campbell and Shiozawa, 1992).

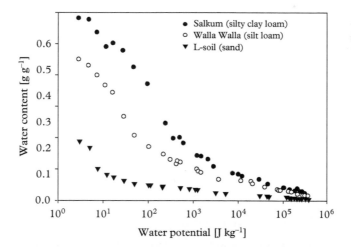

Figure 5.5 *Soil water retention curves for three soils having different texture. The water potential is plotted here as absolute value without the negative sign.*

Water retention varies with soil structural and textural properties. As discussed in Section 5.2.2, the matric forces can be described by capillary and adsorptive forces. Since the water retention curve represents the relationship between the matric potential and the soil water content, the soil water retention curve can be divided into two regions: capillary and adsorptive. Clearly, the ranges of these regions are also dependent on soil properties such as texture and structure (Tuller and Or, 2005). The capillary region is affected by the pore size distribution and by structural properties, while the adsorption region is affected by particle size distribution and specific surface area. Sandy soils have a greater amount of large pores, in a narrow range; therefore the soil water retention curve tends to drop quickly (L-soil). On the other hand, silty and clay soils have a greater amount of mid and small pores, with a broader distribution; therefore the soil water retention curves tend to be smoother and broader (Salkum and Walla Walla).

Soil water retention hysteresis

The relationship between water potential and water content is not unique. It depends on the wetting and drying history of the porous material. The curves in Fig. 5.5 represent the relationship obtained by wetting air-dry soil to the water content shown on the diagram. A different relationship could have been obtained if the soil had been dried to the indicated water content. A still different relationship would have been obtained if the soil had been wet, say to $-20\,\mathrm{J\,kg^{-1}}$, then dried to $-40\,\mathrm{J\,kg^{-1}}$. When the soil wets from air dryness or dries from saturation, the characteristics are called primary wetting or drying curves. The wetting curve always has a lower water content for a given potential than does the drying curve. The characteristics that result from drying a partially wet soil or wetting a partially dry soil are called scanning curves. They lie between the primary wetting and drying loops. An explicit analytical treatment of hysteresis has been worked out and hysteresis is sometimes included in water flow models, but we will not include hysteresis in our models. The reason for the hysteretic behaviour of the water retention curve is the formation of bottlenecks and empty pores in the soil pore network (in particular during wetting), which often determine partial saturation of some soil pores and therefore result in a curve with a different shape from the drying curve.

5.3.2   Soil Water Retention Curve Functions

The soil water retention curve can be characterized by a variety of different mathematical functions. In this book, we present and use three alternative models, which have been extensively tested and are the most commonly used. We also introduce an additional model that is a modification of one of the three models presented.

Campbell model

Campbell (1974) describes the soil moisture characteristic curve by a power-law relation:

$$\theta = \begin{cases} \theta_s \left(\dfrac{\psi_m}{\psi_e} \right)^{-1/b} & \text{if } \psi_m \geq \psi_e \\ \theta_s & \text{if } \psi_m < \psi_e \end{cases} \tag{5.27}$$

where ψ_m [J kg^{-1}] is the water potential, ψ_e [J kg^{-1}] is the air entry potential, θ [m^{3}m^{-3}] is the volumetric water content, θ_s [m^3 m^{-3}] is the saturated volumetric water content and b is a shape parameter related to the pore size distribution of the porous medium. Here the water potential is expressed as an absolute value.

van Genuchten model

An alternative equation commonly used to describe the soil water retention curve is the van Genuchten (1980) equation,

$$S_e(\psi) = \frac{\theta - \theta_r}{\theta_s - \theta_r} = \frac{1}{[1 + (\alpha\psi)^n]^m} \tag{5.28}$$

Solving for θ, can be written as

$$\theta = \theta_r + (\theta_s - \theta_r)\frac{1}{[1 + (\alpha\psi)^n]^m} \tag{5.29}$$

where S_e is the degree of saturation, lying in the range [0, 1], and α, n, m, θ_s and θ_r are fitting parameters. Different restrictions can be imposed on the parameters n and m depending on the shape of the curve. In particular, when only a limited range of water retention values are available (usually in the 'wet' range of the curve), it might be necessary to restrict the parameters n and m. More stable results are generally obtained when the restriction $m = 1 - 1/n$ is implemented for incomplete data sets.

Ippisch–van Genuchten model

Ippisch *et al.* (2006) pointed out that the van Genuchten model, under certain conditions, is problematic when water retention data are used to compute hydraulic conductivity, and they demonstrated that if $n < 2$ or $\alpha h_a > 1$ (where h_a is the air-entry value of the soil, corresponding to the largest pore radius), the van Genuchten–Mualem model predicts erroneous hydraulic conductivities. A detailed description of the van Genuchten–Mualem model, and its limitations, is provided in Chapter 6.

The modified formulation, as proposed by Ippisch *et al.* (2006), has the form

$$S_e = \begin{cases} \dfrac{1}{S_c}[1 + (\alpha\psi_m)^n]^{-m} & \text{if } \psi_m \geq \psi_e \\ 1 & \text{if } \psi_m < \psi_e \end{cases} \tag{5.30}$$

where S_e is the degree of saturation, α, m and n are fitting parameters, and $S_c = [1 + (\alpha\psi_e)^n]^{-m}$ is the water saturation at the air-entry potential ψ_e.

Modified evaluation of the residual water content

As pointed out by Campbell and Shiozawa (1992), the models just described do not correctly represent the soil water retention curve in the dry range and they give poor results at low water contents. Indeed, the concept of the existence of a soil residual water content is an incorrect idea that does not represent experimental data. When accurate measurement of the soil water retention curve in the dry range are performed, as shown in Fig. 5.5, the water retention curve does not tend asymptotically to a 'residual water content', but decreases towards a zero value corresponding to 'oven-dry' matric potential. Campbell and Shiozawa (1992) proposed a model where the 'residual water content' is not a constant, but a decreasing variable as described by the following equation:

$$\theta_r = \theta_{r0} \left(1 - \frac{\ln(\alpha\psi + 1)}{\ln(\alpha\psi_0 + 1)} \right) \tag{5.31}$$

where ψ_0 is the 'oven-dry' matric potential at zero water content, which is about 10^6 J kg^{-1}, and θ_{r0} corresponds to the original value of the 'residual water content' in the van Genuchten (1980) model. This modification improves the characterization of the soil water retention curve, without adding any additional parameters.

Campbell–Ippisch–van Genuchten model

In the present analysis, we have described the two main limitations of the van Genuchten (1980) model in the wet range and in the dry range. Modifications of the model in the wet range were proposed by Ippisch *et al.* (2006) and in the dry range by Campbell and Shiozawa (1992), as described above. Here we are presenting a model where the two modifications are combined and incorporated into the original van Genuchten (1980) model. Therefore the model is an implementation of the Ippisch *et al.* (2006) model, with the modified 'residual water content' proposed by Campbell and Shiozawa (1992). Figure 5.6 shows a water retention curve fitted using the van Genuchten (1980) model, while Fig. 5.7 depicts a curve fitted using the Campbell–Ippisch–van Genuchten model.

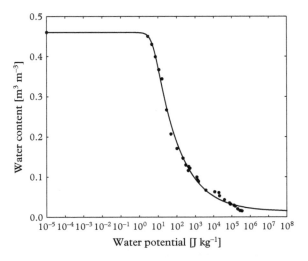

Figure 5.6 *Soil water retention curve for a soil sample fitted using the van Genuchten model.*

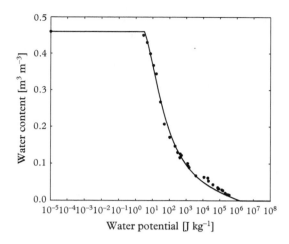

Figure 5.7 *Soil water retention curve for a soil sample fitted using the Campbell–Ippisch–van Genuchten model. Note how the curve reaches a zero value of water content.*

Note how the former reaches an asymptotic value of soil water content at dry water potentials, which does not represent the experimental data, while the latter reaches a zero value of water content, and correctly describes the experimental data.

Parameterization of the soil water retention curve can be obtained by (a) fitting a mathematical model to experimental data using least squares nonlinear fitting algorithms or neural networks, (b) employing inverse methods, which are methods where model parameters are iteratively changed so that a given selected hydrological model approximates the observed response, and (c) using pedotransfer functions, which are regression equations based on the dependence of the soil water retention curves on basic soil properties such as particle size distribution, porosity, bulk density and organic matter. Below we present a Python program for least squares, nonlinear fitting.

Fitting water retention curve models to experimental data

The problem of minimizing a nonlinear function over a space of parameters of the function is usually referred as nonlinear fitting. These problems arise commonly in least squares curve fitting and nonlinear programming. The problem is formulated as

$$\Phi(\beta) = \sum_{i=1}^{m}[y_i - f(x_i, \beta)]^2 \tag{5.32}$$

where there are m empirical data pairs (x_i, y_i). The objective function $\Phi(\beta)$ is minimized by optimizing the parameters β of the model $f(x, \beta)$ so that the sum of the squares of the deviations is minimized.

The Levenberg–Marquardt algorithm (Levenberg, 1944; Marquardt, 1963) is an iterative procedure where the user must provide an initial guess of the parameter vector β. Then, at each iteration, β is replaced by a new approximation $\beta + \delta$. This new parameter set will lead to a new sum of squares. The basis of the algorithm is a linear approximation of the function $f(x, \beta + \delta)$ in the neighbourhood of β. Denoting the Jacobian matrix by $\mathcal{J}$, a Taylor series expansion for small δ leads to

$$f(x, \beta + \delta) \approx f(x, \beta) + \mathcal{J}_i \delta \qquad (5.33)$$

where

$$\mathcal{J}_i = \frac{\partial f(x_i, \beta)}{\partial \beta} \qquad (5.34)$$

is the gradient of f with respect to β. The new objective function is now

$$\Phi(\beta + \delta) = \sum_{i=1}^{m} [y - f(x_i, \beta) - \mathcal{J}_i \delta]^2 \qquad (5.35)$$

Therefore the sought δ that minimizes this new objective function is the solution of the least squares problem. The iteration ends when the function Φ is less than a threshold for the error. One of the limitation of the Levenberg–Marquardt algorithm is that the solution for the best-fit parameters may depend on the initial conditions of the parameter set, and therefore the solution may not be unique.

Numerical implementation

A Python project called `PSP_waterRetentionFitting` was implemented to apply nonlinear fitting to the models described above. Specifically, the user can select among the following functions (or models):

1. Campbell
2. van Genuchten
3. van Genuchten with the restriction $m = 1 - 1/n$
4. Ippisch–van Genuchten
5. Campbell–Ippisch–van Genuchten

The code is written in a modular form, such that the user can easily write a different soil water retention curve model and include it in the program. Moreover, the program modularity allows the user to modify the program and use other equations to fit different data sets of other soil parametric functions. For instance, the Gaussian function for particle size distribution presented in Chapter 2 can be implemented using the present program, to parameterize particle size distribution models.

The program comprises four files:

1. main.py
2. PSP_readDataFile.py
3. PSP_Marquardt.py
4. PSP_waterRetention.py

The main.py file is written to import the experimental data, select the water retention function, call the functions for nonlinear fitting and plot the results. The PSP_readDataFile.py is written to read the experimental data and is described in Appendix A. The PSP_waterRetention.py contains the water retention functions, while the PSP_Marquardt.py implements the equation for least–squares algorithms. To run the program, the user must run the file main.py; then the user will be prompted with the following commands to select the water retention function:

```
1   Campbell
2   van Genuchten
3   van Genuchten with m = 1-1/n restriction
4   Ippisch-van Genuchten
5   Campbell-Ippisch-van Genuchten
Choose model type: 1
```

The program then performs the computations, the results are printed on the screen and a plot is generated. The screen output is

```
Fitting
iterations nr: 23
sum of squared residuals: 0.00345163976933

thetaS =  0.46
AirEntry =  4.21150485478
b =  3.57866322236
```

The first section of the output for the statistical information is the number of iterations and the sum of square residuals. The second section has the water retention parameters. For example, we can run Campbell's equation for the experimental data stored in the file soil.txt, and the saturated water content, the air-entry potential and the parameter b are printed. These values will be used in Chapters 9 and 11 for numerical solutions to water flow.

An example of the fitting results is shown in Fig. 5.8 for the models listed above. The program's output is one graph only for the selected model, but in Fig. 5.8, to show the different shapes of the curves using the different models, we have combined the four outputs. Note, for instance, how the selection of van Genuchten's equation with the restriction $m = 1 - 1/n$ determined a different shape of the curve around the air-entry

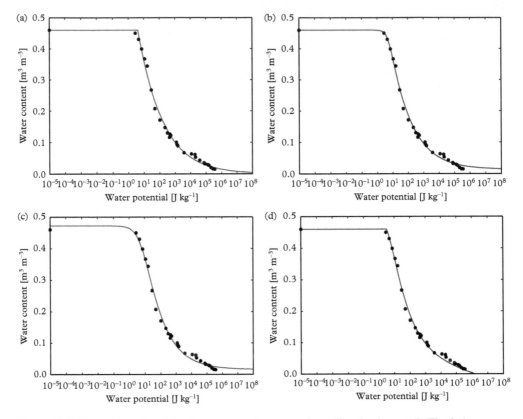

Figure 5.8 *Experimental and fitted water retention curves for a silty clay loam soil. The fitting was performed using (a) Campbell's equation, (b) van Genuchten's equation, (c) van Genuchten's equation with the restriction m = 1 − 1/n and (d) the Campbell–Ippisch–van Genucthen equation.*

potential compared with the curve obtained using van Genuchten's equation with free parameters. Also note that the Campbell and Campbell–Ippisch–van Genuchten models are discontinous at the air-entry potential.

The `main.py` file is shown below. The first lines are used to import modules and files. Then experimental data are read by using the file `PSP_readDataFile.py` The program then has a series of `if` and `elif` to print the choices regarding the selection of the water retention function. The function **Marquardt**() has arguments (`waterRetentionCurve`, `b0`, `bmin`, `bmax`, `waterPotential`, `waterContent`), which are the type of water retention function, the initial, minimum and maximum values of the parameters, and the experimental water potential and water content. The function returns an array of fitted parameters b, which can change in number and type depending on the selected model. The last section of the function **main** comprises the usual `matplotlib` instructions to plot the graph with the results.

```
#PSP_waterRetentionFitting
from __future__ import print_function, division
try: input = raw_input
except: pass

import numpy as np
import matplotlib.pyplot as plt
from PSP_readDataFile import readDataFile
from PSP_Marquardt import *

def main():
    # read experimental values
    myOutput, isFileOk = readDataFile("data/soil.txt", 1, '\t', False)
    if (not isFileOk):
        print('Wrong file: error reading row nr.', myOutput)
        return(False)
    waterPotential = myOutput[:,0]
    waterContent = myOutput[:,1]

    # select water retention curve
    print (CAMPBELL,' Campbell')
    print (VAN_GENUCHTEN,' van Genuchten')
    print (RESTRICTED_VG,' van Genuchten with m = 1-1/n restriction')
    print (IPPISCH_VG,' Ippisch-van Genuchten')
    print (CAMPBELL_IPPISCH_VG,' Campbell-Ippisch-van Genuchten')

    waterRetentionCurve = 0
    while (waterRetentionCurve < CAMPBELL) or \
                        (waterRetentionCurve > CAMPBELL_IPPISCH_VG):
        waterRetentionCurve = float(input("Choose model type: "))
        if (waterRetentionCurve < CAMPBELL) or \
                        (waterRetentionCurve > CAMPBELL_IPPISCH_VG):
            print('wrong choice.')

    # initialize parameters
    thetaS = max(waterContent)
    #thetaR = min(waterContent)
    thetaR = 0.08
    air_entry = 1.0
    Campbell_b = 4.0
    VG_alpha = 1/air_entry
    VG_n = 1.5
    VG_m = 1. - 1./VG_n

    if (waterRetentionCurve == CAMPBELL):
        b0 = np.array([thetaS, air_entry, Campbell_b], float)
        bmin = np.array([thetaS, 0.1, 0.1], float)
```

```
        bmax = np.array([thetaS*1.1, 20., 10.], float)
    elif (waterRetentionCurve == VAN_GENUCHTEN):
        b0 = np.array([thetaS, thetaR, VG_alpha, VG_n, VG_m], float)
        bmin = np.array([thetaS, 0.0, 0.01, 0.01, 0.01], float)
        bmax = np.array([1.0, thetaR, 10., 10., 1.], float)
    elif (waterRetentionCurve == RESTRICTED_VG):
        b0 = np.array([thetaS, thetaR, VG_alpha, VG_n], float)
        bmin = np.array([thetaS, 0.0, 0.01, 1.], float)
        bmax = np.array([1, thetaR, 10., 10.], float)
    elif (waterRetentionCurve == IPPISCH_VG):
        b0 = np.array([thetaS, thetaR, air_entry, VG_alpha, VG_n],
                      float)
        bmin = np.array([thetaS, 0.0, 0.1, 0.01, 1.], float)
        bmax = np.array([1, thetaR, 10., 10., 10.], float)
    elif (waterRetentionCurve == CAMPBELL_IPPISCH_VG):
        b0 = np.array([thetaS, thetaR, air_entry, VG_alpha, VG_n],
                      float)
        bmin = np.array([thetaS, 0.0, 0.1, 0.01, 1.], float)
        bmax = np.array([1, thetaR, 10., 10., 10.], float)

    else:
        print ('wrong choice.')
        return(False)

    print ("\nFitting")
    b = Marquardt(waterRetentionCurve, b0, bmin, bmax,
                  waterPotential, waterContent)

    print ("\nthetaS = ", b[0])
    if (waterRetentionCurve == CAMPBELL):
        print ("AirEntry = ", b[1])
        print ("b = ", b[2])
    elif (waterRetentionCurve == VAN_GENUCHTEN):
        print ("thetaR = ", b[1])
        print ("alpha = ", b[2])
        print ("n = ", b[3])
        print ("m = ", b[4])
    elif (waterRetentionCurve == RESTRICTED_VG):
        print ("thetaR = ", b[1])
        print ("alpha = ", b[2])
        print ("n = ", b[3])
    elif (waterRetentionCurve == IPPISCH_VG):
        print ("thetaR = ", b[1])
        print ("AirEntry = ", b[2])
        print ("alpha = ", b[3])
        print ("n = ", b[4])
    elif (waterRetentionCurve == CAMPBELL_IPPISCH_VG):
        print ("thetaR = ", b[1])
```

```
      print ("AirEntry = ", b[2])
      print ("alpha = ", b[3])
      print ("n = ", b[4])

  myWP = np.logspace(-5, 8, 500)
  myWC = estimate(waterRetentionCurve, b, myWP)

  plt.figure(figsize=(10,8))
  plt.plot(myWP, myWC,'k-')
  plt.plot(waterPotential, waterContent,'ko')

  plt.xscale('log')
  plt.xlabel('Water Potential [J kg$^{-1}$]',
                                  fontsize=20,labelpad=8)
  plt.xticks(size='16')
  plt.ylabel('Water Content [m$^{3}$ m$^{-3}$]',
                                  fontsize=20,labelpad=8)
  plt.tick_params(axis='both', which='major', labelsize=20,pad=8)
  plt.tick_params(axis='both', which='minor', labelsize=20,pad=8)
  plt.yticks(size='16')
  #plt.savefig('waterRetention.eps')
  plt.show()

main()
```

The file `PSP_waterRetention.py` contains the four functions used for fitting the experimental data, as described above. The reader can easily implement a different model and include it in this file. The first lines are written to assign a number to the model, such that the user can then select it. The function **Campbell()** implements the Campbell (1974) model. Note the **if** statement used to discriminate between water potential values larger or smaller than the air-entry value, since this function is discontinuous at the air-entry value. In this code we used the general names water content and water potential within the function name, while the variables were named `theta` and `psi`, respectively. The van Genuchten (1980) model with no restriction imposed on the parameters, is presented as the next function. The van Genuchten (1980) model with the restriction $m = 1 - 1/n$ is implemented, while the Campbell–Ippisch–van Genuchten model is written in the last lines.

```
#PSP_waterRetention.py
from __future__ import division
import numpy as np

CAMPBELL = 1
VAN_GENUCHTEN = 2
RESTRICTED_VG = 3
IPPISCH_VG = 4
CAMPBELL_IPPISCH_VG = 5
```

```
def Campbell(v, psi, theta):
    thetaS = v[0]
    he = v[1]
    Campbell_b= v[2]
    for i in range(len(psi)):
        if psi[i] <= he:
            theta[i] = thetaS
        else:
            Se = (psi[i]/he)**(-1./Campbell_b)
            theta[i] = Se * thetaS

def VanGenuchten(v, psi, theta):
    thetaS = v[0]
    VG_thetaR = v[1]
    VG_alpha = v[2]
    VG_n = v[3]
    VG_m = v[4]
    for i in range(len(psi)):
        Se = 1. / pow(1. + pow(VG_alpha * psi[i], VG_n), VG_m)
        theta[i] = Se * (thetaS - VG_thetaR) + VG_thetaR

def VanGenuchtenRestricted(v, psi, theta):
    thetaS = v[0]
    VG_thetaR = v[1]
    VG_alpha = v[2]
    VG_n = v[3]
    VG_m = 1. - (1. / VG_n)
    for i in range(len(psi)):
        if psi[i] <= 0:
            Se = 0
        else:
            Se = (1. + (VG_alpha * abs(psi[i]))**VG_n)**(-VG_m)
        theta[i] = Se * (thetaS - VG_thetaR) + VG_thetaR

def IppischVanGenuchten(v, psi, theta):
    thetaS = v[0]
    VG_thetaR = v[1]
    he = v[2]
    VG_alpha = v[3]
    VG_n = v[4]
    VG_m = 1. - (1./VG_n)
    VG_Sc = (1. + (VG_alpha * he)**VG_n)**VG_m
    for i in range(len(psi)):
        if (psi[i] <= he):
            Se = 1.0
        else:
            Se = VG_Sc * (1. + (VG_alpha * abs(psi[i]))**VG_n)**(-VG_m)
        theta[i] = Se * (thetaS - VG_thetaR) + VG_thetaR
```

```
def CampbellIppischVanGenuchten(v, psi, theta):
    thetaS = v[0]
    VG_thetaR = v[1]
    he = v[2]
    VG_alpha = v[3]
    VG_n = v[4]
    VG_m = 1. - (1./VG_n)
    VG_Sc = (1. + (VG_alpha * he)**VG_n)**VG_m
    for i in range(len(psi)):
        if (psi[i] <= he):
            Se = 1.0
        else:
            Se = VG_Sc * (1. + (VG_alpha * abs(psi[i]))**VG_n)
                                                **(-VG_m)
        residual = VG_thetaR * (1 -
        ((np.log(VG_alpha*psi[i] + 1.0)/np.log(VG_alpha*(10**6) +
                                                1.0))))
        theta[i] = max(0.0, Se * (thetaS - residual) + residual)
```

The following code, implemented in the file PSP_Marquardt.py, contains the functions needed for the nonlinear fitting algorithm. The variables EPSILON and MAX_ITERATIONS_NR are the two convergence parameters: the precision and the maximum number of iterations. The function **estimate**() is written to pass the values of water content and water potential to the different retention curve models that were selected by the user. The function **Marquardt**() contains the Marquardt (1963) algorithm. The function **Norm**() computes the squared difference between the experimental data y[i] and the estimated data yEst[i], as described by the computation: dy= y[i] - yEst[i] and norm+= dy * dy Indeed, this algorithm is based on the minimization of the sum of the squared differences. The least squares algorithm is coded in the function **LeastSquares**(). Note that in the first and second for loops of this function, the parameters are changed and evaluated again to check if the new set is reducing the sum of the squared differences between the experimental and the fitted data. The final objective is to minimize the sum of the squared differences to a number that is less than the accepted error. Clearly, the choice of the error parameter can significantly change the output. Moreover, this algorithm is subject to error due to the local minimum error.

```
#PSP_Marquardt.py
from __future__ import print_function, division
import numpy as np
from math import sqrt
from PSP_waterRetention import *

EPSILON = 0.00001
MAX_ITERATIONS_NR = 100
```

```
def Marquardt(waterRetentionCurve, v0, vmin, vmax, x, y):
    n = len(v0)
    Lambda0 = 0.01
    vFactor = 2.
    l = np.array([Lambda0]*n)      # damping parameters
    v = np.zeros(n, float)
    for i in range(n): v[i] = v0[i]

    nrIter = 1
    maxDiff = 1.0
    sse = norm(waterRetentionCurve, v0, x, y)

    while (maxDiff > EPSILON) and (nrIter < MAX_ITERATIONS_NR):
        diff = LeastSquares(waterRetentionCurve, l, v, vmin, vmax, x, y)
        maxDiff = max(abs(diff))
        v_new = computeNewParameters(v, vmin, vmax, diff, l, vFactor)
        sse_new = norm(waterRetentionCurve, v_new, x, y)

        if (sse_new < sse):
            sse = sse_new
            for i in range(n): v[i] = v_new[i]
            l /= vFactor
        else:
            l *= vFactor
        nrIter += 1
    print ("iterations nr:", nrIter)
    print ("sum of squared residuals:", sse)
    return(v)

def estimate(waterRetentionCurve, v, Psi):
    waterContent = np.zeros(len(Psi))
    if (waterRetentionCurve == CAMPBELL):
        Campbell(v, Psi, waterContent)
    elif (waterRetentionCurve == VAN_GENUCHTEN):
        VanGenuchten(v, Psi, waterContent)
    elif (waterRetentionCurve == RESTRICTED_VG):
        VanGenuchtenRestricted(v, Psi, waterContent)
    elif (waterRetentionCurve == IPPISCH_VG):
        IppischVanGenuchten(v, Psi, waterContent)
    elif (waterRetentionCurve == CAMPBELL_IPPISCH_VG):
        CampbellIppischVanGenuchten(v, Psi, waterContent)

    return(waterContent)

def computeNewParameters(v, vmin, vmax, diff, l, factor):
    n = len(v)
    v_new = np.zeros(n, float)
    for i in range(n):
```

```
            v_new[i] = v[i] + diff[i]
            if (v_new[i] > vmax[i]):
                v_new[i] = vmax[i]
                l[i] *= factor
            if (v_new[i] < vmin[i]):
                v_new[i] = vmin[i]
                l[i] *= factor
    return(v_new)

def norm(waterRetentionCurve, v, x, y):
    yEst = estimate(waterRetentionCurve, v, x)
    norm = 0
    for i in range(len(x)):
        dy = y[i] - yEst[i]
        norm += (dy*dy)
    return(norm)

def LeastSquares(waterRetentionCurve, l, v, vmin, vmax, x, y):
    n = len(v)
    m = len(x)
    p = np.resize(np.zeros(n, float),(n,m))
    a = np.resize(np.zeros(n, float),(n,n))
    z = np.zeros(n, float)
    g = np.zeros(n, float)
    v1 = np.zeros(n, float)
    diff = np.zeros(n, float)

    for i in range(n): v1[i] = v[i]
    est = estimate(waterRetentionCurve, v, x)

    for i in range(n):
        change = (vmax[i] - vmin[i]) * 0.01
        v1[i] +=   change
        # get a new set of estimates
        yEst = estimate(waterRetentionCurve, v1, x)
        v1[i] -= change
        for j in range(m):
            # compute derivatives
            p[i][j] = (yEst[j] - est[j]) / change

    for i in range(n):
        for j in range(i, n):
            a[i][j] = 0
            for k in range(m):
                a[i][j] = a[i][j] + p[i][k] * p[j][k]
        z[i] = sqrt(a[i][i]) + EPSILON
```

```
for i in range(n):
    g[i] = 0
    for k in range(m):
        g[i] = g[i] + p[i][k] * (y[k] - est[k])
    g[i] = g[i] / z[i]
    for j in range(i, n):
        a[i][j] = a[i][j]  / (z[i] * z[j])

for i in range(n):
    a[i][i] = a[i][i] + l[i]
    for j in range(i+1, n):
        a[j][i] = a[i][j]

for j in range(n-1):
    pivot = a[j][j]
    for i in range(j+1, n):
        mult = a[i][j] / pivot
        for k in range(j+1, n): a[i][k] -= mult * a[j][k]
        g[i] -=  mult * g[j]

diff[n-1] = g[n-1] / a[n-1][n-1]

for i in range(n-2, -1, -1):
    top = g[i]
    for k in range(i+1, n):
        top -= a[i][k] * diff[k]
    diff[i] = top / a[i][i]

for i in range(n):
    diff[i] /=  z[i]

return(diff)
```

Estimated parameters for different textural classes

We present two tables with soil water retention curve parameters. Table 5.3 lists parameters for the 12 textural classes for Campbell's equation, while Table 5.4 lists parameters for van Genuchten's equation and the Campbell–Ippisch–van Genuchten equation.

In spite of the difficulties encountered in producing soil water retention functions from texture data, the benefits, especially for simulation models, of being able to produce one from the other justify additional effort in this area. The relationship we produce here will be based on the definition of geometric mean diameter presented in Chapter 3.

For Campbell's model, we expect ψ_e to decrease (become more negative) and b to increase as the mean pore diameter becomes smaller (note that when $b = 0$, all of the water is held at a single potential, and when b approaches infinity, no change in water content occurs when ψ_m changes). We expect pore size and particle size to be correlated;

Table 5.3 *Soil water retention curve parameters for the Campbell's equation (Rawls et al., 1992)*

Textural class	b	ψ_e [J kg^{-1}]	θ_s [m m^{-3}]
Sand	1.69	−0.73	0.37
Loamy sand	2.11	−0.87	0.38
Sandy loam	3.11	−1.47	0.41
Silt loam	4.74	−2.08	0.43
Loam	4.55	−1.12	0.43
Silt	4.23	−2.6	0.44
Sandy clay loam	4.00	−2.81	0.41
Silty clay loam	6.62	−3.26	0.46
Clay loam	5.15	−2.59	0.45
Sandy clay	5.95	−2.92	0.44
Silty clay	7.87	−3.42	0.51
Clay	7.63	−3.73	0.50

therefore, as the geometric particle diameter d_g increases, ψ_e should become less negative. Rawls *et al.* (1992) presented data from 5000 soil samples. Table 5.3 summarizes some of those data for 12 texture classes. There still exists enormous variation within each texture class, but general patterns do emerge from the table that allow relationships to be developed between texture and hydraulic properties.

Correlating the geometric mean diameter and the air-entry potential gives

$$\psi_e = 0.61 \ln d_g - 3.9 \tag{5.36}$$
$$b = 8.25 - 1.26 \ln d_g \tag{5.37}$$

The geometric mean diameter d_g is obtained by knowledge of sand, silt and clay fractions and applying eqn (2.36).

The parameters presented in Table 5.4 were obtained by a combination of data from the Rosetta software (Schaap *et al.*, 2001) and data presented by Rawls *et al.* (1992). To obtain this table, we tested the parameterization over a large database of soil hydraulic properties data from the Emilia–Romagna region, Italy. This database was therefore used as an independent database for parameter testing and statistical analysis of the parameters. Estimation of van Genucthen parameters from basic soil data can be obtained using the Rosetta software (Schaap *et al.*, 2001).

Table 5.4 *SWR curve parameters for van Genuchten's and modified van Genuchten's equations*

Textural class	α [kg J^{-1}]	n	ψ_e [J kg^{-1}]	θ_r [m m^{-3}]	θ_s [m m^{-3}]
Sand	0.68	1.7	−0.7	0.02	0.38
Loamy sand	0.49	1.5	−1.1	0.03	0.39
Sandy loam	0.39	1.4	−1.5	0.05	0.4
Silt loam	0.14	1.2	−2.6	0.05	0.44
Loam	0.18	1.21	−2.3	0.05	0.42
Silt	0.08	1.24	−2.7	0.04	0.44
Sandy clay loam	0.23	1.22	−2.1	0.05	0.41
Silty clay loam	0.14	1.2	−3.1	0.05	0.46
Clay loam	0.19	1.18	−2.7	0.06	0.45
Sandy clay	0.22	1.18	−2.5	0.07	0.44
Silty clay	0.18	1.16	−3.3	0.06	0.5
Clay	0.17	1.16	−3.3	0.07	0.5

5.3.3 Capacity

The change in water content as a result of a change in water potential is an important property. As described in the discussion of the equation of capillarity, the pressure (potential) change at an interface depends on the curvature of the interface, which is related to the radius of the capillary and the surface tension. Therefore, the water fraction in a given capillary must depend on the change in interfacial properties (i.e. the water potential).

The capacity is the derivative of water content with respect to matric potential:

$$C(\psi_m) = \frac{d\theta}{d\psi_m} \qquad (5.38)$$

If the water content is expressed as a degree of saturation S_e, then the capacity becomes

$$C(\psi_m) = \frac{d\theta}{d\psi_m} = (\theta_s - \theta_r)\frac{dS_e}{d\psi_m} \qquad (5.39)$$

In Campbell's equation, θ_r is not used, so capacity becomes

$$C(\psi_m) = \frac{d\theta}{d\psi_m} = \theta_s\frac{dS_e}{d\psi_m} \qquad (5.40)$$

The capacities for the three water retention equations are as follows:

- for Campbell's equation,

$$C = \frac{d\theta}{d\psi_m} = -\frac{\theta_s}{b\psi_m}\left(\frac{\psi_e}{\psi_m}\right)^{1/b} = \frac{-\theta}{b\psi_m} \qquad (5.41)$$

- for van Genuchten's equation,

$$C = \frac{dSe}{d\psi_m} = -\alpha nm \left[1 + (\alpha\psi_m)^n\right]^{-(m+1)} (\alpha\psi_m)^{n-1} \qquad (5.42)$$

- for the Ippisch–van Genuchten equation,

$$C = \frac{dS_e}{d\psi_m} = -\alpha nm \left[1 + (\alpha\psi_m)^n\right]^{-(m+1)} (\alpha\psi_m)^{n-1} \frac{1}{S_c} \qquad (5.43)$$

5.3.4 Hydrostatic Equilibrium of Soil Water in a Gravitational Field

If water is in hydrostatic equilibrium with the gravitational field, all the water contained in the soil profile must have a total potential equal to zero. If the reference level for the gravitational potential is taken as the surface of the water table, the gravitational potential increases on going towards the surface. To have a total potential equal to zero, the matric potential must decrease by the same amount; therefore the amount of water contained in the soil must decrease from the water table towards the surface (according to the soil water retention curve of that specific soil type).

Figure 5.9 shows the profile of hydrostatic equilibrium of soil water in a gravitational field. The gravitational potential is zero at the surface of the water table and the total potential is zero across the whole profile.

It is possible to compute the total amount of water contained in a soil column in hydrostatic equilibrium with a gravitational field by numerical integration of the water retention curve for a matric potential decreasing from 0 (water table) to the depth of the soil column.

In this example, two water retention curve models (Campbell's and van Genuchten's) were used for a loamy sand to perform the numerical integration over a two-dimensional profile of width 1 m and depth 2 m. The numerical integration is performed with the code Extended methods presented in Appendix B, where the theory for numerical integration is described in detail. The project for computing the amount of water in the soil profile is called `PSP_columnWaterContent.py` and comprises three files:

1. `main.py`
2. `PSP_soil.py`
3. `PSP_integration.py`

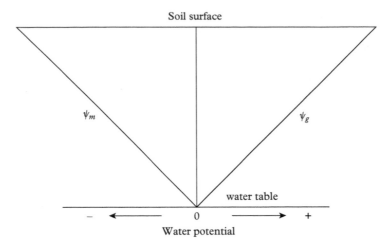

Figure 5.9 *Gravitational (ψ_g) and matric (ψ_m) potential profiles in hydrostatic equilibrium with the water table.*

The file `main.py` (shown below) includes the module needed for the computation and plotting as described in previous chapters. Here we are also importing the module `Image` from the `Python Imaging Library` that we used in Chapter 2, since we are plotting the results of the integration over a picture of a soil profile. The first lines of the **main** function are used to select the water retention model, the depth of the soil profile is defined and the variable `integral` is defined by assigning it the function `qsimp`. Since the integration is performed over the soil column, the units of soil water potential are converted from m to J kg^{-1} (by multiplying by the gravitational constant) when the function `qsimp(soil.waterContent, -waterTableDepth*9.81, 0)` is called. Indeed, the first argument of the function **qsimp** is the type of function to be used for integration and the upper and lower limits of the integral. The integration is therefore performed over matric potential values ranging from 0 to –2 m, but with matric potential in J kg^{-1}.

In the last section of this function, an image called `soilProfile.png` is imported and then a plot is performed over the image itself.

```
#PSP_columnWaterContent
from __future__ import print_function, division
import matplotlib.pyplot as plt
import matplotlib.cm as cm
import Image
import numpy as np
import PSP_soil as soil
from PSP_integration import qsimp

def main():
    choice = 0
    print (soil.CAMPBELL,' Campbell')
```

```
    print (soil.VAN_GENUCHTEN,' van Genuchten')
    while (choice < soil.CAMPBELL) or (choice > soil.VAN_GENUCHTEN):
        choice = float(input("Choose water retention curve: "))
        if (choice < soil.CAMPBELL) or (choice > soil.VAN_GENUCHTEN):
            print('wrong choice.')
    soil.waterRetentionCurve = choice

    waterTableDepth = 2.0
    nrValues = 100
    step = waterTableDepth/nrValues

    integral = qsimp(soil.waterContent, -waterTableDepth*9.81, 0)
    totalWaterContent = integral / 9.81
    print ("\nTotal water content [m^3]:", totalWaterContent)

    x = np.zeros(nrValues+1, float)
    y = np.zeros(nrValues+1, float)
    for i in range(nrValues+1):
        y[i] = step*i
        psi = y[i] - waterTableDepth
        psi *= 9.81
        x[i] = soil.waterContent(psi)

    img = Image.open('soil_column.png').convert("L")
    mgplot = plt.imshow(img, cmap = cm.Greys_r,
            extent=[0,1,waterTableDepth,0])
    plt.plot(x, y, 'k', linewidth='3')
    plt.xlim(0, 1)
    plt.ylim(waterTableDepth, 0)
    plt.title('water content in soil column')
    plt.xlabel('water content [m^3/m^3]')
    plt.ylabel('depth [m]')
    plt.show()

main()
```

The following is the file `PSP_soil.py`, in which Campbell's and van Genuchten's water retention models are implemented:

```
#PSP_soil.py

CAMPBELL = 1
VAN_GENUCHTEN = 2
waterRetentionCurve = 0

thetaS      = 0.46
thetaR      = 0.01
Campbell_he = -4.2
```

```
Campbell_b   = 3.58
VG_alpha     = 0.18
VG_n         = 2.86
VG_m         = 0.11

def waterContent(signPsi):
    if (waterRetentionCurve == CAMPBELL):
        if (signPsi >= Campbell_he):
            return thetaS
        else:
            Se = (signPsi / Campbell_he)**(-1. / Campbell_b)
            return Se*thetaS

    elif(waterRetentionCurve == VAN_GENUCHTEN):
        if (signPsi >= 0.):
            return thetaS
        else:
            Se = 1. / (1. + (VG_alpha * abs(signPsi))**VG_n)**VG_m
            return Se*(thetaS - thetaR) + thetaR
```

The file `PSP_integration.py` implements an algorithm for numerical integration. It is a numerical code that combines the trapezoidal and the Simpson equations, with the idea of increasing the number of evaluated values until a certain precision is fulfilled. A detailed description of this code is provided in Appendix B, where examples of numerical integration are provided.

```
#PSP_integration.py
from __future__ import division

def trapzd(func, a, b, n):
    if (n == 1):
        trapzd.s = 0.5*(b-a)*(func(a) + func(b))
        return trapzd.s
    else:
        it = 1
        for j in range(1, n-1):
            it <<= 1
        tnm = float(it)
        del_ = (b-a)/tnm
        x = a+0.5*del_
        sum = 0.0
        for j in range(1, it+1):
            sum += func(x)
            x += del_
        trapzd.s = 0.5*(trapzd.s + (b-a)*sum/tnm)
        return trapzd.s
```

```
def qsimp(func, a, b):
    EPS = 1.0e-6
    JMAX = 20
    ost=0.0
    os=0.0
    for j in range(1, JMAX+1):
        st = trapzd(func,a,b,j)
        s = (4.0*st-ost)/3.0
        if (j > 5):
            if (abs(s-os) < EPS*abs(os) or (s == 0.0 and os == 0.0)):
                return s
        os=s
        ost=st
```

The result of the program is shown in Fig. 5.10, where the variation of the soil water content across the profile is plotted. The area under the left side of the line is the computed integral. The zero value of the water potential corresponds to the water table surface, as depicted in Fig. 5.9, where the matric potential decreases further above the

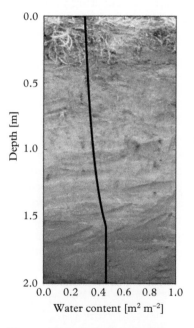

Figure 5.10 *Variation of soil water content in a soil profile as a function of height above a water table. The water is in hydrostatic equilibrium with the gravitational field; therefore all the water in the soil profile has total potential equal to zero.*

water table, while the gravitational potential increases. The solution of the integral is about 750 kg of water in the soil profile ($1 \, \text{m} \times 2 \, \text{m} \times \int \theta(z) \, dz \approx 0.75 \, \text{m}^3$).

5.4   Liquid- and Vapour-Phase Equilibrium

When liquid water is in equilibrium with water in the vapour phase, the water potentials in the two phases are equal. The water potential in the vapour phase can be found by computing the work that would be required to create a unit volume of vapour from free liquid water (Campbell, 1977). The relationship thus obtained between water potential and vapour pressure is

$$\psi = \frac{RT_K}{M_w} \ln \frac{e}{e_s} \qquad (5.44)$$

where ψ is the water potential [J kg^{-1}], R is the gas constant (8.3143 J mol^{-1} K^{-1}), T_K is temperature [K], M_w is the molecular weight of water (0.018 kg mol^{-1}), e is the vapour pressure [Pa] and e_s is the vapour pressure at saturation [Pa]. The ratio e/e_s is the relative humidity h_r. For humidities above about 0.95, eqn (5.44) can be approximated by the first two terms of the series for the logarithm. At $T_K = 293 \, \text{K}$,

$$\psi = 1.37 \times 10^5 (h_r - 1) \qquad (5.45)$$

so a 1 % reduction in relative humidity reduces water potential by 1370 J kg^{-1}. Obviously, humidities in moist soil (soil wetter than about –2000 J kg^{-1}) are always near 1.0. Table 5.2 shows relative humidities of soils over a range of water potentials. The relationship between humidity and water potential is useful in describing vapour-phase water in soils and in providing a means for measuring water potential.

   Equation (5.44) can also be used to compute the water potential of frozen soil. When an ice phase is present in soil, some of the water is frozen and some remains unfrozen. The water potential of the unfrozen water is completely determined by the temperature of the soil. Lowering the temperature lowers the water potential and the unfrozen water content. The saturation vapour pressures $e_s(T)$ of both water and ice can be computed using the equation of Buck (1981):

$$e_s(T) = a \exp\left(\frac{bT}{T + c}\right) \qquad (5.46)$$

where a, b and c are constants for water ($a = 0.611$ kPa, $b = 17.502$ and $c = 240.97 \, °\text{C}$) and for ice ($b_i = 22.452$, $c_i = 272.55$, where the subscript i refers to ice). When ice is present, the ratio of vapour pressures in eqn (5.44) is the ratio of the saturation vapour pressure over ice to the saturation vapour pressure over water. Using eqns (5.46) and (5.44) gives

$$\psi = \frac{RT_K}{M_w} \left(\frac{b_i T}{c_i + T} - \frac{b}{c + T} \right) \tag{5.47}$$

where T is the Celsius temperature (temperature drop below the freezing point) and T_k is the kelvin temperature. For temperatures above about -10 °C. Equation (5.47) is almost linear, with a slope of about $1220\,\mathrm{J\,kg^{-1}\,°C^{-1}}$. Values for freezing-point depression corresponding to various water potentials are given in Table 5.2.

5.5 EXERCISES

5.1. Use the program PSP_travelTimeAnalysis to compute the water content of the soil sample dataTDRSoil.dat. Change the values of bulk density, temperature and solid-phase dielectric permittivity to investigate the effects on the computed water content by using the dielectric mixing model.

5.2. If the temperature of the soil liquid phase increases, does the bulk dielectric permittivity increase or decrease? What effect would this change have on soil water content computation?

5.3. Compute the matric water potential for water in a saturated pore 10 μm in diameter, at 20 °C. The surface tension and liquid density at this temperature are $72.75 \times 10^{-3}\,\mathrm{N\,m^{-1}}$ and $1027.8\,\mathrm{kg\,m^{-3}}$, respectively.

5.4. By using the integration presented in the program PSP_columnWaterContent.py, compute the total amount of soil water in a clay soil for a profile that is 1 m deep. Use the parameters presented in Table 5.4

5.5. By using the two files sand.txt and silt_loam.text, fit Campbell's equation, van Genuchten's equation with no parameter restriction and van Genucthen's equation with the restriction $m = 1 - 1/n$. Print the parameters and discuss the differences obtained by using the different models. Also discuss the differences in the estimated parameters for the two soils.

5.6. Compute the water content at field capacity (assume $\psi_{FC} = -33\,\mathrm{J\,kg^{-1}}$), and permanent wilting point (assume $\psi_{PWP} = -1500\,\mathrm{J\,kg^{-1}}$) using van Genuchten's equation for a soil having the following parameters: $\alpha = 0.15$, $n = 1.35$, $\theta_r = 0.01$ and $\theta_s = 0.46$. The units of water potential are $\mathrm{J\,kg^{-1}}$. Plant available water is assumed to be the difference between these values. Compute the plant available water of the soil.

5.7. Derive the capacity for Campbell's and van Genuchten's equations by computing the derivatives of the two functions with respect to ψ.

6

Steady-State Water Flow and Hydraulic Conductivity

6.1 Forces on Water in Porous Media

Before considering the flow equation, it is helpful to consider the forces that cause water to flow in soil and the forces that retard water flow. The units used to measure water potential are joules per kilogram. A potential gradient therefore has units of newtons per kilogram, or force per unit mass of water. A potential gradient can therefore be thought of as an actual force acting on the water in a porous system.

In steady flow, this force from the water potential gradient is balanced by a retarding force. Reynolds numbers quantify the relation between these two forces. For water flow in porous materials, Reynolds numbers are less than 10^{-3}, indicating that inertial forces are much smaller than viscous forces. The retarding force on the water is therefore viscous drag. The energy in the flowing water is dissipated as heat from friction in the soil. If we can assume that water in porous material behaves as a Newtonian fluid (viscosity independent of rate of shear), then the flux of water through a porous material can be shown to be directly proportional to the potential gradient. This relationship was obtained experimentally by Darcy, and is known as Darcy's law:

$$ f_w = -K\frac{d\psi}{dx} \tag{6.1} $$

where f_w is the water flux density and K is the hydraulic conductivity.

At higher pore flow velocities, deviations from this linear relation occur. The deviations result from the kinetic energy, which determines the formation of domains of rotational flow and the onset of turbulence (Bear, 1972). With increasing flow velocities, these domains of rotational flows increase, until the entire flow is turbulent. Commonly, in soils, such flow velocities are not reached and the flow is laminar.

6.2 Water Flow in Saturated Soils

When water flows in saturated soil, the driving force usually results from gradients in pressure, gravitational or overburden potentials. The hydraulic conductivity of saturated

Soil Physics with Python. First Edition. Marco Bittelli, Gaylon S. Campbell and Fausto Tomei.
© Marco Bittelli, Gaylon S. Campbell and Fausto Tomei 2015. Published in 2015 by Oxford University Press.

soil depends on the size and distribution of pores in the soil and therefore on bulk density, soil texture and soil structure. It is generally assumed to remain constant for a given material and location, although clogging of pores by clay migration can alter hydraulic conductivity markedly in saturated flow experiments. Moreover, in swelling–shrinking clay soils, the volumetric and structural changes due to swelling and shrinking also determine changes in hydraulic conductivity.

If f_w and K are constant in eqn (6.1), then the variables can be separated and the equation integrated from x_1 to x_2:

$$\frac{d\psi}{dx} = -\frac{f_w}{K} \tag{6.2}$$

$$\int_{x_1}^{x_2} \frac{d\psi}{dx}\, dx = \frac{\psi(x_2) - \psi(x_1)}{x_2 - x_1} = -\frac{f_w}{K} \tag{6.3}$$

Rearranging eqn (6.3) leads to

$$f_w = -K\frac{\psi(x_2) - \psi(x_1)}{x_2 - x_1} = -K\frac{\Delta\psi}{\Delta x} \tag{6.4}$$

Using eqn (6.4), the flux density of water in saturated soil can be found if the saturated hydraulic conductivity and the potential difference across a soil column of length x are known. Alternatively, if the flux density of water, the pressure drop across a soil column and the length of the column are known, then K can be determined. In Darcy's law, the empirical parameter depending on material properties is the hydraulic conductivity. If the soil is saturated, this parameter is called the saturated hydraulic conductivity; if it is not saturated, it is called the unsaturated hydraulic conductivity, or simply the hydraulic conductivity. This formulation applies for steady water flow, where there is no water storage and flux is not dependent on location.

6.3   Saturated Hydraulic Conductivity

A number of equations have been derived for predicting the saturated hydraulic conductivity from the pore size distribution. One begins with Poiseuille's equation, which describes the flux of water in a single capillary. Capillaries are interconnected in various ways to represent soil pores. A tortuosity correction is often applied to account for the increased path that must be covered by water moving through the porous material, and the equation is integrated over all pore sizes present in the soil. The result of one such model is (Scheidegger, 1960)

$$K_s = \rho_l \frac{\phi_f r^2}{\eta l} \tag{6.5}$$

where ϕ_f is the total porosity, l is a tortuosity factor, r is a 'mean hydraulic radius' representative of the pore size distribution of the porous material and η is viscosity. A different approach was used by Childs and Collis-George (1950) and Marshall (1958). Since this approach is useful for finding unsaturated as well as saturated conductivity, it will be examined in detail.

Poiseuille's law describes the flow of water in a cylinder of radius R [m] and length L [m]. A hydrostatic pressure difference [N m^{-2}] between points P_1 and P_2 determines a pressure gradient ΔP. Let us now consider a cylinder of radius r smaller than the total cylinder radius R. The hydrostatic force F_h [N] is given by the pressure difference ΔP multiplied by the cross-sectional area πr^2:

$$F_h = \pi r^2 \Delta P \qquad (6.6)$$

The force that acts against the hydrostatic pressure is the viscous drag force F_d, acting along the interior area of the cylinder:

$$F_d = 2\tau \pi r L \qquad (6.7)$$

where τ is the shear stress [N m^{-2}] and $2\pi r L$ is the total interior area of the cylinder. When the flow is steady-state, the two forces are balancing each other and the shear stress can be expressed as

$$\tau = r\frac{\Delta P}{2L} \qquad (6.8)$$

The law of viscosity for momentum transport states that the shear stress is determined by the velocity gradient and the dynamic fluid viscosity η [kg m^{-1} s^{-1}]. Therefore the shear stress of eqn (6.8) can be written as

$$\tau = r\frac{\Delta P}{2L} = -\eta\frac{dv}{dr} \qquad (6.9)$$

The first step in the derivation of Poiseuille's law is to compute the velocity as a function of the internal radius r, therefore integrating eqn (6.9). Rearranging eqn (6.9) to bring the radius on one side and the velocity on the other leads to

$$r\,dr = -\frac{2L\eta}{\Delta P}\,dv \qquad (6.10)$$

The integration is performed from r to the total radius of the cylinder R and from the corresponding velocities $v(r)$ to $v(R)$ (note that on the cylinder wall, the velocity is zero and therefore $v(R) = 0$):

$$\int_r^R r\,dr = -\frac{2L\eta}{\Delta P} \int_{v(r)}^0 dv \tag{6.11}$$

$$\left.\frac{r^2}{2}\right|_r^R = -\left.\frac{2L\eta}{\Delta P}\right|_{v(r)}^0 \tag{6.12}$$

$$\frac{R^2}{2} - \frac{r^2}{2} = -\frac{2L\eta}{\Delta P} - v(r) \tag{6.13}$$

$$v(r) = \frac{\Delta P}{4L\eta}\left(R^2 - r^2\right) \tag{6.14}$$

The velocity distribution shows that the velocity is zero where the radius r is equal to the radius of the cylinder R and it is maximum at the centre of the cylinder, where $r = 0$. This equation describes the well-known parabolic distribution of velocities in a cylinder.

The following code `PSP_Poisseulle.py` is written to compute and plot the velocity vectors in a cylinder. The cylinder and then the velocity vectors are drawn.

```
#PSP_Poisseulle
from visual import *
import numpy as np

scene.background=(1,1,1)
scene.title = ""
l = 5.
R = 1.
D_P = 40.
eta = 1.
# draw cylinder
angles=arange(1.1*pi,2.1*pi,pi/20.)
n = 500
for i in range(n):
 spring=curve(color=(0,1,1), radius=0.06)
 for phi in angles:
    spring.append(pos=(l*(float(i)/float(n)-0.55), R*cos(phi),
                    R*sin(phi)))

for i in range(11):
  x = 0
  y = (float(i)/11.-0.5)*2.*R
  r = np.sqrt(x*x+y*y)
  if (r < R):
   arrow(pos=(0,y,x),axis = (D_P * (R*R-r*r)/(eta*l*4.),0,0),
                 shaftwidth = 0.03, color=(0,0,0))
```

Figure 6.1 shows the output of the program. Note that eqn (6.14) is implemented in the last line, to generate the arrows representing the velocity vectors as functions of

Figure 6.1 *Section of a capillary tube and parabolic distribution of velocities.*

the radius (r). The instructions `shaftwidth = 0.03, color=(0,0,0)` are Visual Python instructions to control the width of the arrow and its colour (0,0,0) for black.

However, for computation of water flow, in capillary tubes, the total volume of water flowing in the cylinder is of interest; therefore, the velocity $v(r)$ must be integrated over the flowing area:

$$Q = \int v(r)\, dA \tag{6.15}$$

The area is the cross-sectional area of the cylinder, which is a circle, and the integration can then be performed in cylindrical coordinates, where the incremental area is given by the change in radius multiplied by the change in angle: $dA = dr\, d\Phi$, where Φ is the angle, ranging from 0 to 2π. A double integration is then performed over the radius and the angle:

$$Q = \int_0^R \int_0^{2\pi} v(r) r\, dr\, d\Phi \tag{6.16}$$

Since the integral over the angle is 2π,

$$Q = 2\pi \int_0^R v(r) r\, dr \tag{6.17}$$

Now the integral can be solved by substituting the definition of velocity from eqn (6.14):

$$Q = \frac{\Delta P}{4L\eta} 2\pi \int_0^R \left(R^2 - r^2\right) r\, dr \tag{6.18}$$

$$Q = \frac{\Delta P \pi}{2L\eta} \left(\int_0^R R^2 r\, dr - \int_0^R r^2 r\, dr\right) \tag{6.19}$$

Since R is constant, it can be moved out of the integral:

$$Q = \frac{\Delta P \pi}{2L\eta} \left(R^2 \int_0^R r\, dr - \int_0^R r^3\, dr\right) \tag{6.20}$$

We now solve the integral:

$$Q = \frac{\Delta P \pi}{2 L \eta} \left(R^2 \left. \frac{1}{2} r^2 \right|_0^R - \left. \frac{1}{4} r^4 \right|_0^R \right) \tag{6.21}$$

$$Q = \frac{\Delta P \pi}{2 L \eta} \left(\frac{1}{2} R^4 - \frac{1}{4} R^4 \right) = \frac{\Delta P \pi}{2 L \eta} \frac{1}{4} R^4 \tag{6.22}$$

Equation (6.22) shows that the flow in a cylinder is proportional to the pressure gradient and the fourth power of the radius and inversely proportional to the dynamic viscosity of the fluid. This equation can be written for an infinitesimal variation of pressure with respect to the cylinder length (dP/dx):

$$Q = \frac{\pi R^4}{8 \eta} \frac{dP}{dx} \tag{6.23}$$

where R is now the total radius of the capillary and x is the axis of flow direction. Considering the discharge over a specific cross-cylinder area, it is possible to define a specific discharge

$$q = \frac{Q}{A} = \frac{\pi R^4}{8 \eta} \frac{1}{\pi R^2} \frac{dP}{dx} = \frac{R^2}{8 \eta} \frac{dP}{dx} \tag{6.24}$$

The term $R^2/8\eta$ is the hydraulic conductivity, which is the product of an intrinsic conductivity given by the pore radius and an extrinsic conductivity given by the liquid properties (viscosity). Equation (6.24) shows that the conductivity is proportional to the square of the pore radius. If the soil is seen as a large number of capillary tubes of varying sizes, then the flux density in soil is the flux density for each capillary size present in the soil multiplied by the area of that size of capillary per unit cross-sectional area of soil. Contributions from each pore class are summed to get the total flux. Since pores are not continuous, some provision must be made for the way pores can fit together. The Childs and Collis-George (1950) model assumes that only pores in a direct sequence contribute to the flux and that flux is always controlled by the smaller of two pores in a sequence.

If a soil column were broken at some arbitrary point as shown in Fig. 6.2, the exposed face could be examined to determine the area of pores having radii between r and $r + dr$. This area can be expressed as

$$\frac{dA}{A} = F(r) \, dr \tag{6.25}$$

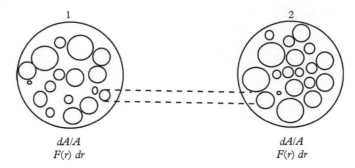

Figure 6.2 *Schematic of the statistical approach for estimation of hydraulic conductivity.*

where $F(r)$ is a pore size distribution function defined such that the total porosity of the soil is

$$\phi_f = \int_0^{r_{max}} F(r) \, dr \qquad (6.26)$$

where r_{max} is the largest capillary tube (or pore). In eqn (6.25), dA/A represents the probability of finding a pore of a given size on one face of the broken soil column. The probability of finding a continuous pore from one face to the other is the product of the probabilities of finding a pore of given size on either face. The contribution to conductivity for pores of radius r and $r + dr$ is therefore

$$\left(\frac{\rho_l r^2}{8\eta}\right)\left(\frac{dA}{A}\right)^2 = \left(\frac{\rho_l r^2}{8\eta}\right) F(r) \, dr \, F(r) \, dr \qquad (6.27)$$

The hydraulic conductivity is the integral over all pore classes, or

$$K = \frac{\rho_l}{8\eta} \int_0^r \int_0^r r^2 F(r) \, dr \, F(r) \, dr \qquad (6.28)$$

The pore size distribution function can be inferred from a moisture retention function if the pores are assumed to be cylindrical so that the capillary rise equation (5.18) gives an estimate of the largest water-filled pore at any given matric potential. Combining eqs (5.18) and (5.27) gives

$$r = \left(\frac{2\gamma}{\rho_l \psi_e}\right)\left(\frac{\theta}{\theta_s}\right)^b \qquad (6.29)$$

Since the pores are assumed to be cylindrical, $F(r) \, dr = d\theta$ is the change in water content associated with draining all pores between radius r and $r + dr$. Substituting for r^2 and $F(r) \, dr$ in eqn (6.28) and integrating gives

$$K = \frac{\gamma^2 \theta_s^2}{2\rho_l \eta \psi_e^2 (2b+1)(2b+2)} \left(\frac{\theta}{\theta_s}\right)^{2b+2} \qquad (6.30)$$

The saturated hydraulic conductivity is the value from eqn (6.30) when $\theta = \theta_s$. Equation (6.30) says that the saturated hydraulic conductivity of a soil is determined by three soil properties: θ_s, ψ_e and b. Of these, ψ_e is the most important. For soils with similar θ_s and b values,

$$K_s \psi_e^2 = \text{constant} \qquad (6.31)$$

This important result is the basis for scaling in studies of variability in hydraulic properties of soils. It says that at a given water content (again assuming constant θ_s and b), variation in ψ_m is directly related to variation in ψ_e, and variation in K is inversely related to ψ_e^2. It is apparent that for water flow and retention calculations, ψ_e is an important soil parameter. We will investigate scaling properties of hydraulic properties in Chapter 7.

6.3.1 Calculating Saturated Conductivity from Soil Texture Data

The relationships between moisture retention and soil texture in Chapter 5 and the result obtained in eqn (6.31) suggest that saturated hydraulic conductivity is related to soil texture. Table 6.1 gives saturated hydraulic conductivity data from Rawls *et al.* (1992)

Table 6.1 *Typical saturated hydraulic conductivities for the soil textural classes*

Textural class	d_g [μm]	ψ_e [J kg^{-1}]	b	K_s [g s m^{-3}][a]		
				Rawls *et al.* (1992)	Eqn (6.32)	Eqn (6.34)
Sand	211.11	−0.7	1.7	5.800	3.697	2.814
Loamy sand	121.77	−0.9	2.1	1.700	2.117	1.994
Sandy loam	61.67	−1.5	3.1	0.720	0.419	0.703
Loam	19.82	−1.1	4.5	0.370	0.978	0.333
Silt loam	10.53	−2.1	4.7	0.190	0.142	0.167
Sandy clay loam	25.16	−2.8	4.0	0.120	0.056	0.196
Clay loam	7.10	−2.6	5.2	0.064	0.072	0.101
Silty clay loam	3.35	−3.3	6.6	0.042	0.035	0.033
Sandy clay	11.36	−2.9	6.0	0.033	0.050	0.055
Silty clay	2.08	−3.4	7.9	0.025	0.030	0.017
Clay	1.55	−3.7	7.6	0.017	0.023	0.017

[a] To convert to values in kg s m^{-3} (or cm s^{-1}), multiply these values by 10^{-3}.

for the various textural classes. There is a general trend towards lower conductivity with smaller mean particle diameter, but the variation within each texture class is large, so one must be careful in using average data like those in Table 6.1.

Correlation of saturated hydraulic conductivity with air entry potentials given in Table 6.1 gives

$$K_s = 0.0013 \, |\psi_e|^{-3} \tag{6.32}$$

which is similar to eqn (6.31), but with a power of 3 instead of 2 on the air-entry potential. Saturated hydraulic conductivity calculated using eqn (6.32) with representative values for silt and clay contents are also shown in Table 6.1. Equation (6.32) was derived by integrating from the smallest to the largest pores, and considers all pores in the soil, even though the small pores have almost nothing to do with saturated hydraulic conductivity. Rawls *et al.* (1992) suggest a different approach that focuses on the largest pores. They suggest computing saturated hydraulic conductivity from the equation

$$K_s = B\phi_e^4 \tag{6.33}$$

where ϕ_e is the porosity between saturation and -33 J kg^{-1} water potential (the largest soil pores). If we assume a power-law water retention curve, this equation becomes

$$K_s = 0.07 \left[\theta_s \left(1 - \left(\frac{\psi_e}{-33} \right)^{1/b} \right) \right]^4 \tag{6.34}$$

where B has been replaced with 0.07, the value that best fits the data in Table 6.1. The values predicted by eqn (6.34) are shown in Table 6.1. Even though this equation has only one free parameter (a value of $\theta_s = 0.5$ was assumed for all textures), it fits the data with only about half the error of eqn (6.32). It also has the advantage that it shows an explicit bulk density dependence (through θ_s) of K_s.

6.4   Unsaturated Hydraulic Conductivity

When the soil desaturates, the driving force for flow becomes the gradients in matric and gravitational potentials. As the largest pores in the soil drain, the hydraulic conductivity is rapidly reduced. This leads to an apparent paradox where doubling the driving force for flow across a soil column by lowering the water potential on one end will actually decrease flow. This occurs because the hydraulic conductivity is reduced more than the driving force is increased. Hydraulic conductivity ranges over many orders of magnitude between saturation and dryness. The dramatic reduction in unsaturated hydraulic conductivity with water content results from the fact that as water content is reduced, the largest pores empty first. According to Poiseuille's equation, small pores conduct water much less readily than large pores (eqn (6.24)). In addition, the path for flow becomes much more tortuous as the soil desaturates. These factors have been successfully incorp-orated in models that give reasonable descriptions of unsaturated hydraulic conductivity as a function of water content (Campbell, 1974).

Campbell model

Based on the theory presented above, Campbell (1974) proposed a formulation to compute the unsaturated hydraulic conductivity. The advantage of this model is that the soil water retention curve proposed by Campbell (1985) is analytically integrable. The unsaturated hydraulic conductivity function is given by

$$
K = \begin{cases} K_s \left(\dfrac{\psi_e}{\psi_m} \right)^{(2+3/b)} & \text{if } \psi_m < \psi_e \\ K_s & \text{if } \psi_m \geq \psi_e \end{cases} \tag{6.35}
$$

where K [kg s m^{-3}] is the unsaturated conductivity and K_s [kg s m^{-3}] is the saturated conductivity.

Mualem–van Genuchten model

Based on the same theory, Mualem (1976) later proposed a model in which the integral was squared and normalized over the saturated water content using the degree of saturation S_e:

$$
K_r(S_e) = S_e^l \left[\frac{\displaystyle\int_0^r r F(r)\, dr}{\displaystyle\int_0^{r_{\max}} r F(r)\, dr} \right]^2 \tag{6.36}
$$

Note that the square in eqn (6.36) comes from eqn (6.27). As described above, by employing the Young–Laplace law, Mualem derived the final formula

$$
K_r(S_e) = S_e^l \left[\frac{\displaystyle\int_0^{S_e} \frac{1}{\psi(S_e)}}{\displaystyle\int_0^1 \frac{1}{\psi(S_e)}} \right]^2 \tag{6.37}
$$

where ψ is the water potential and l is a parameter related to the tortuosity of the pore space. The Mualem model was later combined with a model for the water retention curve (van Genuchten, 1980), to obtain the so-called Mualem–van Genuchten model. The van Genuchten (1980) model was described in Chapter 5 and is given by

$$
S_e(\psi) = \frac{\theta - \theta_r}{\theta_s - \theta_r} = \frac{1}{[1 + (\alpha\psi)^n]^m} \tag{6.38}
$$

where S_e is the degree of saturation, lying in the range $[0, 1]$, and α, n, m, θ_s and θ_r are fitting parameters.

The major limitation of this model is that when the parameters m and n are independent in the van Genuchten (1980) equation, the integral has no analytical solution, and therefore the integration must be performed numerically using 'special' functions. Here we present an algorithm to perform the numerical integration of eqn (6.37). The integration is performed using the incomplete beta function (Press *et al.*, 1992). Equation (6.37) can be written as

$$K(S_e) = K_s S_e^l [I_\xi (p, q)]^2 \tag{6.39}$$

where $p = m + 1/n$, $q = 1 - 1/n$, assuming independent parameters n and m and I_ξ is the incomplete beta function. A detailed descriptions of the derivation from eqn (6.37) to eqn (6.39) is provided by van Genuchten *et al.* (1991) (pp. 13–17).

The following is a Python code, named PSP_unsaturatedConductivity.py, where integration is performed to obtain the unsaturated hydraulic conductivity curve. The parameters of the soil water retention curve (obtained from the nonlinear fitting presented in Chapter 5), α, n, m, θ_r, θ_s and K_s, are read from a text file called soil.txt, where parameters for a silt loam soil are written.

```python
#PSP_unsaturatedConductivity
from PSP_readDataFile import readDataFile
import matplotlib.pyplot as plt
import math
import numpy

NODATA = -9999

def betacf(a, b, x):
    maxNrIterations = 50
    maxEpsilon = 0.0000003
    am = 1
    bm = 1
    az = 1
    bz = 1 - (a+b) * x / (a+1)
    myEpsilon = 1
    m = 1
    while (myEpsilon > (maxEpsilon * abs(az))):
        if (m > maxNrIterations): return (NODATA)
        d = (m * (b - m) * x) / ((a + 2*m -1) * (a + 2*m))
        ap = az + d * am
        bp = bz + d * bm
        d = -((a + m) * (a + b + m) * x) / ((a + 2*m) * (a + 2*m + 1))
        app = ap + d * az
        bpp = bp + d * bz
```

```
        am = ap / bpp
        bm = bp / bpp
        old_az = az
        az = app / bpp
        bz = 1
        m += 1
        myEpsilon = abs(az - old_az)

    return (az)

def incompleteBetaFunction(a, b, x):
    if ((x < 0.) or (x > 1.)):
        return (NODATA)
    if ((x == 0.) or (x == 1.)):
        bt = 0.
    else:
        bt = (math.exp(math.lgamma(a + b) - math.lgamma(a) -
                math.lgamma(b) + a * math.log10(x) + b *
                math.log10(1. - x)))
    if (x < ((a + 1.) / (a + b + 2.))):
        return(bt * betacf(a, b, x) / a)
    else:
        return(1. - bt * betacf(b, a, 1. - x) / b)

def computeConductivity(currentSe, n, m, Ks):
    p = m + 1. / n
    q = 1. - 1. / n
    z = currentSe ** (1. / m)
    myBeta = incompleteBetaFunction(p, q, z)
    if (myBeta == NODATA):
        return (NODATA)
    else:
        return(Ks * currentSe * (myBeta ** 2.))

def main():

    A, isFileOk = readDataFile("soil.txt",1,',', True)
    if ((not isFileOk) or (len(A[0]) != 6)):
        print('warning: wrong soil file.')
        return (False)

    VG_alpha = A[0,0]
    VG_n = A[0,1]
    VG_m = A[0,2]
    VG_thetaR = A[0,3]
```

```
thetaS = A[0,4]
Ks = A[0,5]

A, isFileOk = readDataFile("SWC.txt",1,'\t', False)
if (not isFileOk):
    print('warning: wrong SWC file in row nr.', A+1)
    return (False)

waterPotential = A[:,0]
waterContent = A[:,1]
conductivity = numpy.zeros(len(waterContent))

for i in range(0, len(waterContent)):
    currentSe = (waterContent[i] - VG_thetaR) / (thetaS - VG_thetaR)
    conductivity[i] = computeConductivity(currentSe, VG_n, VG_m, Ks)
    if (conductivity[i] == NODATA):
        print ('Error in compute conductivity')
        return (False)

plt.loglog (waterPotential, conductivity, 'ko')
plt.xlabel('Water Potential [J kg$^{-1}$]',fontsize=14,labelpad=2)
plt.ylabel('Hydraulic Conductivity [kg s$^{-1}$ m$^{-3}$]',
            fontsize=14,labelpad=2)
plt.show()

main()
```

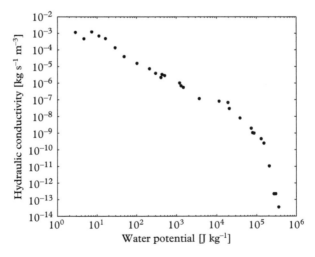

Figure 6.3 *Estimated unsaturated hydraulic conductivity curve as function of water potential obtained from numerical integration for a silt loam soil.*

The numerical integration presented above requires evaluation of special functions such as the beta function. Convergence of these functions is not assured—in particular when the data set is incomplete or scattered. The restriction of $m = 1 - 1/n$ in the van Genuchten (1980) function allows for analytical integration of eqn (6.38):

$$K(h) = \frac{K_s \{1 - (\alpha\psi)^{mn} [1 + (\alpha\psi)^n]^{-m}\}^2}{[1 + (\alpha\psi)^n]^{ml}} \tag{6.40}$$

A feature of this derivation is that because of the relationship between the hydraulic conductivity and the radius, the estimation of hydraulic conductivity is very sensitive to large pores. Since the curve is continuous around the air-entry potential, when the soil water potential $h \to 0$, according to the Young–Laplace law, the radius becomes infinitely large, which is clearly not realistic. Ippisch *et al.* (2006) demonstrated that for values of $n < 2$ and $\alpha h_a > 1$, an air-entry value must be introduced in the van Genuchten (1980) formulation. A detailed theoretical description of these limitations was presented by Ippisch *et al.* (2006), who described and specified the conditions for which the classical van Genuchten–Mualem model leads to wrong predictions of the hydraulic conductivity curve and presented a modified formulation that included an air-entry value. The modified formulation is presented below.

Ippisch–Mualem–van Genuchten model

The modified formulation as proposed by Ippisch *et al.* (2006) has the form

$$S_e = \begin{cases} \dfrac{1}{S_c} [1 + (\alpha\psi)^n]^{-m} & \text{if } \psi \le \psi_e \\[2ex] 1 & \text{if } \psi > \psi_e \end{cases} \tag{6.41}$$

where S_e is the degree of saturation, α, m and n are fitting parameters, and $S_c = [1 + (\alpha\psi_e)^n]^{-m}$ is the water saturation at the air-entry potential ψ_e (the water potential is expressed as a negative number). The resulting hydraulic conductivity using the Mualem model is

$$K = \begin{cases} K_s S_e^l \left[\dfrac{1 - (1 - (S_e S_c)^{1/m})^m}{1 - (1 - S_c^{1/m})^m} \right]^2 & \text{if } S_e < 1 \\[2ex] K_s & \text{if } S_e \ge 1 \end{cases} \tag{6.42}$$

where l is the same parameter as in the original Mualem equation. Ippisch *et al.* (2006) suggested that ψ_e can be obtained from knowledge of the largest pore size, from the water retention curve or by inverse modelling. The air-entry potential could be obtained by letting ψ_e be a fitting parameter during the fitting procedure.

6.5 EXERCISES

6.1. Plot 'typical' hydraulic conductivity for sand, silt loam and clay soil as a function of water potential using data from Table 6.1 and eqn (6.14). Use a log–log scale and plot both saturated and unsaturated conductivity. Use the graphs to determine the water potential at which water would flow from silt loam into sand in a layered profile with infiltration.

6.2. Write a Python function to compute saturated hydraulic conductivity from bulk density, silt fraction and clay fraction. Use your program to show how tillage affects hydraulic conductivity (assuming that tillage decreases the bulk density).

7

Variation in Soil Properties

In each of the chapters so far, physical properties of soil have been discussed. In several instances, relationships have been derived between measured properties and other variables. Implicit in all of this is the assumption that a particular measurement or estimate of a property contains information about the value of that property or of a related property in the entire system represented by that sample at that specific scale. In Chapter 2, we presented the concept of a representative elementary volume, a soil volume representing an average value of a given property, such that the differences among the values of the properties at increasing scales are smaller than a given number ϵ. This approach was used to describe soil physical properties, from the pore scale of observations to the soil sample scale of observations. The concept of bulk density is an example of this 'upscaling' process, where the different densities of the different phases (solid, liquid and gas) are averaged into a value representing the 'bulk' density of the sample. As we pointed out in Chapter 2, this process is not reversible, meaning that when information is combined into an 'average' value, the information about that variable at smaller scales is lost. The dimension of the sample of a representative elementary volume is usually that of the cylindrical rings used for soil sampling, with height and diameter of a few centimetres. Clearly, this concept is limited in its applicability to a range of scales, since at increasing spatial scales, variation of the property in question may again be larger than a given number ϵ, because of soil layering, topographical or geological features. For instance, the average density of a soil volume of $2\,\mathrm{m}^3$, representing a soil profile, may be substantially different from the density of the chosen representative elementary volume, sampled at a given depth.

Moreover, experience shows that even measurements on several 'identical' soil samples of the same size never produce identical results. A single sample or estimate is therefore unlikely to contain complete information about the value of that property for other possible samples, for instance over a given area. How much information then is or can be contained in a given observation or set of observations? The theory of stochastic processes provides a tool for properly describing systems with uncertainty or variation. A stochastic description of a property can be used to determine the information content of data relating to that property. In this chapter, appropriate statistical descriptions of soil physical properties will be given and models will be extended to include uncertainty in input variables.

Soil Physics with Python. First Edition. Marco Bittelli, Gaylon S. Campbell and Fausto Tomei.
© Marco Bittelli, Gaylon S. Campbell and Fausto Tomei 2015. Published in 2015 by Oxford University Press.

7.1 Frequency Distributions

In the language of stochastic process theory, a given soil property is represented as a continuous real random variable. The random variable represents for example all possible values of bulk density or thermal conductivity in a plot, field or other specified area. A description of the random variable is required, based generally on observations of a small fraction of all possible realizations.

A first step in analysing variability in a set of observations of a physical property is to divide the range of the data into equal intervals, determine the number of observations falling within each interval and plot numbers as a function of values of the variate, as in Fig. 7.1. Figure 7.1 is a useful, but bulky summary of our knowledge about the random variable, the bulk density of Pima clay loam. To be useful, the properties of the random variable must be further summarized. Two obvious properties of the frequency distribution are its position and its spread.

Several measures of position are possible. The most useful is the arithmetic mean, which is calculated from

$$\langle x \rangle = \frac{1}{n} \sum_{i=1}^{n} x_i \tag{7.1}$$

where the x_i are members of a set of n observations of the soil property x. Other measures of position are the median and the mode. The median value is the central value,

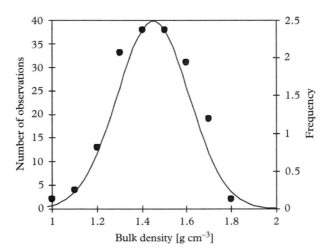

Figure 7.1 *Frequency distribution (points) for 180 bulk density observations on Pima clay loam (Warrick and Nielsen, 1980) and a normal distribution function with mean of 1.45 g cm^{-3} and standard deviation of 0.16 g cm^{-3}.*

which is the value having half the observations smaller and half larger than it. The mode is the value of x at which the frequency distribution is maximum. For symmetrical distributions, such as that in Fig. 7.1, the median, mean and mode coincide. For skew distributions, the median or mode may represent typical values better than the mean. Our discussion will cover only distributions that either are symmetrical, or can be made so by transformation. Properties of skew distributions will therefore not be discussed. The spread, or dispersion, about the mean can also be described in several ways. The most useful is called the variance, and is calculated from

$$s^2 = \frac{1}{n} \sum_{i=1}^{n} (x_i - \langle x \rangle)^2 \tag{7.2}$$

The standard deviation is s, the square root of the variance, and represents the root mean square (rms) deviation of the observations from the mean. The sample mean $\langle x \rangle$ and the variance s^2 can be used to estimate the mean, μ, and the variance, σ^2 of the random variable. It can be shown (Webster, 1977) that best unbiased estimates of the population mean and variance are given by $\mu = \langle x \rangle$ and $\sigma^2 = s^2 n/(n-1)$. The population variance is likely to be larger than the sample variance, because deviations of sample values from μ are likely to be larger than the deviations from $\langle x \rangle$. It is generally more useful to know the mean and variance of the population rather than those of the sample.

The mean and variance of a sample are often combined to form a coefficient of variation. This is given by

$$\mathrm{CV} = \frac{s}{\langle x \rangle} \tag{7.3}$$

The CV can be multiplied by 100 and expressed as a percentage rather than a fraction. This is a convenient way of expressing variation relative to the mean value. It is easily misused, however, especially when the mean is with respect to an arbitrarily chosen zero, as is the case with temperature scales.

Warrick and Nielsen (1980) give a useful table of CVs for several soil properties. Table 7.1 shows averages of a few of their entries.

Table 7.1 *Coefficients of variation for some soil properties*

Property	Mean	Standard deviation	CV-%
Bulk density [Mg m^{-3}]	1.4	0.098	7.0
Saturation water content [m^3 m^{-3}]	0.44	0.047	10.7
Saturated hydraulic conductivity [g s m^{-3}]	0.23	0.25	110

7.2   Probability Density Functions

The frequency distributions for many soil properties approximate the shape of an ideal frequency distribution known as the normal or Gaussian distribution. We have seen the Gaussian distribution in the description of particle size distribution in Chapter 2. For this distribution, the probability of occurrence of the value x is given by

$$p(x) = \left[\frac{1}{\sigma (2\pi)^{1/2}}\right] \exp\left[-\frac{(x-\mu)^2}{2\sigma^2}\right] \tag{7.4}$$

where μ is the mean of x and σ is the standard deviation. Figure 7.1 compares a measured frequency distribution with eqn (7.4), which represents the distribution one might obtain for a large number of samples. Equation (7.4) gives the probability density at x. It is normalized so that

$$\int_{-\infty}^{\infty} p(x)\, dx = 1 \tag{7.5}$$

The area under any part of the probability density function (PDF) is equal to the probability of finding a value of the random variable x in that range. It is useful to keep a few of these probabilities in mind. The area between $\mu - \sigma$ and $\mu + \sigma$ is 0.68, the area from $\mu - 2\sigma$ and $\mu + 2\sigma$ is 0.95 and that from $\mu - 3\sigma$ and $\mu + 3\sigma$ is 0.997. Thus, if the random variable representing a given soil property has a normal probability density function, a mean μ and a variance σ^2, then 68 % of the values will lie within ±1 standard deviation of the mean and 95 % within two standard deviations. This provides useful information for determining how much a given sample or group of samples can tell us about the mean of the population or how many samples are needed to reduce uncertainty to a given value. The relationship between uncertainty of an estimate of the mean and uncertainty in the individual observations is

$$\sigma_{(x)} = \frac{\sigma}{\sqrt{n}} \tag{7.6}$$

where n is the number of samples in the mean. This equation can be used to determine the number of samples needed to estimated mean bulk density to within ±0.1 Mg m^{-3} with 95 % certainty. First set $2\sigma_{(x)} = 0.1$ Mg m^{-3} $= 2\sigma/n^{1/2}$. From Table 7.1, $\sigma = 0.098$ Mg m^{-3}, so $n = 4\sigma^2/(0.1)^2 = 3.81$; four samples should therefore be sufficient.

7.3   Transformations

If the frequency distribution for a soil property follows a normal probability density function, then its position and dispersion are easily and conveniently described by the arithmetic mean and the variance. Probabilities of occurrence for various values of the property are easily found. The most useful of the transformations for soil physical properties is the log transform. In Chapter 2, we have seen how the particle size distribution can be described by a Gaussian or bimodal Gaussian distribution when the particle diameter is log-transformed. The frequency distribution of the transformed data is symmetric, with a mean

$$\langle \ln x \rangle = \frac{1}{n} \sum \ln x \tag{7.7}$$

and a variance

$$\sigma^2 = \frac{1}{n-1} \sum (\ln x_i - \langle \ln x \rangle)^2 \tag{7.8}$$

The mean calculated in eqn (7.7) is the geometric mean discussed in Chapter 3 and can also be calculated from

$$\bar{g} = \left(\prod_{i=1}^{n} x_i \right)^{1/n} \tag{7.9}$$

where $\prod$ is the product operator: $x_1 x_2 x_3 x_4 \ldots x_n$ (Webster, 1977). The geometric mean $\bar{g} = \exp(\langle \ln x \rangle)$ is preferable to the arithmetic mean for representing log-normally distributed data. This can be seen by considering a hypothetical set of values [0.1, 0.1, 0.1, 0.1, 10] (the last value was from a core with a wormhole in it). The arithmetic mean is 2.08. The geometric mean is 0.25, a number much more representative of the data set. The reciprocal transform also deserves mention. For this, the mean is

$$\left\langle \frac{1}{x} \right\rangle = \left(\frac{1}{n} \right) \sum \frac{1}{x_i} \tag{7.10}$$

The reciprocal of $\langle 1/x \rangle$ is known as the harmonic mean. For instance, stomatal resistances of plant leaves have a skew distribution, but their reciprocals (conductances) are normally distributed. One would therefore want to do reciprocal transforms on stomatal resistance data to compute means or perform statistical tests.

7.4 Spatial Correlation

All of the derivations so far have assumed that individual observations are independent and uncorrelated. This is likely to be a poor assumption for soil data, since samples taken in close proximity to each other are generally more similar than more widely spaced samples. Data can be tested for spatial correlation using one of three related functions: the autocovariance function C, the autocorrelation function r or the semivariogram function γ. All are calculated from sets of measured values of the soil property x, with measurements separated in space. Measurements in a grid pattern are best for computing these functions, since such data allow us to determine whether the functions are direction dependent.

The autocovariance for separation (or lag) k is calculated from

$$C_k = \frac{1}{n-k-1} \sum_{i=1}^{n-k} (x_i - \langle x \rangle)(x_{i+k} - \langle x \rangle) \tag{7.11}$$

For lag 0 ($k = 0$), $C_0 = \sigma^2$. The autocovariance function is the set of values of C_k produced for all possible values of k.

The autocorrelation is the ratio of the autocovariance at lag k to the total variance or autocovariance at lag zero:

$$r_k = \frac{C_k}{C_0} = \frac{C_k}{\sigma^2} \tag{7.12}$$

The autocorrelation is related to the correlation coefficient used in linear regression analysis. A linear regression of x_i on x_{i+k} would produce an estimate of a slope and intercept, and a correlation coefficient. The square of the correlation coefficient gives the fraction of the total variation in x_i that can be explained by variation in x_{i+k}. The autocorrelation for lag k gives the same information. The set of values for all possible k's produces the autocorrelation function. The semivariance is calculated from

$$\gamma_k = \frac{0.5}{n-k-1} \sum_{i=1}^{n-k} (x_i - x_{i+k})^2 \tag{7.13}$$

For a stationary random process (one for which the means and variances of all possible samples are equal),

$$\gamma_k = C_0 - C_k \tag{7.14}$$

The semivariance is zero at zero lag, and with increasing lag, approaches σ^2. Some measure of the distance over which to expect significant correlation would be useful. Russo and Bresler (1981) used the term 'integral scale' to describe such a measure. This is a term taken from the atmospheric turbulence literature (Lumley and Panofsky, 1964) and is defined as

$$\mathcal{J} = \sqrt{\int_0^\infty r_k \, k \, dk} \tag{7.15}$$

Russo and Bresler (1981) fitted empirical equations to r_k data to find values for the integral scale. The integral scales they found typically were about half the distance required for r_k to go to zero or for γ_k to reach its maximum value. They found integral scales for saturated water content to be 40–80 m, with the integral scale for conductivity being about half that.

The next step is to ask what it means if data are correlated. Russo and Bresler (1982) provide answers to this question, but at a level beyond that of this book. Their results and those of other similar studies can be summarized in more qualitative terms. Two questions might be considered. One would be how correlation affects our knowledge of adjacent, unsampled values, and the other would be how correlation affects our knowledge of characteristics of the population. For uncorrelated data, the best estimate of a value at an unsampled point is the population mean. Samples from any location are equally valuable in estimating the population mean, and an estimate of the uncertainty in the mean is given by eqn (7.6). For correlated data, a measurement taken near the location of an unsampled point should better represent the value at that point than does the mean. This is the basis of the mapping method known as kriging (Burgess and Webster, 1980) where unknown values of a property are computed as weighted averages of sampled values, with weights being proportional to correlations. On the other hand, two points separated by less than two integral scales provide less information about population properties than do samples separated by more than two integral scales. Efficient sampling for population properties therefore requires that sample locations be separated by several integral scales.

7.5   Approaches to Stochastic Modelling

The models presented in previous chapters have been deterministic; that is, for a given set of input values, the output is completely determined. In this chapter, descriptions of variability for the input values have been developed. We will now show how variability in input parameters relates to variability in model output. The PDF and its properties have been used to describe the soil characteristics that are model inputs. The next step is to make the model produce a PDF of output values. Once the output PDF is known, the mean, variance and other statistical moments of that property can be found.

In Section 7.4, data were transformed using log and reciprocal functions, and we noted that these transforms produced new PDFs. Other models can be thought of as transforms of the PDF as well. A model changes input variables into output variables and input PDFs into output PDFs. A deterministic model can therefore become a stochastic model by treating input parameters and variables as stochastic processes.

The simplest transforms, from a stochastic model point of view, are linear transformations. For these, the output PDF has the same shape as the input PDFs. If the inputs are Gaussian random processes, then the output PDF will be Gaussian. For a Gaussian process, the PDF is described by the mean and variance. If y is a function of variables a, b and c,

$$y = f(a, b, c) \tag{7.16}$$

then the expected value of y is obtained from

$$\langle y \rangle = f(\langle a \rangle, \langle b \rangle, \langle c \rangle) \tag{7.17}$$

If a, b and c are uncorrelated, then

$$\sigma_y^2 = \left(\frac{\partial y}{\partial a}\right)^2 \sigma_a^2 + \left(\frac{\partial y}{\partial b}\right)^2 \sigma_b^2 + \left(\frac{\partial y}{\partial c}\right)^2 \sigma_c^2 \qquad (7.18)$$

As an example, consider the error in gas-filled porosity ϕ_g that would result from measurement error in bulk density ρ_b and water content w. Combining eqns (2.20) and (2.23) gives

$$\phi_g = 1 - \rho_b \left(\frac{1}{\rho_s} - \frac{w}{\rho_l}\right) \qquad (7.19)$$

Assume the ρ_s and ρ_l are accurately known. The expected value for ϕ_g is the value obtained by calculation from eqn (7.19) using mean values of ρ_b and w. If errors in ρ_b and w are uncorrelated, then

$$\sigma_\phi^2 = \left(\frac{\rho_b}{\rho_l}\right)^2 \sigma_l^2 - \left(\frac{1}{\rho_s} - \frac{w}{\rho_l}\right)^2 \sigma_\rho^2 \qquad (7.20)$$

This equation is evaluated using mean values of ρ_b and w.

One serious pitfall of this approach is the assumption that the errors are uncorrelated. It is likely that ρ_b and w are calculated from the same set of measurements. If, for example, the measured soil mass were too low, that would decrease the calculated bulk density and increase the calculated water content.

Nonlinear transforms, such as those for moisture characteristics or hydraulic conductivity functions, often produce log-transformed PDFs. Water content has a Gaussian PDF. Water potential and hydraulic conductivity have log-normal distributions. Equations (7.17) and (7.18) can still be used to calculate means and variances, but these values are more difficult to interpret than for linear transforms.

If the mean and variance of a Gaussian process are known for an input variable of a nonlinear transform, the PDF of the output can be generated using a technique called Monte Carlo simulation. For this technique, input variables are drawn from populations having the specified mean and variance.

7.5.1 Scaling Methods

Because of the heavy computing requirements and the difficulty of visualizing the effects of variability in several input variables at once, it is sometimes useful to use scaled variables. This concept was introduced by Miller and Miller (1956) and is called *Miller scaling*. The Miller scaling method derives scaling laws between porous media samples that differ in their characteristic length scale only and are geometrically similar, as depicted in Fig. 7.2.

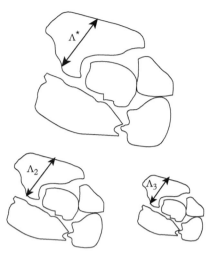

Figure 7.2 *Similarity for three scales.*

The main idea is that the geometry of the pore space is the same, but the media differ in their characteristic length scales, Λ. A reference characteristic length scale (identified by Λ^* in the figure) must be defined. For practical purposes, a mean diameter of the particle or pore size distribution is often used, but other measures could be used. To apply the Miller concept of similarity for derivations of hydraulic properties, some assumptions are necessary:

1. The medium is homogeneous, isotropic and permanent, and hence independent of position, orientation and time, when described macroscopically.
2. The liquid phase has uniform surface tension, contact angle, viscosity and density.
3. The medium is assumed to be in the *Richards* regime; this means that both the liquid and the gas are connected. Hence there are neither isolated bubbles nor isolated drops.

To find the relation between the scaling factor and the corresponding hydraulic properties, we assume that a given soil, S_1, is at a reference state with water content θ^*, matric potential ψ^*, conductivity K^*, characteristic length scale Λ^* and curvature of the water–air interface r_m^*. Another soil, S_2, is in a geometrically similar state, defined by a scaling factor κ. Based on the hydraulic property values of soil S_1, the corresponding values for soil S_2 can be obtained from Table 7.2.

While the first relations are quite obvious, it is necessary to discuss the scaling of the matric potential and hydraulic conductivity. The pressure at the air–water interface is proportional to the radius of curvature:

$$P = \frac{2\gamma}{r_m} \tag{7.21}$$

Table 7.2 *Miller scaling*

Soil	S_1	S_2
Scaling factor	1	κ
Characteristic length	Λ^*	$\Lambda = \kappa \Lambda^*$
Curvature of the water-air interface	r_m^*	$r_m = \kappa r_m^*$
Water content	θ^*	$\theta = \theta^*$
Matric potential	ψ^*	$\psi = \kappa^{-1}\psi^*$
Conductivity	K^*	$K = \kappa^2 K^*$

where γ is the surface tension [N m^{-1}]; therefore the scaling of the pressure leads to

$$P = \frac{2\gamma}{r_m} = \frac{2\gamma}{\kappa r_m^*} = \kappa^{-1} P^* \tag{7.22}$$

Hence the matric potential scales inversely proportional to the scaling factor:

$$\psi(\theta) = \kappa^{-1}\psi^*(\theta) \tag{7.23}$$

By using a dimensional analysis of Richards' equation (Richards' equation will be described in the following chapters), it is possible to show that the scaling approach can be applied to the hydraulic conductivity:

$$K(\theta) = \kappa^2 K^*(\theta) \tag{7.24}$$

The hydraulic properties described by the Miller and Miller (1956) approach are therefore specified by providing a reference point where they are computed and a characteristic length.

The limitation of the Miller assumption is that the soil must have constant porosity of the heterogeneous soil, since there is only one scaling factor κ. Some authors have presented different scaling methods, but at the price of increasing the number of parameters.

The Miller characteristic length can be chosen to obtain a spatial distribution of the media that is homogeneous at the macroscopic scale but heterogeneous at the microscopic length. This objective can be accomplished by generating the characteristic length Λ as a realization of a stationary random space function, having a distribution function and an autocovariance function. This stochastic approach is useful since it can generate probability distributions of variables such as the soil water retention curve or the hydraulic conductivity.

Roth (1995) investigated fields that have certain statistical properties and are characterized by the scaling parameter κ, where the scaling parameter is used for

Miller–Miller-type scaling. It is assumed that the logarithm of the scaling factor,

$$f = \log\left(\frac{\Lambda}{\Lambda^*}\right) \tag{7.25}$$

has the following properties: (a) it is normally distributed, (b) it has variance $\mathrm{var}(f) = \sigma_f^2$, (c) it has an expectation $\langle f \rangle = 0$ and (d) it has autocovariance $C(\vec{h}) = E[f(\vec{r})f(\vec{r} + \vec{h})]$.

To generate a realization f_0 of a stationary random space f with these properties, the *Wiener–Khinchin* theorem can be used (van Kampen, 1981). This theorem states that the Fourier transform of the autocovariance function C is equal to the spectrum S of f_0, which is defined as $S(\nu) = |\tilde{f}_0(\nu)|^2$:

$$S = \mathcal{F}[C] \tag{7.26}$$

which defines a relation between the autocovariance of a random function and the amplitude of the different realizations. Since the autocovariance is the covariance of the variable against a time-shifted version of itself, a specific realization can be obtained by assigning fixed values of phases $\phi(\nu)$ to the frequency modes. The variable $\phi(\nu)$ is a fast-changing variable that makes the phases of adjacent frequency intervals mutually independent. A realization $\tilde{f}_0$ can be defined by

$$\tilde{f}_0(\nu) = [\mathcal{F}[C]]^{(1/2)} \exp[i\phi(\nu)] \tag{7.27}$$

For the numerical simulations, the function is needed only at discrete points. Therefore eqn (7.27) can be modified with the use of a random variable Ω, which simplifies eqn (7.27) and leads to

$$\tilde{f}_0(\nu_k) = [\mathcal{F}[C]]^{(1/2)} \exp(i\Omega) \tag{7.28}$$

where Ω is a uniformly distributed random variable on the interval $[0, 2\pi]$. The fast Fourier transform of $\tilde{f}_0$ finally leads to a discrete realization f_0 in position space.

Robin *et al.* (1993) gave a detailed analysis of this method, where they presented an algorithm, based on Fourier transforms, to generate pairs of random fields. As pointed out by Roth (1995), this method is advantageous for studying the influence of autocovariance models on phenomena relevant to soil physics, such as flow and transport, because it allows the generation of fields having identical large structure but different autocovariances. Two models for the autocovariance are used: a Gaussian model

$$C(x) = \sigma^2 \exp\left(-\frac{\pi}{4}\frac{|\chi|^2}{\iota^2}\right) \tag{7.29}$$

and an exponential model

$$C(x) = \sigma^2 \exp\left(-\frac{|\chi|}{\iota}\right) \tag{7.30}$$

where $|\chi| = \sqrt{x^2 + y^2}$ and ι is the correlation length, which is the characteristic length for the microscopic scale. In the program, these two models are implemented in the function **statisticalFunction()**. The realizations for the two models are identical, with equal generating sequences of Ω in eqn (7.28) for both fields. The difference between the two models is that the exponential model is more irregular with respect to the Gaussian model, since the former is not differentiable.

7.6   Numerical Implementation

We now present the project called `PSP_MillerMiller`. The project comprises three files:

1. `main.py`
2. `PSP_MillerMiller.py`
3. `PSP_soil.py`

The file `main.py` contains the function **main** with instructions to prompt the selection of the water retention and the autocovariance model. The function **plotHydraulic-Properties()** calls a function written in the file `MillerMilller.py`, to plot the soil water retention and the hydraulic conductivity curves, defined by different characteristic lengths $\sigma - \mu$, σ and $\sigma + \mu$. The variation in characteristic lengths is then defined by variation around the mean.

The number of grid cells (n) and the distance between two grid cells (dx, dy) respectively are initialized to generate the two-dimensional field as shown below. The values of the correlation lengths in the x and y directions are also defined: (lx, ly). The **main** function contains two calls to the functions **plotThetaField()** and **plotConductivityField()**. Then the function **heterogeneousField()** is called. This function is used to generate the heterogeneous field as described above. It is written in the file `MillerMilller.py`.

```
#main.py
from __future__ import print_function, division
from PSP_MillerMiller import *

def main():
    choice = 0
    print (CAMPBELL,' Campbell')
    print (VAN_GENUCHTEN,' van Genuchten')
    while (choice < CAMPBELL) or (choice > VAN_GENUCHTEN):
        choice = float(input("Choose water retention curve: "))
        if (choice < CAMPBELL) or (choice > VAN_GENUCHTEN):
            print('wrong choice.')
    waterRetentionCurve = choice
```

```
    sigma = 1.0
    plotHydraulicProperties(waterRetentionCurve, sigma)

    choice = 0
    print (GAUSSIAN,' Gaussian model ')
    print (EXPONENTIAL,' exponential model')
    while (choice < GAUSSIAN) or (choice > EXPONENTIAL):
        choice = float(input("Choose autocovariance model: "))
        if (choice < GAUSSIAN) or (choice > EXPONENTIAL):
            print('wrong choice.')
    funcType = choice

    n = 128
    dx = dy = 0.01
    lx = ly = 0.1
    f = heterogeneousField(n, dx, dy, lx, ly, funcType)
    f *= (sigma*sigma)
    plt.imshow(f,cmap=plt.gray()), plt.axis('off'), plt.title(''),
             plt.show()

    plotThetaField(waterRetentionCurve, f, n)
    plotConductivityField(waterRetentionCurve, f, n)
main()
```

The second file PSP_soil.py is used to define the class Csoil that is used to define the soil properties. This file will also be used in following chapters. The water retention variables are now objects of the class Csoil, which will be called by different routines. Here the Campbell (1985) and van Genuchten (1980) equations are implemented as described in Chapter 5.

```
#PSP_soil.py
CAMPBELL = 1
VAN_GENUCHTEN = 2


class Csoil:
    def __init__(self):
        self.thetaS = 0.46
        self.thetaR = 0.02
        self.Ks = 1.e-3
        self.Mualem_L = 0.5

        self.Campbell_he = -4.2
        self.Campbell_b = 3.6

        self.VG_he = 3.8
        self.VG_alpha = 0.29
```

```
        self.VG_n = 1.33
        self.VG_m = 1.-(1./self.VG_n)

def hydraulicConductivity(waterRetentionCurve, soil, signPsi):
    if (waterRetentionCurve == CAMPBELL):
        if (signPsi >= soil.Campbell_he):
            K = soil.Ks
        else:
            K = soil.Ks * (soil.Campbell_he / signPsi)**(2.+ 3./
                            soil.Campbell_b)
    elif (waterRetentionCurve == VAN_GENUCHTEN):
        if (signPsi >= 0.):
            K = soil.Ks
        else:
            Se = 1. / (1. + (soil.VG_alpha * abs(signPsi))**
                        soil.VG_n)**soil.VG_m
            K = soil.Ks * Se**soil.Mualem_L *
                (1.-(1.-Se**(1./soil.VG_m))**soil.VG_m)**2
    return(K)

def waterContent(waterRetentionCurve, soil, signPsi):
    if (waterRetentionCurve == CAMPBELL):
        if (signPsi >= soil.Campbell_he):
            return soil.thetaS
        else:
            Se = (soil.Campbell_he / signPsi)**(1. / soil.Campbell_b)
            return Se * soil.thetaS
    elif(waterRetentionCurve == VAN_GENUCHTEN):
        if (signPsi >= 0.):
            return soil.thetaS
        else:
            Se = 1. / (1. + (soil.VG_alpha * abs(signPsi))**
                    soil.VG_n)**soil.VG_m
            return Se*(soil.thetaS - soil.thetaR) + soil.thetaR
```

As described above, to generate spatial fields of heterogeneities, autocovariance were generated using stationary random space functions, by employing Fourier transforms as stated in the Wiener–Khinchin theorem. This computation is written in the function **heterogeneousField()**. In this function, the main variables used for generating the fields are described. The autocovariance can be chosen as prompted to the user. Then the function **statisticFunction()** performs the computation. If $\text{Type} = 1$, the Gaussian model is selected, while if $\text{Type} = 2$, the exponential model is selected. In the function **heterogeneousField()**, the realizations of the function f are generated. The random angle $\phi(\nu)$ of eqn (7.28) is generated by using the function **random** available from numpy. The variable r is assigned the statistical function by the instruction and then the Fourier transform of the statistical function is computed. Note that the entries of the array `fft_f[i][j]` are complex numbers. The Fourier transform is implemented in the module numpy.

```
#PSP_MillerMiller.py
import numpy as np
import matplotlib.pyplot as plt
from PSP_soil import *

GAUSSIAN = 1
EXPONENTIAL = 2

def statisticalFunction(x, y, funcType):
    if (funcType == GAUSSIAN):
        return np.exp(-0.25*np.pi*(x*x + y*y))
    elif (funcType == EXPONENTIAL):
        return np.exp(-np.sqrt(x*x + y*y))

def heterogeneousField(n, dx, dy, lx, ly, funcType):
    f = np.array([[complex(0.,0.)]*int(n+1)]*int(n+1))
    phi = np.zeros((n+1)**2)
    for i in range((n+1)**2):
        phi[i]=2*np.pi*np.random.random_sample()
    for i in range (int(n/2)+1):
        for j in range (int(n/2)+1):
            r = statisticalFunction(i*dx/lx, j*dy/ly, funcType)
            f[-i +int(n/2)][-j+int(n/2)] = complex(r,0.)
            f[+i +int(n/2)][-j+int(n/2)] = complex(r,0.)
            f[-i +int(n/2)][+j+int(n/2)] = complex(r,0.)
            f[+i +int(n/2)][+j+int(n/2)] = complex(r,0.)
    fft_f = np.fft.fft2(f);
    for i in range(n+1):
        for j in range(n+1):
            r = pow(2.*np.real(fft_f[i][j])**2,0.25);
            fft_f[i][j] = complex(r*np.cos(phi[i*(n+1)+j]),
                                  r*np.sin(phi[i*(n+1)+j]))
    f   = np.real(np.fft.ifft2(fft_f))
    mean_f = np.mean(f)
    std_f = np.std(f, dtype=np.float64)
    f   = (f - mean_f) / std_f
    return(f)

def plotHydraulicProperties(waterRetentionCurve, sigma):
    soil = Csoil()
    soil_minus_sigma = Csoil()
    soil_plus_sigma = Csoil()
```

```python
    soil_minus_sigma.Campbell_he /= np.exp(-sigma*sigma)
    soil_minus_sigma.VG_alpha *= np.exp(-sigma*sigma)
    soil_minus_sigma.Ks *= np.exp(-2*sigma*sigma)
    soil_plus_sigma.Campbell_he /= np.exp(sigma*sigma)
    soil_plus_sigma.VG_alpha *= np.exp(sigma*sigma)
    soil_plus_sigma.Ks *= np.exp(2*sigma*sigma)

    n = 100
    psi = -np.logspace(-2, 5, n)
    theta = np.zeros(n); theta_minus = np.zeros(n);
            theta_plus = np.zeros(n)
    K = np.zeros(n); K_minus = np.zeros(n); K_plus = np.zeros(n)
    for i in range(n):
        theta[i] = waterContent(waterRetentionCurve, soil, psi[i])
        K[i] = hydraulicConductivity(waterRetentionCurve,
                        soil, psi[i])
        theta_minus[i] = waterContent(waterRetentionCurve,
                        soil_minus_sigma, psi[i])
        K_minus[i] = hydraulicConductivity(waterRetentionCurve,
                        soil_minus_sigma, psi[i])
        theta_plus[i] = waterContent(waterRetentionCurve,
                        soil_plus_sigma, psi[i])
        K_plus[i] = hydraulicConductivity(waterRetentionCurve,
                        soil_plus_sigma, psi[i])

    plt.subplot(1,2,1)
    plt.plot(-psi, theta_minus, '--k', label='-$\sigma$')
    plt.plot(-psi, theta, '-k', label='$\mu$')
    plt.plot(-psi, theta_plus, ':k', label='+$\sigma$')
    plt.xlabel('$\psi$ [J kg$^{-1}$]',labelpad=8,fontsize=20)
    plt.ylabel('$\\theta$ [m$^3$ m$^{-3}$]',labelpad=8,fontsize=20)
    plt.xscale('log'), plt.legend()
    plt.tick_params(axis='both', which='major', labelsize=14,pad=8)

    plt.subplot(1,2,2)
    plt.plot(-psi, K_minus*3600., '--k', label='-$\sigma$')
    plt.plot(-psi, K*3600., '-k',label='$\mu$')
    plt.plot(-psi, K_plus*3600., ':k', label='+$\sigma$')
    plt.xscale('log'), plt.yscale('log')
    plt.xlabel('$\psi$ [J kg$^{-1}$]',labelpad=8,fontsize=20)
    plt.ylabel('K [kg s$^{-1}$ m$^{-3}$]',labelpad=8,fontsize=20)
    plt.tick_params(axis='both', which='major', labelsize=14,pad=8)
    plt.legend()
    plt.tight_layout()
    plt.show()

def plotThetaField(waterRetentionCurve, f, n):
    soil = Csoil()
```

```
    currentSoil = Csoil()
    theta = np.resize(np.zeros(((n+1)**2)*3),(3,n+1,n+1))
    psi = [-0., -15., -100.]
    for i in range(len(psi)):
        for k in range(n+1):
            for l in range(n+1):
                currentSoil.Campbell_he = soil.Campbell_he /
                                           np.exp(f[k,l])
                currentSoil.VG_alpha = soil.VG_alpha * np.exp(f[k,l])
                theta[i,k,l] = waterContent(waterRetentionCurve,
                                      currentSoil, psi[i])

        plt.subplot(1, len(psi), i+1),
        plt.imshow(theta[i], cmap=plt.gray(), vmin = 0.0, vmax = 0.5)
        plt.axis('off'), plt.title(r'$\theta$
                              ($\psi$='+str(psi[i])+')')
        cbar = plt.colorbar(orientation='horizontal',
                        ticks = [0.0,0.5], shrink=0.8,
                        aspect = 7., pad = 0.05)
        cbar.set_ticklabels(['%.1f' %(0.0), '%.1f' %(0.5)])
    plt.show()


def plotConductivityField(waterRetentionCurve, f, n):
    soil = Csoil()
    currentSoil = Csoil()
    K = np.resize(np.zeros(((n+1)**2)*3),(3,n+1,n+1))
    psi = [-0., -15., -100.]
    for i in range(len(psi)):
        for k in range(n+1):
            for l in range(n+1):
                currentSoil.Campbell_he = soil.Campbell_he /
                                           np.exp(f[k,l])
                currentSoil.VG_alpha = soil.VG_alpha * np.exp(f[k,l])
                currentSoil.Ks = soil.Ks * np.exp(2*f[k,l])
                K[i,k,l] = hydraulicConductivity(waterRetentionCurve,
                                            currentSoil, psi[i])

        plt.subplot(1, len(psi), i+1),
        plt.imshow(np.log(K[i])/np.log(10.), cmap=plt.gray(),
                            vmin = -10, vmax = -1)
        plt.axis('off'), plt.title(r'$\log_{10}$K ($\psi$='
                                    + str(psi[i])+')')
        cbar = plt.colorbar(orientation='horizontal',
                            ticks = [-10,-8,-6,-4,-2],
                            shrink=0.8, aspect = 7., pad = 0.05)
        cbar.set_ticklabels([-10,-8,-6,-4,-2])
    plt.show()
```

The first example of the Miller–Miller concept is applied to the hydraulic properties described by the Campbell and Mualem–van Genuchten parameterization.

Figure 7.3 depicts the Campbell hydraulic functions for three different sandy soils, defined by different characteristic lengths $\sigma - \mu$, σ and $\sigma + \mu$. The variation in characteristic lengths is then defined by a variation around the mean.

In Fig. 7.3(a), the solid line is the mean μ and the dotted line is the characteristic length $\sigma + \mu$; therefore a large value of the characteristic length corresponds to a coarser soil. For the coarser soil, for the same value of water potential, the soil can hold less water and therefore has a lower value of water content. On the other hand, the dashed line represents $\sigma - \mu$; therefore a smaller characteristic length corresponds or a finer soil, which at the same value of water potential has a higher value of water content. At a zero value of the water potential ($\psi = 0$), all the soils have the same value of water content, because the scaling does not change the total porosity. These results are qualitatively in agreement with the experimental relationships between particle size distribution and soil water retention curve (Campbell and Shiozawa, 1992).

The effect of scaling on the hydraulic conductivity (Fig. 7.3(b)) shows that at a zero value of the water potential ($\psi = 0$) the value of hydraulic conductivity is different among

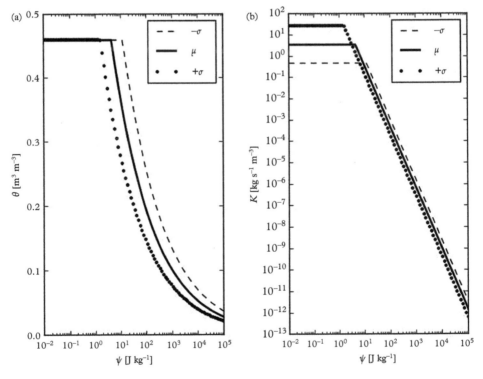

Figure 7.3 *Hydraulic properties for simulation of Miller similarity. The solid line is the average value (μ), while the dotted and dashed lines are the $\pm$ standard deviations σ.*

the three different soils. The hydraulic conductivity is higher for coarse sand (dotted line) and lower for finer sand (dashed line). These results are also qualitatively in accordance with experimental results, where at drier water potentials the coarser textures conductivity drops below the finer, as has been previously noted.

Overall, the Miller–Miller scaling shows an important feature of heterogenous soils, namely that the hydraulic structure of the media depends on the matric potential (Roth, 1995) and that the scaling in characteristic length reflects these features and is in accordance with experimental results.

7.6.1   Statistical Fields of Hydraulic Properties

As described above, to generate spatial fields of heterogeneities, autocovariances were generated using stationary random space functions, by employing Fourier transforms as stated in the Wiener–Khinchin theorem. Figure 7.4 shows three examples of a realization of f_0 obtained by the autocovariance exponential model for soil water content, at three different levels of soil water potential, while Fig. 7.5 shows three examples of a realization of f_0 obtained by the autocovariance Gaussian model.

Figure 7.6 and Fig. 7.7 depict examples of the hydraulic conductivity for the exponential and Gaussian models at three different levels of soil water potential.

The fields were obtained for values of $\sigma^2 = 1$ and correlation length $\iota = 0.1$ m. The plots represent vertical and horizontal (two-dimensional) values of the variables $\theta_{m,n}$ and $K_{m,n}$. The variables are described as discrete grid points, with x and y separation determined by the variables $dx = dy$ in the program.

The water content field (at a corresponding water potential of -15 J kg^{-1}) is characterized by islands of higher water content and it is non-isotropic. A similar structure is preserved at lower values of soil water content (water potential of -100 J kg^{-1}), but the field is less heterogenous. The hydraulic conductivity fields are plotted on a logarithmic scale and, because of the highly nonlinear dependence of K and ψ_m, the range of K

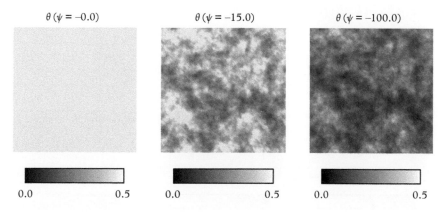

Figure 7.4 *Heterogeneous fields of soil water content for the exponential model at three different level of soil water potential.*

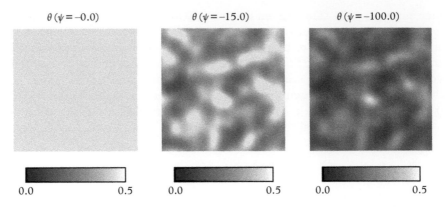

Figure 7.5 *Heterogeneous fields of soil water content for the Gaussian Model at three different levels of soil water potential.*

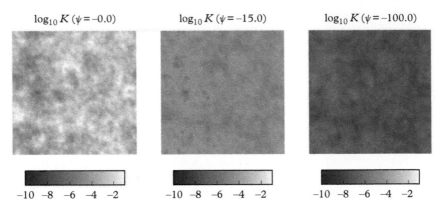

Figure 7.6 *Heterogeneous fields of hydraulic conductivity for the exponential model at three different levels of soil water potential.*

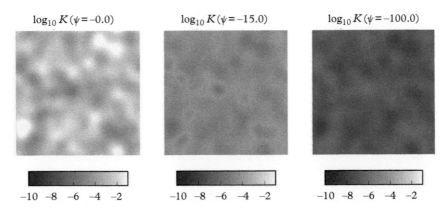

Figure 7.7 *Heterogeneous fields of hydraulic conductivity for the Gaussian Model at three different levels of soil water potential.*

is much larger. The exponential field leads to more small-scale structure in both the θ and K fields compared with the Gaussian fields. This is due to the higher irregularity of random functions with exponential autocovariance compared with Gaussian autocovariance. In general, as pointed out by Roth (1995), the spatial structures of the hydraulic variables are rather insensitive to the magnitude of heterogeneity for $\sigma^2 < 0.5$.

Under the assumption that Miller-similar media represent the structural features of natural soils, this technique allows for reproduction of a variety of structural properties due to heterogenous distribution at small scales of properties, such as particle size distribution, and therefore their effects on the hydraulic properties.

7.7 EXERCISES

7.1. Use the program `PSP_MillerMiller` to produce statistical fields for the exponential and Gaussian model at different levels of water potential. Discuss the results.

7.2. Use the program `PSP_MillerMiller` to produce water retention and hydraulic conductivity curves. Change the value of the standard deviation and discuss the results.

8

Transient Water Flow

In Chapter 6, we have seen that the flux of water through a porous medium is proportional to the gradient in water potential and to the hydraulic conductivity, which depends on the material and fluid properties:

$$f_w = -K\frac{d\psi}{dx} \tag{8.1}$$

where f_w is the water flux density [kg m^{-2} s^{-1}], $d\psi/dx$ is the water potential gradient that drives the flow, K is the hydraulic conductivity [kg s m^{-3}], ψ is the water potential [J kg^{-1}] and x is the space dimension [m]. A formulation for application in three dimensions can be written as

$$\vec{f_w} = -\overline{K}\nabla \cdot \psi \tag{8.2}$$

where the overbar here indicates that the hydraulic conductivity is a tensor and ∇ is the vector differential operator.

This formulation applies for steady water flow, where there is no water storage and flux is not dependent on location. However, steady water flow describes only a limited subset of conditions occurring in natural soils. A typical example of steady water flow is groundwater flow. The unsaturated (vadose) zone is characterized by cycles of soil wetting and drying determined by a variety of processes such as infiltration, evaporation and plant water uptake, where water content of soil changes with time. Steady conditions may occur in the unsaturated zone, but they are usually short-lived. In this chapter, we will discuss the solution of transient water flow problems.

8.1 Mass Conservation Equation

To formulate a transient mass transport problem, it is necessary to invoke the mass conservation equation. The law of conservation of mass states that the mass of a closed system will remain constant over time. This law can be applied to water flow, where, during flow, mass must be locally preserved. Let us consider a cube of dimensions Δx, Δy and Δz (Fig. 8.1). The continuity equation states that the change in water content

Soil Physics with Python. First Edition. Marco Bittelli, Gaylon S. Campbell and Fausto Tomei.
© Marco Bittelli, Gaylon S. Campbell and Fausto Tomei 2015. Published in 2015 by Oxford University Press.

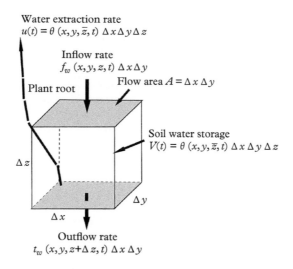

Figure 8.1 *Mass conservation for a cubic volume.*
Modified from Jury et al. (1991).

in the volume at a given time must be equal to the difference between the inflow and outflow rates f_w. The continuity equation may also contain a sink or source term. In the case of a sink (water extraction), it is the water removed from the volume (as indicated in the figure). This term is often used to represent the extraction of water by plant roots. The source term represents an input of water. An example is subsoil irrigation in agriculture, where the irrigation lines are buried below ground, and therefore the mass balance equation must include a source term to account for additional input of water into the soil volume.

8.2   Water Flow

In process-based models, the movement of water is usually described by Richards' equation, which is the application of the mass conservation equation to water flux. Richards' equation is a mass balance equation (as described above) applied to water flow, with the change in water content over time on the left-hand side and the difference in flow rate into an infinitesimal volume on the right-hand side. In one dimension, it is

$$\rho_l \frac{\partial \theta}{\partial t} = \frac{\partial}{\partial z}\left[K(\psi_m)\left(\frac{\partial \psi_m}{\partial z} + g\hat{\mathbf{z}} \right)\right] \qquad (8.3)$$

Richards' equation can also be written with the vector differential operator ∇:

$$\rho_l \frac{\partial \theta}{\partial t} = \nabla \cdot \left[K(\psi_m)\left(\nabla \psi_m + g\hat{\mathbf{z}} \right)\right] \qquad (8.4)$$

where θ [m³ m⁻³] is the volumetric water content, ψ_m [J kg⁻¹] the matric potential, K [kg s m⁻³] the hydraulic conductivity, t time [s], g the gravitational acceleration [9.8 m s⁻²], ρ_l the liquid water density [kg m⁻³] and $\hat{z}$ the unit vector in the vertical direction, taken positively upwards. This form is also referred as the *mixed* form of Richards' equation, because water content and matric potential occur simultaneously.

Richards' equation can also be expressed in terms of *water potential* only (ψ-based form) or *water content* only (θ-based form). The ψ-based form is obtained by applying the chain rule of calculus, where the derivative of the water content with respect to time on the left-hand side of the equation can be written as

$$\frac{d\theta}{dt} = \frac{d\theta}{d\psi}\frac{d\psi}{dt} = C(\psi)\frac{d\psi}{dt} \tag{8.5}$$

The capacity $C = d\theta/d\psi$ has been described in Chapter 5. Therefore, by using the capacity, eqn (8.4) can be expressed in the ψ-based form

$$\rho_l C(\psi_m)\frac{\partial \psi_m}{\partial t} = \nabla \cdot \left[K(\psi_m)\left(\nabla\psi_m + g\hat{z}\right)\right] \tag{8.6}$$

The θ-based form of Richards' equation is obtained by employing the water diffusivity $D = K/C$:

$$\frac{\partial \theta}{\partial t} = \nabla \cdot \left[D(\theta)\left(\nabla\theta + g\hat{z}\right)\right] \tag{8.7}$$

Equations (8.3), (8.6) and (8.7) are second-order nonlinear partial differential equations. The strong nonlinearity of Richards' equation is due to the highly nonlinear dependence of the hydraulic properties such as hydraulic conductivity, capacity and diffusivity on water potential and water content.

8.3   Infiltration

Infiltration is the process by which liquid water enters soil. Through redistribution, water that entered the soil during infiltration is redistributed within the soil. Both infiltration and redistribution profoundly affect the soil water balance. If water were applied to the surface of a uniform column of air-dry soil and the rate of infiltration of water into the soil column were measured as a function of time, the infiltration rate would be shown to decrease with time. These features are depicted in Fig. 8.2. The infiltration rate is high initially, but decreases with time to a constant value. If a similar experiment were done with a column of moist soil, similar results would be obtained, but the initial rate would be lower. If infiltration were into a horizontal, rather than a vertical, column, once again similar results would be obtained, except that the infiltration rate would decrease towards zero, rather than the constant, non-zero value for a vertical column like the one depicted in Fig. 8.2. If water content were measured at several times during infiltration, the water content profiles in the column would be similar. Two features of these profiles

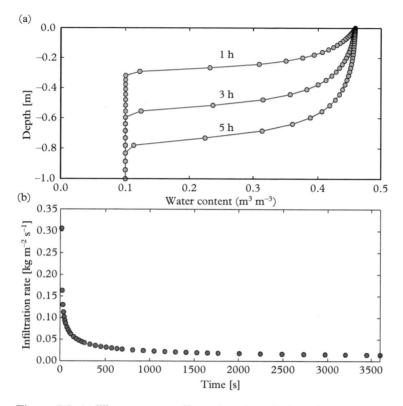

Figure 8.2 *(a) Water content profiles at three times during infiltration.*
(b) Infiltration rate as a function of time.

should be noted. First, there is a zone of almost constant water content extending from the soil surface to the apparent boundary between wet and dry soil. This zone is called the transmission zone. A visible wetting front or boundary between wet and dry soil exists at the lower end of the transmission zone. This sharp front is the result of the sharp decrease in hydraulic conductivity with water content that is characteristic of unsaturated porous materials.

A quantitative analysis of infiltration is obtained by solving the differential equation that describes water potential in soil as a function of position and time. The observations described above are qualitatively consistent with predictions one might make from Darcy's law:

$$\text{vertical flow:} \quad f_w = -K\frac{d\psi_m}{dz} + Kg \tag{8.8}$$

$$\text{horizontal flow:} \quad f_w = -K\frac{d\psi_m}{dx} \tag{8.9}$$

Equation (8.8) is for vertical flow, with the second term on the right-hand side being the gravitational component of the water potential gradient. Equation (8.9) is for horizontal flow. As water first infiltrates a vertical column, $d\psi_m/dz$ is large, so the flux is large. As the length of the transmission zone increases, the absolute value of $d\psi_m/dz$ decreases. When the transmission zone is very long, the matric potential gradient becomes negligible, so the flux becomes $f_w = Kg$. Since the transmission zone is near saturation, K is near K_s. For horizontal infiltration, f_w approaches zero as the matric potential gradient approaches zero. The fact that infiltration rate is initially lower in moist soil is explained by the reduced matric potential gradient in moist soil. These qualitative descriptions can be formalized in a more quantitative way that allows for quantification of infiltration rates, through an infiltration model.

Philip's model

The model most frequently referred to is by Philip (1957). In a one-dimensional soil profile, the infiltration rate v_I is defined as

$$v_I = \frac{dI}{dt} \tag{8.10}$$

where I is the cumulative infiltration [kg m^{-2}]:

$$I = \int_0^t v_I \, dt \tag{8.11}$$

The cumulative infiltration corresponds to the increment in soil water content in the soil volume. For a simple case of infiltration into a horizontal column, it is written as

$$I = \int_{\theta_{\text{low}}}^{\theta_{\text{up}}} x \, d\theta \tag{8.12}$$

where the integration is performed between two limits of water content, here denoted by θ_{low} and θ_{up}. For specific initial and boundary conditions, Philip analytically solved the θ-based form of Richard's equation (8.7). He transformed the partial differential equation into an ordinary differential equation by assuming a relationship between space and time, where $x = \chi \sqrt{t}$, allowing for substitution of the variables x and t with χ.

Using this transformation of variables, the variable x in the integral (8.12) can be written as

$$I = \int_{\theta_{\text{low}}}^{\theta_{\text{up}}} \chi t^{1/2} \, d\theta = t^{1/2} \int_{\theta_{\text{low}}}^{\theta_{\text{up}}} \chi \, d\theta = S_I \sqrt{t} \tag{8.13}$$

where S_I is the value of the integral, called *sorptivity*. Applying this definition to the infiltration velocity, which is the derivative of I with respect to time, we obtain

$$v_I = \frac{d(S_I \sqrt{t})}{dt} = \frac{S_I}{2\sqrt{t}} \tag{8.14}$$

so the infiltration velocity decreases inversely with the square root of time. The constant of proportionality is the sorptivity

$$S_I = \int_{\theta_{\text{low}}}^{\theta_{\text{up}}} \chi \, d\theta \tag{8.15}$$

This solution is valid for infiltration into a horizontal column. Infiltration into a vertical column implies that the integral of the total water content entering the soil surface must be equal to the changes in soil water content in the soil column below plus the water leaving the soil column at the bottom (mass balance). These differences are due to the fact that the water potential in a partially saturated water column is at a value of matric potential, but the gravitation potential decreases towards the bottom boundary with a water potential gradient equal to g.

For infiltration into a vertical column, the definition of cumulative infiltration is

$$I = -\int_{\theta_{\text{low}}}^{\theta_{\text{up}}} z \, d\theta + K_{\theta_{\text{low}}} t \tag{8.16}$$

The negative sign in front of the integral is due to the fact that as depth increases, water content decreases, causing the integral to become negative. Now it is possible to perform the variable substitution proposed by Philip (1957):

$$I = t^{1/2} - \chi \, d\theta + t \left(K_{\theta_{\text{low}}} t + \int_{\theta_{\text{low}}}^{\theta_{\text{up}}} (-\chi) \, d\theta \right) + t^{3/2} \left(\int_{\theta_{\text{low}}}^{\theta_{\text{up}}} (-\chi) \, d\theta \right) + \cdots \tag{8.17}$$

where the integral of the different elements of the series depends on the initial soil condition and soil type. They can therefore be written as

$$I = At^{1/2} + Bt + Ct^{3/2} + \cdots + Mt^{m/2} \tag{8.18}$$

where A is equal to the sorptivity for a horizontal column, while the other terms are constant for a specific problem and soil. For relatively short times, the series in eqn (8.18) converges rapidly. Philip (1957) proposed that for the majority of cases, the two terms are sufficient to provide a robust solution

$$I = S_I t^{1/2} + Bt \tag{8.19}$$

and

$$v_I = \frac{S_I}{2t^{1/2}} + B \tag{8.20}$$

The infiltration process, as described by this model, is strongly affected by the value of S_I for small values of t, while for large values of t, the series in eqn (8.18) does not converge. On the other hand, a variety of experiments confirm the proportionality of the process to $\sqrt{t}$.

8.3.1   Green–Ampt Model

A simpler approach was taken much earlier by Green and Ampt (1911). Their solution gives the most important features of the infiltration process, so it will be examined in detail.

During horizontal infiltration, an observable wetting front or boundary moves through the soil. The water content (and therefore the water potential) at the wetting front is almost constant during infiltration, as long as the profile is uniform. If the distance from the soil–water boundary to the wetting front is x_f and the potentials at the boundary and wetting front are ψ_i and ψ_f, respectively, then the infiltration rate (or flux at the boundary) is

$$f_{wi} = -K\frac{d\psi_m}{dx} = -\overline{K}\frac{\psi_f - \psi_i}{x_f} \tag{8.21}$$

where $\overline{K}$ is the mean conductivity of the transmission zone.

The amount of water stored in the soil per unit time is the change in water content of the soil from its initial dry condition to the average water content of the transmission zone, multiplied by the rate of advance of the wetting front. To satisfy continuity, all of the water flowing in must be stored by advance of the wetting front, so

$$\overline{K}\frac{\psi_i - \psi_f}{x_f} = \rho_l \Delta\theta \frac{dx_f}{dt} \tag{8.22}$$

Here ρ_l is the density of water, $\Delta\theta = (\theta_i + \theta_f)/2 - \theta_o$, θ_i, θ_f and θ_o are the volumetric water contents at the inflow, wetting front and unwetted soil, and t is time. Separation of variables and integration gives the position of the wetting front as a function of time:

$$x_f = \sqrt{2\overline{K}\frac{(\psi_i - \psi_f)t}{\rho_l \Delta\theta}} \tag{8.23}$$

Equation (8.23) indicates that the distance to the wetting front is directly proportional to the square root of time. The infiltration rate is obtained by combining eqns (8.21) and (8.23) to obtain

$$f_{wi} = \sqrt{\rho_l \Delta\theta \overline{K}\frac{(\psi_i - \psi_f)}{2t}} \tag{8.24}$$

showing that the infiltration rate is directly proportional to $\Delta\theta^{1/2}$ and $\overline{K}^{1/2}$ and inversely proportional to $t^{1/2}$. Equation (8.24) can be integrated over time to find the cumulative infiltration. The result of this integration is

$$I = \sqrt{2\rho_l \Delta\theta \overline{K} t (\psi_i - \psi_f)} \tag{8.25}$$

The Green–Ampt equation can also be applied to vertical infiltration, but the result is somewhat less satisfying. The most important result is that, for long times, f_{wi} approaches a constant value.

8.3.2   Infiltration into Layered Soils

The analysis so far has been concerned only for infiltration into uniform soils. Such soils exist mainly in carefully prepared laboratory columns. In the field, processes of soil development, tillage, illuviation, etc. produce layering. When infiltrating water encounters a boundary between layers, the infiltration rate generally decreases. The infiltration rate decreases for either sand or clay layers. The decrease with a clay layer is expected, because water infiltrates clay less readily than coarser-textured materials, so water slows down as it enters the clay, decreasing $\psi_i - \psi_f$ and thus decreasing infiltration rate. The saturated hydraulic conductivity of sand is higher than that of clay, while the unsaturated hydraulic conductivity of sand is lower than the unsaturated hydraulic conductivity of finer-textured materials. The boundary must therefore be wet almost to saturation before the hydraulic conductivity of the sand increases sufficiently to start carrying the water away. This increase in water content at the boundary decreases $\psi_i - \psi_f$ and decreases f_{wi}. The infiltration rate recovers after water enters the sand, but does not recover for the clay. Perched water tables are therefore common over clay layers, but are not possible over coarse-textured materials. In this chapter (and in the numerical solution), we have included the option of setting up the simulation for a soil having horizons with different properties. The user can select any number of soil horizons.

8.4   Numerical Simulation of Infiltration

As a first example, we will show a one-dimensional solution for infiltration. In the following chapters, we will present a three-dimensional solution. Many of the techniques used to simulate water movement in unsaturated soil are similar to those used for heat flow. The soil profile can be represented by a network of conductors and capacitors as described for heat flow, and the network problem can be solved to determine how water potential and water content change with time at each node in the network. The complicating factor compared with the heat flow problem is that for the unsaturated flow problem, the hydraulic conductivity and water capacity are functions of the dependent variable (water content or potential) and the problem is highly nonlinear. The methods used to solve water flow problems are therefore more difficult than those used to solve heat flow problems. Methods that have been used to solve these nonlinear equations include the following:

1. Linear methods. These are used in most numerical simulations of soil water flow, but they require very small time increments during infiltration and redistribution and arbitrary assumptions about the element hydraulic conductivities. With adaptive time steps, these solutions can be improved, but they still require a high number of approximations.

2. Integral transform methods. These require only the lumped parameter assumption (that the continuum of soil and water can be adequately represented by a network of capacitors and resistors).

Both methods will be discussed so that the reader can see the connection between the gas and heat flow models described previously and the water flow models of this chapter. Then the programs will be presented to show how to implement them. In the following chapters, methods to solve these problems in three dimensions will also be presented.

Method 1 will be used with the different options for the hydraulic properties described for the soil water retention and hydraulic conductivity (van Genuchten, Campbell and Ippisch–van Genuchten), while method 2 will be used only with Campbell's equation. The reason for this choice is the following: method 2 is solved using a Newton–Raphson method (discussed below). The Newton–Raphson method is an efficient algorithm to find the root of a function. When it is applied to water flow problems, it is used to find the root of the mass balance equation. This method requires computation of the derivatives of the function with respect to x. When it is applied to the mass balance equation, the derivatives are partial derivatives of the water potential and their computation may be cumbersome. Since Campbell's equation for the soil water retention and hydraulic conductivity curve is simple, the partial derivatives of the mass balance equation have an analytical solution and can be implemented in the code. When the van Genuchten and Mualem–van Genuchten equations are used, the partial derivatives with respect to the water potential do not have a simple analytical solution and the formulation of the Newton–Raphson method is not straightforward.

8.4.1   Linear Methods

The numerical procedure to solve Richards' equation in one dimension is identical to that used for solving Fourier's law for heat flow, described in Chapter 5. The numbering of conductances, capacitances, depths, etc. is the same as that shown in Fig. 4.3. The equation for water balance at a node is

$$\frac{\rho_l \overline{C}_i \left(\psi_i^{j+1} - \psi_i^j \right)(z_{1+1/2} - z_{1-1/2})}{\Delta t} = \frac{\overline{K}_i \left(\overline{\psi}_{i+1} - \overline{\psi}_i \right)}{(z_{i+1} - z_i)} - \frac{\overline{K}_{i-1} \left(\overline{\psi}_i - \overline{\psi}_{i-1} \right)}{(z_i - z_{i-1})} + u_i \qquad (8.26)$$

where $\overline{C}_i$ is the node water capacity, $\overline{K}_i$ is the element hydraulic conductivity and ψ is the node water potential. The overbar denotes a mean value over the time step and/or element length, u is a sink–source term and the superscript j indicates time. The water potentials are defined as with heat flow:

$$\overline{\psi}_i = \eta \overline{\psi}_i^{j+1} + (1 - \eta) \overline{\psi}_i^j \qquad (8.27)$$

where η is a weighting factor ranging between 0 and 1. While any value in this range can be used, the nonlinearity of water flow problems makes $\eta = 1$ the best choice in almost all cases. Hereinafter, $\eta = 1$ will be used. The solution is therefore fully implicit, with the value of the water potential being chosen at the end of the time step.

Ideally, C and K are appropriate mean values of the capacity and conductivity over a time step. The correct values, however, are not easily obtained, since each depends partially on the unknown potential at the end of the time step. Two approaches are normally used to estimate C and K. One is to use small time steps. A second approach is to make two or more computations, the first based on values for C and K at time j, with subsequent estimates made using the average of the known potential at the beginning of the time step and the estimated potential at the end of the time step.

The source term in eqn (8.26), u, combines all of the inputs and losses of water from the node that are not explicit in eqn (8.26). These might include water extraction by roots, gravitational flux, and condensation or evaporation. In most infiltration problems, only the gravitational flux is included, so

$$u = g(\overline{K}_{i-1} - \overline{K}_i) \tag{8.28}$$

Other sink terms can be included, such as water extraction by plant roots.

As we have shown in the example of gas flow, eqn (8.26) is written for each node in a simulated soil profile and then solved for the unknown ψ_i^{j+1}'s. The matrix of coefficients is tridiagonal. For four nodes, we can write

$$\begin{vmatrix} b(1) & c(1) & 0 & 0 \\ a(2) & b(2) & c(2) & 0 \\ 0 & a(3) & b(3) & c(3) \\ 0 & 0 & a(4) & b(4) \end{vmatrix} \begin{vmatrix} pn(1) \\ pn(2) \\ pn(3) \\ pn(4) \end{vmatrix} = \begin{vmatrix} d(1) \\ d(2) \\ d(3) \\ d(4) \end{vmatrix} \tag{8.29}$$

where

$$c(i) = a(i+1) = -\eta K(i) \tag{8.30}$$

$$b(i) = \eta[K(i-1) + K(i)] + cp(i) \tag{8.31}$$

and

$$d(i) = (1-\eta)K(i-1)p(i-1) + \{cp(i) - (1-\eta)[K(i-1) + K(i)]\}\, p(i)$$
$$+ (1-\eta)K(i)p(i+1) + u(i) \tag{8.32}$$

The potentials at the beginning and end of the time step are $p(i)$ and $pn(i)$. The conductance of each element is the element hydraulic conductivity divided by the element length:

$$k(i) = \frac{\overline{K}}{z_{i+1} - z_i} \tag{8.33}$$

The element hydraulic conductivity is computed using the node water potentials and the hydraulic properties functions described in Chapter 6. Initially, these are computed using $p(i)$, but, if iteration is used, subsequent estimates are made using an estimated average potential over the time step. Various schemes have been used for computing element hydraulic conductivity from the values at the nodes. These include arithmetic means, geometric means and weighted arithmetic means, as described in Chapter 5. The storage term $cp(i)$ is given by

$$cp(i) = \rho_l \overline{C}_i \frac{z_{i+1} - z_{i-1}}{2\Delta t} \tag{8.34}$$

where Δt is the time step and $\overline{C}_i$ is the water capacity $d\theta/d\psi_m$ at the node. The capacities for the three models of water retention curves were presented in eqns (5.41) to (5.43).

The computation of the derivative $d\theta/d\psi$ can be performed at the beginning of the time step, but this value will not represent a correct value for the whole time step if, during the time step, there are important variations of the two variables θ and ψ. In fact, numerical tests show that significant over- or under-estimation of water fluxes are experienced, depending on the increasing or decreasing behaviour of the derivative during the time step.

To address this problem, in the code the analytical derivative (5.41) or (5.42) is first computed at the beginning of the time step, while at the end of each approximation the derivative is computed by the incremental ratio using the new values of θ and ψ obtained at each iteration. The final value that is used is an intermediate value between the first and the last value. The advantage of this method is that it is possible to obtain a value of the ratio also in transition between unsaturated and saturated conditions. This approach was proposed by Bittelli *et al.* (2010) to simulate water flow in both saturated and unsaturated conditions.

8.4.2   Boundary Conditions

Either flux or potential boundary conditions can be specified at the top and bottom of the profile. At the bottom, the potential can be set to a known value. If there is a water table, this can be zero. If not, it can be set as a free drainage boundary condition, which is a unit vertical water potential gradient. This boundary condition must be used when the water table is far below the domain of simulation. At the top of the profile, the surface may be at saturation, in which case the potential is set to the air-entry potential ψ_e. If water is applied at a rate below the infiltration capacity of the soil, the potential does not go to the air-entry value, so a flux can be specified. This can be done by adding the flux to the source term at node 1. In the numerical examples, the user will be able to choose different boundary conditions and investigate their effects on the water flow computation.

8.4.3   Cell-Centred Finite Volume Method

The cell-centred finite volume method, a linear method, was introduced in Chapter 4 for the solution of the heat flow equation. Here we are applying it to a one-dimensional solution, while in the following chapters the solution will be applied to a three-dimensional domain. The mass balance equation can be written as

$$\frac{\partial(\theta)}{\partial t} = \nabla(f_w) + u \tag{8.35}$$

where f_w is the water flux, θ is the volumetric water content and u is the sink–source term. This general equation is solved adopting two different laws to describe fluxes within the soil and fluxes on the soil surface (surface runoff). Here we describe only the solution of water flux within the soil, while in the following chapters we will describe the solution of water fluxes on the soil surface (runoff). The solution of this equation is based on its integration over the total volume V_t of the grid cell at each node:

$$\int_{V_t} \frac{\partial(\theta)}{\partial t} = \int_{V_t} \nabla(f_w) + \int_{V_t} u \tag{8.36}$$

where V_t is the total available volume at a grid cell. Gauss' theorem states that the integral of $\nabla(f_w)$ over the total volume of the grid cell can be replaced by the fluxes over the boundary of the grid cell, leading to

$$\int_{V_t} \frac{\partial(V_t\theta)}{\partial t} = \int_{\partial V_t} f_w + \int_{V_t} u \tag{8.37}$$

This equation states that the change in total water mass within one grid cell is equal to the water fluxes across its boundaries into the cell and to the water sources and sinks within the grid cell. The term V_t is included to couple the subsurface and surface flows, as described in the following chapters.

If the simulation domain is approximated by a grid of nodes, eqn (8.37) is equivalent to the mass balance equation for the volume surrounding each node:

$$V_t\frac{\partial\theta}{\partial t} = \sum_{j=1}^{n} F_{ij} + u_i \qquad \forall i \neq j \tag{8.38}$$

where θ is the amount of stored water in the volume surrounding the node, F_{ij} is the flux between the ith and jth nodes, and u_i is sink–source term. We can write a system of equations for all nodes, with the unknowns being the total water potentials ψ_t.

The flux F_{ij} is described by Darcy's law in finite difference form:

$$F_{ij} = -K_{ij}A_{ij}\frac{\psi_{ti} - \psi_{tj}}{L_{ij}} \tag{8.39}$$

where A_{ij} is the interfacial area between nodes i and j, L_{ij} is the distance between the two nodes, ψ_{ti} is the total hydraulic head at node i and K_{ij} is the internode conductivity. For subsurface nodes, the nodal mass balance can be written as

$$V_i \left(\frac{d\theta}{\psi_{ti}}\right)_i \frac{\partial \psi_{ti}}{\partial t} = \sum_{j=1}^{n} \zeta_{ij} K(\theta)_{ij} (\psi_{tj} - \psi_{ti}) + u_i \tag{8.40}$$

where V_i is the soil bulk volume, θ_i is the soil water content, K is the hydraulic conductivity, ψ_{ti} is the water potential of node i, ψ_{tj} is the water potential of node j and ζ_{ij} is the ratio of bulk exchange area between nodes i and j over the internode distance.

8.4.4 Hydraulic Conductivity

When using linear methods, without lumped parameters, a very important aspect of the numerical solution is the choice of the mean used to compute the hydraulic conductivity of elements. This choice is particularly important when the simulation involves large gradients in water potential between two nodes, such as infiltration into a very dry soil. In this condition, a node that is very wet (high water potential) is adjacent to a node that is very dry (very low water potential). The estimated unsaturated conductivity for the wet node is high (very close to saturation), while it is very low for the dry node. These abrupt changes in hydraulic parameters often cause numerical errors and instabilities. Therefore, since the unsaturated hydraulic conductivity curve spans several orders of magnitude, the choice of the value between the two elements is very important. We defined three options: the logarithmic, the harmonic and the geometric mean. The logarithmic mean is given by (Carlson, 1972)

$$\bar{K} = \frac{K_1 - K_2}{\ln K_1 - \ln K_2} \qquad \text{for } K_1 \neq K_2 \tag{8.41}$$

the harmonic mean is given by

$$\bar{K} = \frac{2}{1/K_1 + 1/K_2} \tag{8.42}$$

and the geometric mean is given by

$$\bar{K} = \sqrt{K_1 K_2} \tag{8.43}$$

We ran several numerical and experimental tests and found that the logarithmic mean provided the most stable and reliable results.

8.4.5 Integral Transform Methods

The linear methods just outlined are extensively used to simulate soil water movement, but their success is based on the use of a powerful computer that can cope with short time steps and many solution nodes and on arbitrarily chosen weighting factors to determine element hydraulic conductances and capacitances.

Some methods follow that are suitable for less powerful computers and that avoid arbitrary choices for element conductance by linearizing the problem before putting it in difference form. Element conductances are therefore constant, as they were in the heat flow calculation. Gardner (1958) and Campbell (1985) suggested using the 'matric flux potential' as a driving force for flow. The matric flux potential is defined as

$$\Phi = \int_{-\infty}^{\psi} K \, d\psi \tag{8.44}$$

If the $K(\psi)$ relationship given by eqn (6.35) is assumed, then, for $\psi < \psi_e$,

$$\Phi = K \frac{\psi_m}{1-n} \tag{8.45}$$

where $n = 2 + 3/b$. The values of K and n are always positive, ψ_m is negative, $n > 1$, so Φ is always positive. At air entry,

$$\Phi_e = K_s \frac{\psi_e}{1-n} \tag{8.46}$$

so Φ ranges from zero for dry soil to Φ_e for saturated soil. The use of Φ as the 'driving force' in the flow equation, rather the ψ or θ, results in a linear equation for steady flow:

$$f_w = -\frac{d\Phi}{dz} \tag{8.47}$$

Note that the hydraulic conductivity is no longer present in the flow equation, removing the highly nonlinear dependence $K(\psi)$. The variables here are easily separated to obtain the difference equation

$$f_{wi} = -\frac{\Phi_{i+1} - \Phi_i}{z_{i+1} - z_i} \tag{8.48}$$

The matric flux potential can then be combined with the continuity equation (8.3) to give

$$\rho_l C \frac{\partial \Phi}{\partial t} = \frac{\partial^2 \Phi}{\partial z^2} \tag{8.49}$$

where $C = d\theta/d\Phi$.

Equation (8.45) can be combined with eqn (5.27) to find the relationship between matric flux potential and water content:

$$\Phi = \Phi_e \left(\frac{\theta}{\theta_s}\right)^{b+3} \tag{8.50}$$

Differentiating eqn (8.50) gives the capacity

$$C = \frac{\theta}{(b+3)\Phi} \tag{8.51}$$

Equation (8.49) can now be put into a difference form similar to eqn (8.26), but with Φ for the driving potential, unity for the conductivity and C calculated from eqn (8.51).

The matric flux potential is a useful driving potential, since it results in constant conductances for the soil elements. It does have limitations, however. The most serious is that, like water content, it is not continuous across boundaries between materials having different properties. This is evident from eqn (8.45). Note, however, that flux through an element is linearly related to the product $K\psi$, eqn (8.45). We could therefore rewrite eqn (8.49) as

$$f_{wi} = -\frac{K_{i+1}\psi_{i+1} - K_i\psi_i}{(1-n)(z_{i+1} - z_i)} \tag{8.52}$$

This equation can be derived somewhat differently to make its meaning clearer. In the lumped-parameter models, steady flow within each element is assumed. For steady flow conditions, eqn (8.47) can be integrated analytically between any two nodes. If eqn (6.35) is assumed to describe $K(\psi)$, the result of the integration is eqn (8.52).

We can also use eqn (8.52) to derive an analytically correct value for the element conductivity. This is

$$\overline{K} = -\frac{K_{i+1}\psi_{i+1} - K_i\psi_i}{(1-n)(\psi_{i+1} - \psi_i)} \tag{8.53}$$

which is the correct conductivity to use in eqn (8.26). We can either solve for the Φ_i, using eqn (8.49) and the associated capacity and conductivity variables, or solve for the ψ_i, using eqn (8.52). For infiltration, numerical methods are most stable if the matric flux potential is used. Evaporation, plant water uptake and redistribution work as well, or better, using the standard matric potential. We will develop numerical solutions using both potentials, but generally will use the latter throughout the remainder of the book.

The use of the matric flux potential linearizes the steady flow equation and allows it to be integrated for each element, but the transient flow problem is still nonlinear, since $C(\Phi)$ and $C(\psi)$ both vary with the dependent variable. The solution method must therefore be one that solves nonlinear equations.

8.4.6   Newton–Raphson Method

The Newton–Raphson iterative method is often used to solve nonlinear equations. The method can be illustrated using a nonlinear function of the variable x, where the value or values of x for which $f(x) = 0$ are to be found. For some arbitrary value of x, say $x^{(t)}$, $f(x)$ will have some value, $F^{(t)}$, which usually will not be zero. The function $f(x^t)$ at the point x^t can be replaced by its tangent. Then the root of the tangent is computed, which is used as the new approximation for the root. The root of the tangent is given by

$$x^{(t+1)} = x^{(t)} - \frac{f(x^{(t)})}{f'(x^{(t)})} \tag{8.54}$$

where $f'(x^{(t)})$ is the derivative. Note that $x^{(t+1)}$ is closer to the actual root with respect to $x^{(t)}$ and it becomes the next x value at which to evaluate $f(x)$. The iteration goes on until it is assured that $x^{(n)}$ is an accurate approximation of the solution (assuming that the iteration converges to a solution). Note that n is not a power, but represents the number of iterations. The difference between $x^{(t)}$ and the true solution, $x^{(n)}$, can be approximated as

$$\frac{\partial F}{\partial x}\bigg|_x^o \cdot (x^o - x^1) = F^o \tag{8.55}$$

where x^o is the old value, x^1 is the new value and both $F = f(x)$ and the derivative $f'(x)$ are evaluated at x^o. Figure 8.3 depicts a graphical example of the Newton–Raphson method.

For simultaneous equations (an n-dimensional problem), the solution process is similar to that outlined for a single variable, but matrix algebra is used for the solution. F and $x^{t+1} - x^t$ become vectors. The coefficient matrix is made up of the partial derivatives of each F with respect to each x. This is called the Jacobian matrix. The new set of Δx's is found for each iteration by Gaussian elimination on the Jacobian and back-substitution, as explained in Appendix B. If $F = f(x_1, x_2, x_3)$, then values of x_1, x_2, x_3 for which $F = 0$ can be found by iteratively solving

$$\begin{vmatrix} \dfrac{\partial F_1}{\partial x_1} & \dfrac{\partial F_1}{\partial x_2} & \dfrac{\partial F_1}{\partial x_3} \\[2mm] \dfrac{\partial F_2}{\partial x_1} & \dfrac{\partial F_2}{\partial x_2} & \dfrac{\partial F_2}{\partial x_3} \\[2mm] \dfrac{\partial F_3}{\partial x_1} & \dfrac{\partial F_3}{\partial x_2} & \dfrac{\partial F_3}{\partial x_3} \end{vmatrix} \begin{vmatrix} x_1^{t+1} - x_1^t \\[2mm] x_2^{t+1} - x_2^t \\[2mm] x_3^{t+1} - x_3^t \end{vmatrix} = \begin{vmatrix} F_1 \\[2mm] F_2 \\[2mm] F_3 \end{vmatrix} \tag{8.56}$$

The partial derivatives and F's are evaluated at the x^t and the equations are solved for the x^{t+1}. These are then used to re-evaluate the F's and partial derivatives and solved

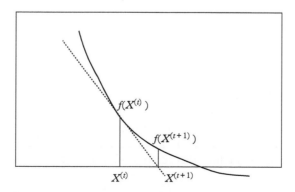

Figure 8.3 *Example of the Newton–Raphson method.*

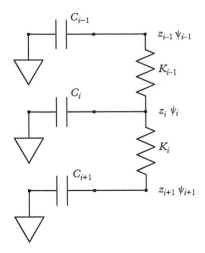

Figure 8.4 *Nodes and potentials for the Newton–Raphson calculation.*

again. Convergence is determined by checking the F's to see if they are sufficiently close to zero.

To apply this method to solution of the infiltration problem, the mass balance for node i must first be written. The symbols are shown in Fig. 8.4. The ψ's represent matric potentials, the K's are conductances, and the f's are fluxes. The equations are written in terms of matric potentials, but could be written using matric flux potentials. The mass balance for node i is .

$$F_i = f_i - f_{i-1} + u_i - u_{i-1} + \rho_l \frac{(\theta_i^{j+1} - \theta_i^j)(z_{i+1} - z_{i-1})}{2\Delta t} \tag{8.57}$$

where f_i, u_i and θ_i are all nonlinear functions of ψ_i. We need to find values for ψ that will make $F_i = 0$ for $i = 1, \ldots, m$, the total number of nodes. It is important to note that the values for the potential that force F to zero at every node are those that ensure mass balance at every node. The gravitational flux is $u_i = gk_i$, where k_i is the conductivity at node i. Using this with eqn (8.49), we obtain

$$F_i = \frac{K_i\psi_i - K_{i-1}\psi_{i-1}}{(1-n)(z_i - z_{i-1})} - \frac{K_{i+1}\psi_{i+1} - K_i\psi_i}{(1-n)(z_{i+1} - z_i)} - g(K_i - K_{i-1}) + \frac{\rho_l(\theta_i^{j+1} - \theta_i^j)(z_{i+1} - z_{i-1})}{2\Delta t} \tag{8.58}$$

Here the K's are hydraulic conductivities calculated from the corresponding node water potentials at the most recent iteration. The solution is therefore a backward difference one. To find the Jacobian, we take the derivatives with respect to each potential:

$$\frac{\partial F_i}{\partial \psi_i} = \frac{K_i}{z_i - z_{i-1}} + \frac{K_i}{z_{i+1} - z_i} + \frac{\rho_l(z_{i+1} - z_{i-1})\theta_i}{2b\psi_i\Delta t} - \frac{ngK_i}{\psi_i} \tag{8.59}$$

$$\frac{\partial F_i}{\partial \psi_{i-1}} = \frac{K_{i-1}}{z_i - z_{i-1}} + \frac{ngK_{i-1}}{\psi_{i-1}} \tag{8.60}$$

$$\frac{\partial F_i}{\partial \psi_{i+1}} = \frac{K_{i+1}}{z_{i+1} - z_i} \tag{8.61}$$

Using eqns (8.59) – (8.61) to form a set of equations like (8.56) gives the set of equations that is solved to find new estimates of ψ at the end of the time step.

Boundary conditions are easily applied in the Newton–Raphson solution scheme. For infiltration, the upper boundary will have either a constant flux or a constant potential. For the constant-flux condition, the flux is added as a source to the F equation at node 1. For a constant potential, ψ_i^{j+1} is set at the start of the time step and assumed not to change during the time step. Since the Newton–Raphson method computes changes in ψ to bring F to zero, if F_1 and $\partial F/\partial \psi_i$ are set to zero before the Gaussian elimination, the value for ψ_1 will stay constant from iteration to iteration. The lower boundary condition is usually a constant potential or a free drainage, and never becomes part of the solution.

8.5   Numerical Implementation

Here we present a series of files containing the code for one-dimensional solution of vertical infiltration and redistribution using three different solvers:

1. Cell-centred finite volume
2. Finite difference with matric potential solved with the Newton–Raphson method
3. Integral transform with matric flux potential solved with the Newton–Raphson method

The program allows one to select among these three options. Different hydraulic properties can be used as described in Chapter 5. The second and third methods are implemented only for the Campbell parameterization of the soil water retention curve and hydraulic conductivity, since the Newton–Raphson method requires the Jacobian. The Campbell parameterization allows for a straightforward derivation of the Jacobian, while the van Genuchten parameterization does not. Also, as was demonstrated by Ippisch et al. (2006), the validity limits of the van Genuchten–Mualem model cause unreliable numerical solutions at high water content.

The soil used for the simulation is the same as was used in Chapter 5 for the soil water retention curve fitting. The project `PSP_infiltrationRedistribution1D` for infiltration and redistribution comprises seven files:

1. `main.py`
2. `PSP_grid.py`
3. `PSP_readDataFile.py`
4. `PSP_soil.py`

5. `PSP_Infiltration1D.py`

6. `PSP_ThomasAlgorithm.py`

7. `soil.txt`

8.5.1 Main

The main file `main.py` contains the calls to the other file `PSP_infiltration1D.py`, where the solvers for the numerical solution are implemented. This file (described below) calls the file used to build the grid (`PSP_grid.py`) and the file where the soil data structure is implemented (`PSP_soil.py`). The soil hydraulic parameters are saved in the file `soil.txt`, which are the parameters for the Campbell, van Genuchten and modified van Genuchten equations. The file `soil.txt` is organized in rows and columns. Each row corresponds to a soil layer, while the data organized by columns are the soil hydraulic function parameters (the reader should open the file to see how it is structured). In this example, the soil is homogenous and therefore there is only one layer of 1 m depth. However, in the following chapters, numerical solutions for layered soils will also be presented, and the file will have multiple rows, each corresponding to a soil layer.

The selection of the models, printed to the screen, are prompted to the user for selection of the hydraulic properties model. The user is asked to select the lower boundary condition, free drainage or constant potential, and the total length of the simulation. Then the results are plotted and cumulative time, incremental time step, number of iterations per time step and cumulative infiltration are printed.

An example of the iterative prompt is shown below. First the user is asked to select the hydraulic function, in this case the Campbell formulation, and then to select the numerical scheme (in this case cell-centred finite volume). The initial degree of saturation of the soil profile is 0.3. The lower boundary condition is free drainage and the number of simulation hours is 5 h.

```
1   Campbell
2   van Genuchten with m = 1-1/n restriction
3   van Genuchten modified
Select water retention curve: 1

1   Cell-Centered Finite Volume
2   Matric Potential with Newton-Raphson
3   Matric Flux Potential with Newton-Raphson
Select solver: 1

Initial degree of saturation ]0-1]: 0.3

1: Free drainage
2: Constant water potential
Select lower boundary condition:1

Nr of simulation hours: 5
```

The three numerical schemes were described above. The upper boundary condition is a constant-potential boundary condition. The model also allows for setting up other upper boundary conditions, such as atmospheric boundaries, which are simulated by applying either a prescribed water potential or a prescribed flux. The implementation of atmospheric boundary conditions will be presented in Chapter 15. The lower boundary condition can be chosen as free drainage or constant water potential. The free-drainage boundary condition is a unit vertical water potential gradient boundary condition that can be implemented in the form of a variable-flux boundary condition. The constant potential corresponds to setting the last node at a fixed value of water potential. The program is as follows:

```python
#PSP_main.py
from __.future__ import print_function, division

import matplotlib.pyplot as plt
import PSP_infiltration1D as inf

def main():
    isSuccess, soil = inf.readSoil("soil.txt")
    if not isSuccess:
        print("warning: wrong soil file.")
        return

    print (inf.CAMPBELL,' Campbell')
    print (inf.RESTRICTED_VG,' van Genuchten with m = 1-1/n
            restriction')
    print (inf.IPPISCH_VG,' Ippisch-van Genuchten')
    funcType = int(input("Select water retention curve: "))

    if (funcType == inf.CAMPBELL):
        print()
        print (inf.CELL_CENT_FIN_VOL,' Cell-Centered Finite Volume')
        print (inf.NEWTON_RAPHSON_MP,' Matric Potential with
                Newton-Raphson')
        print (inf.NEWTON_RAPHSON_MFP,' Matric Flux Potential with
                Newton-Raphson')
        solver = int(input("Select solver: "))
    else:
        solver = inf.CELL_CENT_FIN_VOL

    myStr = "]" + format(soil.VG_thetaR, '.2f')
    myStr += ", " + format(soil.thetaS, '.2f')
    myStr += "] initial water content (m^3 m^-3):"
    thetaIni = inf.NODATA
    print()
    while ((thetaIni <= soil.VG_thetaR) or (thetaIni > soil.thetaS)):
        thetaIni = float(input(myStr))
```

```
inf.initializeWater(funcType, soil, thetaIni, solver)

# [J kg^-1] upper boundary condition
ubPotential = inf.airEntryPotential(funcType, soil)

print()
print ("1: Free drainage")
print ("2: Constant water potential")
boundary = int(input("Select lower boundary condition:"))
if (boundary == 1):
    isFreeDrainage = True
else:
    isFreeDrainage = False

# hours of simulation
simulationLenght = int(input("\nNr of simulation hours:"))

endTime = simulationLenght * 3600
maxTimeStep = 600
dt = maxTimeStep / 10
time = 0
sumInfiltration = 0
totalIterationNr = 0

plt.ion()
f, myPlot = plt.subplots(2, figsize=(8, 8), dpi=80)
myPlot[0].set_xlim(0, 0.5)

myPlot[1].set_xlim(0, simulationLenght * 3600)
myPlot[1].set_ylabel("Infiltration Rate [kg m$^{-2}$ s$^{-1}$]")
myPlot[1].set_xlabel("Time [s]")

while (time < endTime):
    dt = min(dt, endTime - time)
    if (solver == inf.CELL_CENT_FIN_VOL):
        success, nrIterations, flux = inf.cellCentFiniteVolWater
            (funcType, soil, dt, ubPotential, isFreeDrainage,
            inf.LOGARITHMIC)
    elif (solver == inf.NEWTON_RAPHSON_MP):
        success, nrIterations, flux = inf.NewtonRapsonMP(funcType,
                    soil, dt, ubPotential, isFreeDrainage)
    elif (solver == inf.NEWTON_RAPHSON_MFP):
        success, nrIterations, flux = inf.NewtonRapsonMFP(funcType,
                    soil, dt, ubPotential, isFreeDrainage)
    totalIterationNr += nrIterations

    if success:
        for i in range(inf.n+1):
```

```
                inf.oldTheta[i] = inf.theta[i]
            sumInfiltration += flux * dt
            time += dt

            print("time =", int(time), "\tdt =", int(dt),
                    "\tIter. =", int(nrIterations),
                    "\tInf:", format(sumInfiltration, '.3f'))
            myPlot[0].clear()
            myPlot[0].set_xlim(0, 0.5)
            myPlot[0].set_xlabel("Water content [m$^3$ m$^{-3}$]")
            myPlot[0].set_ylabel("Depth [m]")
            myPlot[0].plot(inf.theta, -inf.z, 'r-')
            myPlot[0].plot(inf.theta, -inf.z, 'yo')
            myPlot[1].plot(time, flux, 'ro')
            plt.draw()

            if (float(nrIterations/inf.maxNrIterations) < 0.25):
                    dt = min(dt*2, maxTimeStep)

        else:
            dt = max(dt / 2, 1)

            for i in range(inf.n+1):
                inf.theta[i] = inf.oldTheta[i]
                if (solver == inf.NEWTON_RAPHSON_MFP):
                    inf.psi[i] = inf.MFPFromTheta(soil, inf.theta[i])
                else:
                    inf.psi[i] = inf.waterPotential(funcType, soil,
                                                    inf.theta[i])
            print ("dt =", dt, "No convergence")

    print("nr of iterations per hour:", totalIterationNr /
            simulationLenght)
    plt.ioff()
    plt.show()
main()
```

The algorithm has an adaptive time step. If the ratio between the number of iterations and the maximum number of iterations is less than 0.25, then the selected time step is the minimum value between $2t$ and the maximum time step. Otherwise, the time step is the maximum value between $dt/2$ and 1. The algorithm also has the option of using a linear or a geometric grid, which is generated by the same functions used for solution of gas flow in Chapter 3. The file PSP_grid.py contains the algorithm.

8.5.2 Soil Functions

The file PSP_soil.py is written to define the soil hydraulic properties. For numerical solutions of water flow, the functions are different formulations of the water retention curve, hydraulic conductivity, capacity and means of hydraulic conductivities. In the following chapters, the file PSP_soil.py will also contain additional soil properties.

The first lines are written to define variables that will be used to select between different hydraulic functions. The following lines are written to define variables used to select the numerical scheme and then those to choose the type of mean used for computation of hydraulic conductivity across different elements. The class Csoil is defined. Note here the use of object-oriented programming (OOP) through the use of a class. The reader should read the section on OOP in Appendix A for details about OOP in Python. The function **readSoil**( ) reads the data contained in the file soil.text, by employing the function **readDataFile**, which has been used and described in previous chapters. The function **readSoil**( ) reads all the rows of the file soil.txt and to each row a horizon is associated that is an instance of the class Csoil. The user can therefore include as many horizons as needed. The function returns soil[ ], which is a list of instances. In the example presented here, the list has length = 3. The function **getHorizonIndex** returns the horizon index (given the depths). The horizon index begins at zero; therefore, in this case, they are 0,1 and 2. There is a control to be sure that there are no holes or missing depth intervals. The total depth of the soil profile is the lower depth of the last horizon.

The selection of the air-entry potential is performed in the function **airEntryPotential**( ). When the fitting of water retention data is done with the Campbell and the modified van Genuchten equations the air-entry values may be different and so two sets of parameters can be read from the file soil.txt; therefore different values are used in the numerical solution. The following functions are written to obtain relationships between variables, for instance to convert from degree of saturation to volumetric water content, or to define the water retention curve for the different models and obtain the degree of saturation.

```
#PSP_soil.py
from PSP_readDataFile import *
from math import sqrt, log

g = 9.8065

NODATA = -9999.

CAMPBELL = 1
RESTRICTED_VG = 2
IPPISCH_VG = 3
VAN_GENUCHTEN = 4

CELL_CENT_FIN_VOL = 1
NEWTON_RAPHSON_MP = 2
NEWTON_RAPHSON_MFP = 3

LOGARITHMIC = 0
HARMONIC = 1
GEOMETRIC = 2
```

```
class Csoil:
    upperDepth = NODATA
    lowerDepth = NODATA
    Campbell_he = NODATA
    Campbell_b = NODATA
    CampbellMFP_he = NODATA
    Campbell_b3 = NODATA
    VG_alpha = NODATA
    VG_n = NODATA
    VG_m = NODATA
    VG_he = NODATA
    VG_alpha_mod = NODATA
    VG_n_mod = NODATA
    VG_m_mod = NODATA
    VG_Sc = NODATA
    VG_thetaR = NODATA
    Mualem_L = NODATA
    thetaS = NODATA
    Ks = NODATA


def readSoil(soilFileName):
    mySoil = []
    A, isFileOk = readDataFile(soilFileName, 1, ',', False)
    if ((not isFileOk) or (len(A[0]) < 12)):
        print("error: wrong soil file.")
        return False, mySoil

    for i in range(len(A)):
        horizon = Csoil()
        horizon.upperDepth = A[i,0]
        horizon.lowerDepth = A[i,1]
        horizon.Campbell_he = A[i,2]
        horizon.Campbell_b = A[i,3]
        horizon.Campbell_n = 2.0 + (3.0 / horizon.Campbell_b)
        horizon.VG_he = A[i,4]
        horizon.VG_alpha = A[i,5]
        horizon.VG_n = A[i,6]
        horizon.VG_m =  1. - (1. / horizon.VG_n)
        horizon.VG_alpha_mod = A[i,7]
        horizon.VG_n_mod = A[i,8]
        horizon.VG_m_mod =  1. - (1. / horizon.VG_n_mod)
        horizon.VG_Sc =((1.+ (horizon.VG_alpha_mod*
        abs(horizon.VG_he))**horizon.VG_n_mod)**(-horizon.VG_m_mod))
        horizon.VG_thetaR = A[i,9]
        horizon.thetaS = A[i,10]
        horizon.Ks = A[i,11]
        horizon.Mualem_L = 0.5
        horizon.CampbellMFP_he = (horizon.Ks * horizon.Campbell_he
```

```
            / (1.0 - horizon.Campbell_n))
        horizon.Campbell_b3 = ((2.0 * horizon.Campbell_b + 3.0)
            / (horizon.Campbell_b + 3.0))
        mySoil.append(horizon)
    return True, mySoil

def getHorizonIndex(soil, depth):
    for index in range(len(soil)):
        if ((depth >= soil[index].upperDepth)
        and (depth < soil[index].lowerDepth)):
            return(index)
    lastHorizon = len(soil)-1
    if depth >= soil[lastHorizon].lowerDepth:
        return lastHorizon
    else:
        return(-1)

def airEntryPotential(funcType, soil):
    if (funcType == CAMPBELL):
        return(soil.Campbell_he)
    elif (funcType == IPPISCH_VG):
        return(soil.VG_he)
    elif (funcType == RESTRICTED_VG):
        return(0)
    else:
        return(NODATA)

def waterPotential(funcType, soil, theta):
    psi = NODATA
    Se = SeFromTheta(funcType, soil, theta)
    if (funcType == RESTRICTED_VG):
        psi = (-(1./soil.VG_alpha)*
            ((1./Se)**(1./soil.VG_m) - 1.)**(1./soil.VG_n))
    elif (funcType == IPPISCH_VG):
        psi = -((1./soil.VG_alpha_mod)*
            ((1./(Se*soil.VG_Sc))**(1./soil.VG_m_mod)-1.)**(1./soil.VG_n_mod))
    elif (funcType == CAMPBELL):
        psi = soil.Campbell_he * Se**(-soil.Campbell_b)
    return(psi)

def SeFromTheta(funcType, soil, theta):
    if (theta >= soil.thetaS): return(1.)
    if (funcType == CAMPBELL):
        Se = theta / soil.thetaS
    else:
        Se = (theta - soil.VG_thetaR) / (soil.thetaS - soil.VG_thetaR)
    return (Se)
```

```
def thetaFromSe(funcType, soil, Se):
    if (funcType == RESTRICTED_VG) or (funcType == IPPISCH_VG):
        theta = (Se * (soil.thetaS - soil.VG_thetaR) + soil.VG_thetaR)
    elif (funcType == CAMPBELL):
        return(Se * soil.thetaS)
    return(theta)


def degreeOfSaturation(funcType, soil, psi):
    if (psi >= 0.): return(1.)
    Se = NODATA
    if (funcType == IPPISCH_VG):
        if (psi >= soil.VG_he): Se = 1.
        else:
            Se = (1./soil.VG_Sc) * pow(1.+pow(soil.VG_alpha_mod
                * abs(psi), soil.VG_n_mod), -soil.VG_m_mod)
    elif (funcType == RESTRICTED_VG):
        Se = 1 / pow(1 + pow(soil.VG_alpha * abs(psi), soil.VG_n),
                    soil.VG_m)
    elif (funcType == CAMPBELL):
        if psi >= soil.Campbell_he: Se = 1.
        else: Se = pow(psi / soil.Campbell_he, -1. / soil.Campbell_b)
    return(Se)


def thetaFromPsi(funcType, soil, psi):
    Se = degreeOfSaturation(funcType, soil, psi)
    theta = thetaFromSe(funcType, soil, Se)
    return(theta)


def hydraulicConductivityFromTheta(funcType, soil, theta):
    k = NODATA
    if (funcType == RESTRICTED_VG):
        Se = SeFromTheta(funcType, soil, theta)
        k = (soil.Ks * pow(Se, soil.Mualem_L) *
            (1. -pow(1. -pow(Se, 1./soil.VG_m), soil.VG_m))**2)
    elif (funcType == IPPISCH_VG):
        Se = SeFromTheta(funcType, soil, theta)
        num   = (1. - pow(1. -
        pow(Se * soil.VG_Sc, 1./ soil.VG_m_mod), soil.VG_m_mod));
        denom = (1. - pow(1. -
        pow(soil.VG_Sc, 1./ soil.VG_m_mod), soil.VG_m_mod));
        k = soil.Ks * pow(Se, soil.Mualem_L) * pow((num / denom), 2.)
    elif (funcType == CAMPBELL):
        psi = waterPotential(funcType, soil, theta)
        k = soil.Ks * (soil.Campbell_he / psi)**soil.Campbell_n
    return(k)
```

```
#-----------------------------------------------
# dTheta/dH = dSe/dH (Theta_s - Theta_r)
#-----------------------------------------------

def dTheta_dPsi(funcType, soil, psi):
    airEntry = airEntryPotential(funcType, soil)
    if (psi > airEntry): return 0.0

    if (funcType == RESTRICTED_VG):
        dSe_dpsi = (soil.VG_alpha * soil.VG_n * (soil.VG_m
        * pow(1. + pow(soil.VG_alpha * abs(psi), soil.VG_n),
        -(soil.VG_m + 1.)) *
        pow(soil.VG_alpha * abs(psi), soil.VG_n - 1.)))
        return dSe_dpsi * (soil.thetaS - soil.VG_thetaR)
    elif (funcType == IPPISCH_VG):
        dSe_dpsi = (soil.VG_alpha_mod * soil.VG_n_mod * (soil.VG_m_mod
                * pow(1. +
                  pow(soil.VG_alpha_mod * abs(psi), soil.VG_n_mod),
                -(soil.VG_m_mod + 1.)) *
                pow(soil.VG_alpha_mod * abs(psi), soil.VG_n_mod - 1.)))
        dSe_dpsi *= (1. / soil.VG_Sc)
        return dSe_dpsi * (soil.thetaS - soil.VG_thetaR)
    elif (funcType == CAMPBELL):
        theta = soil.thetaS * degreeOfSaturation(funcType, soil, psi)
        return -theta / (soil.Campbell_b * psi)

def MFPFromTheta(soil, theta):
    return (soil.CampbellMFP_he *
    (theta / soil.thetaS)**(soil.Campbell_b + 3.0))

def MFPFromPsi(soil, psi):
    return (soil.CampbellMFP_he *
    (psi / soil.Campbell_he)**(1.0 - soil.Campbell_n))

def thetaFromMFP(soil, MFP):
    if (MFP > soil.CampbellMFP_he):
        return(soil.thetaS)
    else:
        return(soil.thetaS *
        (MFP / soil.CampbellMFP_he)**(1.0/(soil.Campbell_b + 3.0)))

def hydraulicConductivityFromMFP(soil, MFP):
    k = soil.Ks * (MFP / soil.CampbellMFP_he)**soil.Campbell_b3
    return(k)

def dTheta_dH(funcType, soil, H0, H1, z):
    psi0 = H0 + g*z
    psi1 = H1 + g*z
```

```
    if (abs(psi1-psi0) < 1E-5):
        return dTheta_dPsi(funcType, soil, psi0)
    else:
        theta0 = thetaFromPsi(funcType, soil, psi0)
        theta1 = thetaFromPsi(funcType, soil, psi1)
        return (theta1 - theta0) / (psi1 - psi0)

def meanK(meanType, k1, k2):
    if (meanType == LOGARITHMIC):
        if (k1 != k2):
            k = (k1-k2) / log(k1/k2)
        else:
            k = k1
    elif (meanType == HARMONIC):
        k = 2.0 / (1.0 / k1 + 1.0 / k2)
    elif (meanType == GEOMETRIC):
        k = sqrt(k1 * k2)
    return k
```

8.5.3 Solvers

The file `PSP_infiltration1D.py` contains the solvers, where the different numerical schemes are written. The first lines import the functions for creating the grid, the solutions of the Thomas algorithm and the soil data. The Thomas algorithm was described in Chapter 3 and is given in detail in Appendix B. The following lines define the density of water, the area between nodes, the maximum number of iterations and the tolerance for the mass balance. Note that for a one-dimensional solution the area between elements will not be important, but this variable will be used for the three-dimensional solution presented later. The vectors for the numerical solution are initialized by the numpy arrays definition. Descriptions of each one are given in the comments. The depth of each layer in the computational grid is defined by the function **grid.geometric()**, which is the function to build a geometric grid, written in the file `PSP_grid.py`. Here a geometric grid is used; the function **grid.linear()** defines a linear grid. Then the three numerical solutions are written in the functions **cellCentFiniteVolWater()**, **NewtonRapsonMP()** and **NewtonRapsonMFP()**.

```
#PSP_infiltration1D
from __future__ import print_function, division

import PSP_grid as grid
from PSP_ThomasAlgorithm import ThomasBoundaryCondition
from PSP_soil import *

waterDensity = 1000.
area = 1
```

```
maxNrIterations = 100
tolerance = 1e-5
n = 50
hor = np.zeros(n+2, int)
z = np.zeros(n+2, float)
zCenter = np.zeros(n+2, float)
vol = np.zeros(n+2, float)
a = np.zeros(n+2, float)
b = np.zeros(n+2, float)
c = np.zeros(n+2, float)
d = np.zeros(n+2, float)
dz = np.zeros(n+2, float)
psi = np.zeros(n+2, float)
dpsi = np.zeros(n+2, float)
theta = np.zeros(n+2, float)
oldTheta = np.zeros(n+2, float)
C = np.zeros(n+2, float)
k = np.zeros(n+2, float)
u = np.zeros(n+2, float)
du = np.zeros(n+2, float)
f = np.zeros(n+2, float)
H = np.zeros(n+2, float)
H0 = np.zeros(n+2, float)

def initializeWater(funcType, soil, se_0, solver):
    global z
    # vector depth [m]
    lastHorizon = len(soil)-1
    z = grid.geometric(n, soil[lastHorizon].lowerDepth)
    vol[0] = 0
    for i in range(n+1):
        dz[i] = z[i+1]-z[i]
        if (i > 0): vol[i] = area * dz[i]
    for i in range(n+2):
        zCenter[i] = z[i] + dz[i]*0.5

    if (solver == CELL_CENT_FIN_VOL):
        for i in range(n+1):
            dz[i] = zCenter[i+1]-zCenter[i]

    #initial conditions
    psi[0] = 0
    for i in range(1, n+2):
        hor[i] = getHorizonIndex(soil, zCenter[i])
        theta[i] = thetaFromSe(funcType, soil[hor[i]], se_0)
        oldTheta[i] = theta[i]
        if (solver == NEWTON_RAPHSON_MFP):
            psi[i] = MFPFromTheta(soil[hor[i]], theta[i])
```

```
            k[i] = hydraulicConductivityFromMFP(soil[hor[i]], psi[i])
        else:
            psi[i] = waterPotential(funcType, soil[hor[i]], theta[i])
            k[i] = (hydraulicConductivityFromTheta(funcType,
                                         soil[hor[i]], theta[i]))
        H[i] = psi[i] - zCenter[i]*g

def NewtonRapsonMP(funcType, soil, dt, ubPotential, isFreeDrainage):
    #apply upper boundary condition
    airEntry = airEntryPotential(funcType, soil[0])
    psi[1] = min(ubPotential, airEntry)
    oldTheta[1] = thetaFromPsi(funcType, soil[0], psi[1])
    theta[1] = oldTheta[1]

    if (isFreeDrainage):
        psi[n+1] = psi[n]
        theta[n+1] = theta[n]
        k[n+1] = k[n]

    nrIterations = 0
    massBalance = 1
    while ((massBalance > tolerance) and (nrIterations <
           maxNrIterations)):
        massBalance = 0
        for i in range(1, n+1):
            k[i] = hydraulicConductivityFromTheta(funcType,
                                         soil[hor[i]], theta[i])
            u[i] = g * k[i]
            du[i] = -u[i] * soil[hor[i]].Campbell_n / psi[i]
            capacity = dTheta_dPsi(funcType, soil[hor[i]], psi[i])
            C[i] = (waterDensity * vol[i] * capacity) / dt

        for i in range (1, n+1):
            f[i] = ((psi[i+1] * k[i+1] - psi[i] * k[i])
                    / (dz[i] * (1 - soil[hor[i]].Campbell_n))) - u[i]
            if (i == 1):
                a[i] = 0
                c[i] = 0
                b[i] = k[i] / dz[i] + C[i] + du[i]
                d[i] = 0
            else:
                a[i] = -k[i-1] / dz[i-1] - du[i-1]
                c[i] = -k[i+1] / dz[i]
                b[i] = k[i] / dz[i-1] + k[i] / dz[i] + C[i] + du[i]
                d[i] = f[i-1] - f[i] + (waterDensity * vol[i]
                                    * (theta[i] - oldTheta[i])) /dt
                massBalance += abs(d[i])
```

```
ThomasBoundaryCondition(a, b, c, d, dpsi, 1, n)

for i in range(1, n+1):
        psi[i] -= dpsi[i]
        psi[i] = min(psi[i], airEntry)
        theta[i] = thetaFromPsi(funcType, soil[hor[i]], psi[i])
    nrIterations += 1

    if (isFreeDrainage):
        psi[n+1] = psi[n]
        theta[n+1] = theta[n]
        k[n+1] = k[n]

    if (massBalance < tolerance):
        flux = -f[1]
        return True, nrIterations, flux
    else:
        return False, nrIterations, 0

def cellCentFiniteVolWater(funcType, soil, dt, ubPotential,
                            isFreeDrainage, meanType):
    #apply upper boundary condition
    airEntry = airEntryPotential(funcType, soil[0])
    psi[0] = min(ubPotential, airEntry)
    theta[0] = thetaFromPsi(funcType, soil[0], psi[0])
    theta[1] = theta[0]
    psi[1] = psi[0]

    if (isFreeDrainage):
        psi[n+1] = psi[n]
        H[n+1] = psi[n+1] - zCenter[n+1]*g
        theta[n+1] = theta[n]
        k[n+1] = k[n]

    sum0 = 0
    for i in range(1, n+1):
        H0[i] = psi[i] - zCenter[i]*g
        H[i] = H0[i]
        sum0 += waterDensity * vol[i] * theta[i]

    massBalance = sum0
    nrIterations = 0
    while ((massBalance > tolerance) and (nrIterations <
                                    maxNrIterations)):
        for i in range(1, n+1):
            k[i] = hydraulicConductivityFromTheta(funcType,
                                        soil[hor[i]], theta[i])
```

```
    capacity = dTheta_dH(funcType, soil[hor[i]], H0[i], H[i],
                         zCenter[i])

    C[i] = (waterDensity * vol[i] * capacity) / dt

f[0] = 0
for i in range(1, n+1):
    f[i] = area * meanK(meanType, k[i], k[i+1]) / dz[i]

for i in range(1, n+1):
    a[i] = -f[i-1]
    if (i == 1):
        b[i] = 1
        c[i] = 0
        d[i] = H0[i]
    elif (i < n):
        b[i] = C[i] + f[i-1] + f[i]
        c[i] = -f[i]
        d[i] = C[i] * H0[i]
    elif (i == n):
        b[n] = C[n] + f[n-1]
        c[n] = 0
        if (isFreeDrainage):
            d[n] = C[n] * H0[n] - area * k[n] * g
        else:
            d[n] = C[n] * H0[n] - f[n]* (H[n]-H[n+1])

ThomasBoundaryCondition(a, b, c, d, H, 1, n)

newSum = 0
for i in range(1, n+1):
    psi[i] = H[i] + g*zCenter[i]
    theta[i] = thetaFromPsi(funcType, soil[hor[i]], psi[i])
    newSum += waterDensity * vol[i] * theta[i]

if (isFreeDrainage):
    psi[n+1] = psi[n]
    theta[n+1] = theta[n]
    k[n+1] = k[n]

if (isFreeDrainage):
    massBalance = abs(newSum - (sum0 + f[1]*(H[1]-H[2])*dt
                               - area*k[n]*g*dt))
else:
    massBalance = abs(newSum - (sum0 + f[1]*(H[1]-H[2])*dt
                               - f[n]*(H[n]-H[n+1])*dt))
nrIterations += 1
```

```
    if (massBalance < tolerance):
        flux = f[1]*(H[1]-H[2])
        return True, nrIterations, flux
    else:
        return False, nrIterations, 0

# Infiltration simulation using Matric Flux Potential def
NewtonRapsonMFP(funcType, soil, dt, ubPotential, isFreeDrainage):
    #apply upper boundary condition
    airEntry = airEntryPotential(funcType, soil[0])
    ubPotential = min(ubPotential, airEntry)
    psi[1] = MFPFromPsi(soil[0], ubPotential)
    oldTheta[1] = thetaFromMFP(soil[0], psi[1])
    theta[1] = oldTheta[1]
    k[1] = hydraulicConductivityFromMFP(soil[0], psi[1])
    psi[0] = psi[1]
    k[0] = 0.0

    if (isFreeDrainage):
        psi[n+1] = psi[n]
        theta[n+1] = theta[n]
        k[n+1] = k[n]

    nrIterations = 0
    massBalance = 1
    while ((massBalance > tolerance) and (nrIterations <
                                     maxNrIterations)):
        massBalance = 0
        for i in range(1, n+1):
            k[i] = hydraulicConductivityFromMFP(soil[hor[i]], psi[i])
            capacity = theta[i] / ((soil[hor[i]].Campbell_b + 3.0)
                              * psi[i])
            C[i] = waterDensity * vol[i] * capacity / dt
            u[i] = g * k[i]
            f[i] = (psi[i+1] - psi[i]) / dz[i] - u[i]
            if (i == 1):
                a[i] = 0
                c[i] = 0
                b[i] = 1.0/dz[i] + C[i] + g * soil[hor[i]].Campbell_b3
                                  * k[i] / psi[i]
                d[i] = 0
            else:
                a[i] = -1.0/dz[i-1] -g * soil[hor[i-1]].Campbell_b3
                                  * k[i-1] / psi[i-1]
                c[i] = -1.0/dz[i]
                b[i] = (1.0/dz[i-1] + 1.0/dz[i] + C[i] +
                g * soil[hor[i]].Campbell_b3 * k[i] / psi[i])
```

```
        d[i] = f[i-1] - f[i] + (waterDensity * vol[i]
                               * (theta[i] - oldTheta[i]) / dt)
        massBalance += abs(d[i])

    ThomasBoundaryCondition(a, b, c, d, dpsi, 1, n)

    for i in range(1, n+1):
        psi[i] -= dpsi[i]
        psi[i] = min(psi[i], soil[hor[i]].CampbellMFP_he)
        theta[i] = thetaFromMFP(soil[hor[i]], psi[i])

    if (isFreeDrainage):
        psi[n+1] = psi[n]
        theta[n+1] = theta[n]
        k[n+1] = k[n]

    nrIterations += 1

if (massBalance < tolerance):
    flux = -f[1]
    return True, nrIterations, flux
else:
    return False, nrIterations, 0
```

8.5.4 Results

The results of the simulation are shown below. We ran two different simulations to show the effect of soil horizons having different soil properties. The first simulation is for a uniform profile and the second for a profile displaying three horizons.

Uniform profile

The soil selected for the uniform profile is a silt loam. The program prints on screen the time [s], the time step [s], the number of iterations for each time step needed to reach convergence [number s^{-1}] and the cumulative infiltration [kg m^{-2}]. The last line prints the average number of iterations per hour [number h^{-1}].

```
Nr of simulation hours:2
time = 60    dt = 60    Iter. = 55   Inf: 9.365
time = 120   dt = 60    Iter. = 68   Inf: 13.728
time = 180   dt = 60    Iter. = 57   Inf: 17.052
time = 240   dt = 60    Iter. = 29   Inf: 19.853
time = 300   dt = 60    Iter. = 28   Inf: 22.343
time = 360   dt = 60    Iter. = 19   Inf: 24.621
time = 480   dt = 120   Iter. = 53   Inf: 28.611
time = 600   dt = 120   Iter. = 40   Inf: 32.211
time = 720   dt = 120   Iter. = 9    Inf: 35.630
```

```
time = 960    dt = 240   Iter. = 97   Inf: 41.546
time = 1200   dt = 240   Iter. = 58   Inf: 46.881
time = 1440   dt = 240   Iter. = 42   Inf: 51.811
time = 1680   dt = 240   Iter. = 28   Inf: 56.421
time = 1920   dt = 240   Iter. = 27   Inf: 60.782
time = 2160   dt = 240   Iter. = 19   Inf: 64.948
time = 2640   dt = 480   Iter. = 95   Inf: 72.695
time = 3120   dt = 480   Iter. = 69   Inf: 80.002
time = 3600   dt = 480   Iter. = 42   Inf: 86.966
time = 4080   dt = 480   Iter. = 36   Inf: 93.643
time = 4560   dt = 480   Iter. = 30   Inf: 100.085
time = 5040   dt = 480   Iter. = 32   Inf: 106.327
time = 5520   dt = 480   Iter. = 24   Inf: 112.401
time = 6120   dt = 600   Iter. = 38   Inf: 119.779
time = 6720   dt = 600   Iter. = 35   Inf: 126.976
time = 7200   dt = 480   Iter. = 18   Inf: 132.633
nr of iterations per hour: 524.0
```

While the computation is running, a plot is shown, as depicted in Fig. 8.5. Since the upper boundary condition is a constant potential, set at the air-entry value, this

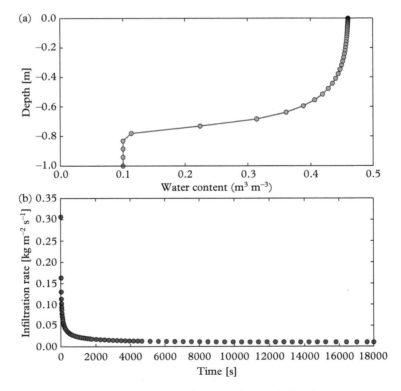

Figure 8.5 *(a) Redistribution of soil water following infiltration.*
(b) Infiltration rate as a function of time.

simulation resembles a layer of ponding water on a dry soil. The initial water content is $0.1 \, \text{m}^3 \, \text{m}^{-3}$. This condition is usually a difficult one to simulate because of the development of high water potential gradients at the wetting front.

Layered profile

The soil selected for the layered profile has an A horizon with a coarser (less clay) texture. This is typical of A horizons where clays are leached. The depth is 20 cm. The illuviated B horizon has a finer texture. Usually, B horizons are zones where clay, soluble salts and/or iron accumulate. The soil properties selected are for a clay loam and the depth is 50 cm. Finally, the C horizon consists of parent material, such as glacial till or lake sediments, that shows little or no alteration. The texture of the C horizon is sand, having a depth of 30 cm, for a total depth of 1 m. The soil properties for the three soils are listed in Tables 5.3 and 5.4, while the values for saturated hydraulic conductivity are listed in Table 6.1.

The simulation was for 24 h to show the effects of the soil horizons on the soil water content distribution. While the computation is running, a plot is shown, as depicted in Fig. 8.6. The transition between the soil layers is very clear. These effects are better

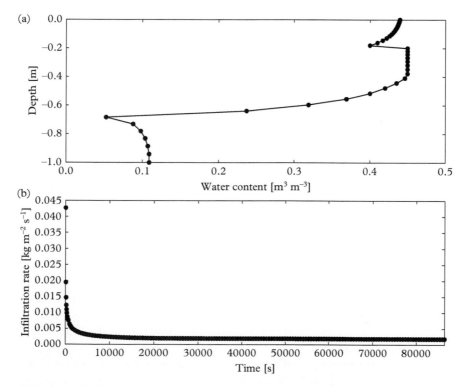

Figure 8.6 *(a) Redistribution of soil water following infiltration for a three-horizon soil. (b) Infiltration rate as a function of time.*

visualized by running the program and observing the dynamics of the system. Observe how the wetting front in the B horizon encounters the layer of coarse sand. The pores in the clay above are much smaller than those in the sand layer below. The large pores in the sands cannot hold water at the same water potential that exists in the wetting front, so water does not move readily into the sand. However, as the soil above the sand becomes very wet, the water eventually moves into the sand. This behaviour is typical of field conditions where the C horizon is a glacial till or a sediment deposition of coarse material.

The opposite situation is a layer of clay in a uniform soil (a clay pan). These layers may restrict rooting and are also the reason for the formation of perched water tables. When excess water is provided to the soil and a clay layer is present (such as a frangipan or a hardpan layer), the downward water movement is prevented and perched water tables are formed. The user can invert the position of the layer in the file soil.txt and simulate the accumulation of water over the clay layer.

Finally, a comparison of three models for the same hydraulic functions and boundary conditions shows a similar value of cumulative infiltration. For a 5 h numerical simulation, the cumulative infiltration was 247.3, 248 and 248.5 $kg\,m^{-2}\,s^{-1}$ for the cell-centered finite volume method, the finite difference method with matric potential and the integral transform method with matric flux potential, respectively. However, the corresponding computational times were 268.3, 290 and 26.4 iter/h. The integral transform methods, because of the linearization of the nonlinear problem, provide much faster and more stable solutions compared with the linear methods. Similar results were obtained for a variety of tests, using different boundary conditions and simulation times.

....

8.6 EXERCISES

8.1. Use the program PSP_infiltrationRedistribution1D with different hydraulic property functions and numerical schemes. Investigate the effect on the cumulative infiltration and on the total number of iterations.

8.2. Using the program PSP_infiltrationRedistribution1D, select a hydraulic property function and a numerical scheme. Using the same settings, change the value of the soil initial water content and investigate the effect on the cumulative infiltration. How do the profiles of soil water content differ?

8.3. Using the program PSP_infiltrationRedistribution1D, select a hydraulic property function and a numerical scheme. Using the same settings, change the value of the lower boundary condition and investigate the effect on the total cumulative infiltration.

8.4. Modify the program such that there are two layers: 0.5 m of sandy soil at the top and clay soil at the bottom. What do you expect for the infiltration process. Run the program and compare the results.

8.5. Repeat Exercise 8.4, but invert the two layers.

9

Triangulated Irregular Network

In Chapter 8, we discussed one-dimensional solution of water flow. For practical applications, it is often necessary to perform multidimensional simulations (2D and 3D). In three-dimensional hydrological models, accurate descriptions of land relief are essential, since terrain affects surface runoff, infiltration, subsurface lateral flow, evaporation and plant transpiration. Indeed, plant growth is affected by aspect and slope, because the spatial distribution of solar radiation (i.e. shadowing due to relief) affects photosynthetic activity and soil water content distribution. Moreover, snow accumulation and melting depend on topography as well, since snow melting in particular depends on solar radiation and therefore is strongly affected by topography and geomorphological features of the catchment (Bittelli *et al.*, 2010).

In general, as the model domain increases in size, the resolution of the terrain representation must decrease to allow efficient model simulation. This is especially true when the model is used on personal computers rather than on clusters of computers assembled in parallel for large computational projects. To decrease the computational time, it is necessary to decrease the number of nodes or volume elements used in the simulation. This task is commonly performed with algorithms that reduce the number of nodes while still preserving the essential features of the surface. In this chapter, we discuss how to reduce the number of computational elements in a digital terrain model (DTM) by using a representation of the surface as a triangulated irregular network (TIN).

9.1 Digital Terrain Model

A DTM is a data structure of rasterized elevation data of the terrain. A raster consists of a matrix of cells (or pixels) organized into rows and columns where each cell contains a value representing specific information, in our case, the elevation. Figure 9.1 shows a DTM for the hilly area of Fontanafredda (Piedmont region, Northern Italy). This DTM will be used as an example for the application of the triangulation program. The term digital surface model (DSM) refers to a data structure representing the land surface and including objects on the surface, such as trees or houses, whereas the DTM refers to the surface of the terrain only. To remove the objects from a DSM to obtain a DTM, several techniques can be used, including land survey or computer algorithms. The term digital

Soil Physics with Python. First Edition. Marco Bittelli, Gaylon S. Campbell and Fausto Tomei.
© Marco Bittelli, Gaylon S. Campbell and Fausto Tomei 2015. Published in 2015 by Oxford University Press.

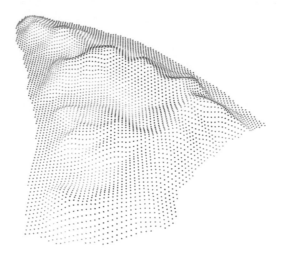

Figure 9.1 *Digital terrain model of Fontanafredda vineyard (Piedmont, Italy), cell size = 20 m, dimension 196 × 207 = 40572 cells.*

elevation model (DEM) refers more generally to both DSMs and DTMs. Usually, the data contained in a DTM are geo-referenced data. An elevation is geo-referenced when its location, in terms of map projections or coordinate systems, has been established. For discussion of different geo-referencing systems, the user should refer to a geography or cartography text (e.g. Slocum *et al.*, 2009).

9.1.1   Data Structure

The data structure used to represent a DTM can be stored in different file formats. We choose the ArcGIS floating-point grid format by ESRI (ESRI, 2011), which is a binary floating-point file with a `.flt` extension and an ASCII header file with a `.hdr` extension. The binary file contains each data point, while the header file includes the following metadata:

1. `nrows, ncols`: the number of rows and columns
2. `xllcorner, yllcorner`: the (x, y) position [m] of the lower left corner of the grid
3. `cellsize`: the size [m] of every cell of the grid
4. `NODATA`: the value that identifies the empty zones of the grid.

We choose this format because of its wide use and simplicity. It allows us to read the DTM data with a few lines of code. The data are stored in a simple grid of floating-point values, with each value per grid element occupying only 4 bytes of memory. The x and y positions of the lower left corner of the grid (`xllcorner, yllcorner`) are geo-referenced, but it is not specified with respect to which datum and UTM zone. The UTM zone of the example presented here is number 32 and the reference ellipsoid is WGS84. The module `PSP_gis.py` implements the functions necessary for reading an ESRI floating-point file and will be discussed in Section 9.6.

9.2   Triangulated Irregular Network

A TIN is a data structure used for the representation of a surface. TINs are made up of a set of irregularly distributed nodes, with three-dimensional coordinates, connected in a network of non-overlapping triangles. TIN are often used to represent terrains. The advantage of using TINs instead of DTMs is that they can reduce the number of nodes and the computational time, maintaining an accurate representation of the terrain. Moreover, DTMs are usually based on a square geometry, while the triangular structure allows for accommodation of more irregular geometries typical of natural terrains. Usually a TIN is generated by applying a triangulation algorithm to an existing regular grid; however, it can be applied to regular or irregular objects of various geometries. A TIN is commonly created by using a Delaunay triangulation, which will be described in detail in Section 9.9. There is a large body of literature on the subject of triangulation and TIN generation (Jay, 1991; Cheng *et al.*, 2012).

9.3   Numerical Implementation

The project is called `PSP_triangulation.py` and comprises the following modules:

1. `main.py`
2. `PSP_triangulation.py`
3. `PSP_gis.py`
4. `PSP_RamerDouglasPeucker.py`
5. `PSP_triangle.py`
6. `PSP_Delaunay.py`
7. `PSP_refinement.py`
8. `PSP_utility.py`
9. `PSP_color.py`
10. `PSP_visual3D.py`

Because of the complexity and length of the algorithm, we divided the program into modules, each performing a specific task. The `main.py` module calls the main functions needed by the program. The `PSP_triangulation.py` module implements the whole triangulation procedure. The `PSP_gis.py` module (where GIS stands for geographic information system) contains the function to open the DTM and defines various properties of DTM needed for the triangulation procedure.

`PSP_RamerDouglasPeucker.py` implements a recursive procedure used to identify and reduce the numbers of boundary points. `PSP_triangle.py` contains a variety of functions used to define geometrical properties of triangles and other geometrical objects. `PSP_Delaunay.py` contains the Delaunay algorithm. `PSP_refinement.py` contains the refinement functions for the triangulation both on the elevation and the angles. `PSP_utility.py` contains utilities for computation and file management. `PSP_color.py` defines the colours used by the plotting algorithms. `PSP_visual3D.py` is used to plot DTM and TIN in three dimensions. Each of these files will now be described in detail.

9.4   Main

The triangulation procedure begins with the **main** module (the first line is described in Appendix A):

```
from __future__ import print_function
from PSP_triangulation import *

def main():
    #open DTM
    header, dtm = openDTM()
    pointListDTM = []
    for col in range(header.nrCols):
        for row in range(header.nrRows):
            if dtm[col, row] != header.flag:
                p = getPointFromColRow(header, dtm, col, row)
                pointListDTM.append(p)
    header.nrPixels = len(pointListDTM)
    visual3D.initialize(header)
    visual3D.drawDTM(pointListDTM, header)
    print ("number of DTM points:", len(pointListDTM))

    #create TIN
    quality = 0.5
    angleThreshold = pi/4.
    isInitialPartition = False
    pointList, triangleList = triangulate(header, dtm,
                        quality, angleThreshold, isInitialPartition)

    visual3D.drawAllSurfaces(triangleList, header)
    print ("number of triangles:", len(triangleList))
```

```
      #write TIN
      writeTIN(triangleList, pointList, header, dtm,
               "data/vertices.csv", "data/triangles.csv",
               "data/neighbours.csv")
      print("End.")
main()
```

main.py imports the PSP_triangulation.py module, which imports all the other modules, such that main.py can call all the principal functions of the program (nested import). At the beginning, the program opens the DTM using the function **openDTM** defined in the module PSP_gis.py. The program then scans the matrix of the DTM, to identify the cells that have a valid value and append each of these points into the array pointListDTM. This is used by the function **drawAllPointsDTM** to draw the DTM in a window by using VisualPython. This function will be described in the module PSP_visual3D.py. The program Visual Python is described in Appendix A.

After having visualized the DTM, main.py defines the principal parameters of the triangulation algorithm (quality, angleThreshold and isInitialPartition) and passes these parameters, along with the structure of the DTM, to the function **triangulate** of the module PSP_triangulation.py.

The function **triangulate** returns the arrays pointList and triangleList, which are passed to the function **drawAllSurfaces** written to visualize the generated TIN using VisualPython. Finally, the function **writeTIN** of the PSP_utility.py module writes the generated TIN as three lists of vertices, triangles and neighbours of triangles, each saved as comma-separated files (.csv). Note that these data will be used in later chapters as input to create the grid used for three-dimensional solutions of water flow.

9.5   Triangulation

The module PSP_triangulation.py implements the function **triangulate**, which is the main function for the triangulation procedure. This function returns the arrays pointList and triangleList, which are passed to the function **drawAllSurfaces** that visualizes the generated TIN using VisualPython.

The parameters used by the triangulation algorithm are

- quality
- angleThreshold
- isInitialPartition
- partitionStep
- thresholdZ
- thresholdRamer
- maxBoundarySide

- `minArea`
- `randomFactor`

The first three parameters are defined in the **main** function. The other parameters are defined in the **triangulate** function to make them dependent on the DEM size and the key parameter `quality`. Therefore the user can change only the first three parameters in the **main** module, and this automatically changes all the others. On the other hand, the user could independently change all parameters, defining a specific value by modifying the code.

The parameter `quality` ranges between 0 and 1, with 1 as the best possible quality. A value smaller than 1 can be used to speed up the computation and reduce the number of generated triangles. `angleThreshold` [radians] defines the minimum accepted angle, to avoid having triangles with very small angles. `isInitialPartition` is a Boolean value that specifies if the user wants to set an initial regular partition of the DTM points or not. This option is important for DTMs having very flat areas, to avoid the generation of very large triangles. Therefore, if the user is applying this algorithm to an area presenting both valleys and hilly or mountainous areas, we suggest that the initial partition option be employed. Otherwise, if the area is in a hilly or mountainous area, it may not be necessary. `thresholdZ` is the threshold on the elevation difference (in meter) used for the triangulation refinement on the elevation. It expresses the maximum accepted elevation difference, while `thresholdRamer` is the threshold for the perpendicular distance used in the boundary selection, as described below. The parameter `maxBoundarySide` is the maximum allowed length [m] of a single side on the DTM boundary. `minArea` is the minimum accepted area [m^2] that a triangle can have. The next two parameters have an effect only if `isInitialPartition` is set to True. `randomFactor` is a variability factor (in the range 0–1) to define if the initial partition is regular or not (0 = the partition is totally regular and 1 = maximum variability). `partitionStep` is the average distance (in number of pixels) between two points of the initial partition. The defined parameters are indicative conditions. The algorithm attempts to meet all conditions, but some triangles of the final set may not meet all the conditions. The scheme to obtain the triangulation is as follows:

- Select the key points defining the boundary of the DTM and store them in the set `pointList`.
- Define an initial partition (if the user selects this option) and add these points to the set.
- Run the Delaunay triangulation algorithm on this set of points to obtain a first triangulation.
- Perform a refinement of the triangulation on the elevation.
- Perform a refinement of the triangulation based on the triangle angles.

Because of the wide variety and heterogeneities of possible geomorphological features, the user is advised to try different options for the parameters described and visually analyse the generated TIN. An inspection should be made to see if the landscape is correctly represented and if all of the geomorphological features are preserved and well described. The `PSP_triangulation.py` module is as follows:

```python
#PSP_triangulation.py
from __future__ import print_function, division
from PSP_RamerDouglasPeucker import getDtmBoundary
import PSP_visual3D as visual3D
from PSP_refinement import *
from PSP_Delaunay import *
from copy import copy

def triangulate(header, dtm, quality, angleThreshold,
                isInitialPartition):
    invQuality = 1.0 / quality
    step = max(1, round(invQuality))     # [pixel] computation step
    nrPixelsSide = sqrt(header.nrPixels)
    # [pixel] average distance between two points of the initial
        partition
    partitionStep = round(invQuality * sqrt(nrPixelsSide))

    # [m] maximum height difference for refinement
    thresholdZ = invQuality * header.dz / 100.0

    # [m] maximum perpendicular distance for boundary
    thresholdRamer = thresholdZ * invQuality**2

    #[m] maximum distance for boundary
    maxBoundarySide = partitionStep * header.cellSize / 2.0

    # [m^2] minimum area of triangles
    minArea = invQuality * invQuality * header.cellSize**2

    # [0-1] fraction variability for partition
    randomFactor = 0.33

    pointList = getDtmBoundary(header, dtm, thresholdRamer,
                              maxBoundarySide)
    flagMatrix = initializeFlagMatrix(header, dtm, pointList, step)

    visual3D.drawAllPoints(pointList, header)

    if isInitialPartition:
        partitionPointList = addPartitionPoints(pointList, header,
                    dtm, partitionStep, randomFactor, flagMatrix, step)
        visual3D.drawAllPoints(pointList, header)
    else:
        partitionPointList = []

    pointList = sortPointList(pointList)
```

```
(pointList, triangleList =
firstTriangulation(pointList, partitionPointList, header, dtm))
visual3D.drawAllTriangles(triangleList, header)

(pointList, triangleList, flagMatrix = refinementZ(pointList,
                  triangleList, header, dtm, thresholdZ,
                  minArea, flagMatrix, step))

(pointList, triangleList, flagMatrix = refinementAngle(pointList,
                  triangleList, header, dtm, angleThreshold,
                  minArea, flagMatrix, step))

return (pointList, triangleList)
```

9.6   GIS Functions

All the geographic functions and the necessary data structures are defined in the
`PSP_gis.py` module. The module imports numpy, `math` and the module `re` (regular
expression). The `re` module provides regular expression operation, which is a sequence
of characters that forms a search pattern used in pattern matching with strings. It allows
for a search for a specific pattern in a string. In particular, we used the function **split** to
separate a text line into an array of values by defining a list of possible separators (for
instance the tab separator).

To make this program usable in both 2.x and 3.x versions of Python, we implemented
an `if` statement, to select the correct version of the `tkinter` dialog windows, using the
`sys` module. Since the `tkinter` implementation differs between Python 2.x and 3.x,
the system must detect the version and apply the correct implementation. `tkinter` is
employed to prompt the dialog box used to select the DTM file name. Then the class
`Cheader` is defined. This class contains the metadata necessary to define the geographic
position of the area of interest and other useful information. The function **openDTM** is
written to open the DTM files (the floating data point .flt and the header file .hdr). This
function calls **loadEsriHeader** and **loadEsriBinary** to perform this task. Specifically,
loadEsriHeader utilizes the function **split** described above to read header files having
different separators between records (spaces or tablature) and it copies the metadata
into a class `Cheader`. The function **loadEsriBinary** loads an entire binary file into a
matrix by using a numpy function called **fromfile**. The resulting matrix is organized in
columns and rows (as in the Fortran language), as indicated by the capital letter (F) in
the parameter `order` of the function. We use this format because the ESRI binary files
are organized in the same way. The function **computeMinMaxGrid** is used to scan all
the DTM data to find the maximum and minimum elevation (such as to define a degree
of magnification for a better visualization) and store it.

In the triangulation code, it will be necessary to scan the values in the grid, both on its
column and row structure, and on the (x, y) position. Therefore, management functions
for both input types are implemented. In particular, the functions **isOutOfGridColRow**

and **isOutOfGridXY** check if a point (defined by columns and rows, or by *x* and *y*) is outside the limits of the grid, which we pass as a header.

The function **getValueFromXY** is used to get the value *z* of a point having coordinates (x, y). A similar function is **getValueFromColRow**, which returns the value of the DTM on the corresponding columns and rows. The function **getColRowFromXY** returns the corresponding column and row, given a position (x, y). Finally, the function **getPointfromColRow** returns the coordinate (x, y, z) of the centre point of the cell, defined by its column and row (cell-centred). The program `PSP_gis.py` is as follows:

```
#PSP_gis.py
from __future__ import print_function, division
from copy import copy
import numpy as np
from math import *
from re import split


NODATA = -9999
NOLINK = -1


import sys
if sys.version_info < (3, 0):
    from tkFileDialog import askopenfilename
else:
    from tkinter.filedialog import askopenfilename

class Cheader():
    nrRows = 0
    nrCols = 0
    nrPixels = 0
    xllCorner = NODATA
    yllCorner = NODATA
    cellSize = 0
    flag = NODATA
    zMin = NODATA
    zMax = NODATA
    dz = NODATA
    magnify = NODATA

def openDTM():
    options = {}
    options['defaultextension'] = '.flt'
    options['filetypes'] = [('ESRI raster files', '.flt')]
    options['initialdir'] = 'data/'
    options['title'] = 'Open DTM'
    fileName = askopenfilename(**options)
    if (fileName != ""):
```

```
        header = loadEsriHeader(fileName[:-4] + ".hdr")
        dtm = loadEsriBinary(fileName, header)
        return(header, dtm)

def loadEsriHeader(fileName):
    header = Cheader()
    file = open(fileName, "r")
    txtLine = file.read()
    #separators: newline, one or more spaces, tab
    values = split('\n| +|\t', txtLine)
    i = 0
    while (i < len(values)):
        tag = values[i].upper()
        if (tag == "NCOLS"):
            header.nrCols = int(values[i+1])
        elif (tag == "NROWS"):
            header.nrRows = int(values[i+1])
        elif (tag == "XLLCORNER"):
            header.xllCorner = float(values[i+1])
        elif (tag == "YLLCORNER"):
            header.yllCorner = float(values[i+1])
        elif (tag == "CELLSIZE"):
            header.cellSize = float(values[i+1])
        elif ((tag == "NODATA") or (tag == "NODATA_VALUE")):
            header.flag = float(values[i+1])
        i += 2
    return(header)

def loadEsriBinary(fileName, header):
    print ("load DTM data...")
    myFile = open(fileName, "rb")
    grid = np.fromfile(myFile, dtype=np.float32).reshape(
                header.nrCols, header.nrRows, order='F')
    computeMinMaxGrid(header, grid)
    return(grid)

def computeMinMaxGrid(header, grid):
    isFirstValue = True
    for col in range(header.nrCols):
        for row in range(header.nrRows):
            z = grid[col, row]
            if (z != header.flag):
                if (isFirstValue):
                    header.zMin = z
                    header.zMax = z
                    isFirstValue = False
                else:
                    header.zMin = min(header.zMin, z)
                    header.zMax = max(header.zMax, z)
```

```
          header.dz = header.zMax - header.zMin
          size = header.cellSize * sqrt(header.nrRows * header.nrCols)
          ratio = size / header.dz
          header.magnify = max(1., min(6., ratio/6.))

      def isOutOfGridColRow(header, col, row):
          if ((row < 0) or (row >= header.nrRows)
          or (col < 0) or (col >= header.nrCols)):
              return(True)
          else:
              return(False)


      def isOutOfGridXY(header, x, y):
          if ((x < header.xllCorner) or (y < header.yllCorner)
          or (x >= (header.xllCorner + header.cellSize * header.nrCols))
          or (y >= (header.yllCorner + header.cellSize * header.nrRows))):
              return(True)
          else:
              return(False)


      def getColRowFromXY(header, x, y):
          row = (header.nrRows -
          floor((y - header.yllCorner) / header.cellSize)-1)
          col = floor((x - header.xllCorner) / header.cellSize)
          return col, row;


      #get value with check on boundary limits
      def getValueFromColRow(header, grid, col, row):
          if (isOutOfGridColRow(header, col, row)):
              return header.flag
          else:
              return grid[col, row]


      #get value with check on boundary limits
      def getValueFromXY(header, grid, x, y):
          if (isOutOfGridXY(header, x, y)):
              return header.flag
          else:
              col, row = getColRowFromXY(header, x, y)
              return grid[col, row]


      def getPointFromColRow(header, grid, col, row):
          x = header.xllCorner + (col + 0.5) * header.cellSize
          y = header.yllCorner + (header.nrRows - row - 0.5) *
                      header.cellSize
          z = NODATA
```

```
    if not isOutOfGridColRow(header, col, row):
        if grid[col, row] != header.flag:
            z = grid[col, row]
    return np.array([x,y,z])

def isTrue(header, booleanGrid, col, row):
    if isOutOfGridColRow(header, col, row):
        return(False)
    else:
        return booleanGrid[col][row]

def distance2D(v1, v2):
    dx = fabs(v1[0] - v2[0])
    dy = fabs(v1[1] - v2[1])
    return sqrt(dx*dx + dy*dy)

def distance3D(v1, v2):
    dx = fabs(v1[0] - v2[0])
    dy = fabs(v1[1] - v2[1])
    dz = fabs(v1[2] - v2[2])
    return sqrt(dx*dx + dy*dy + dz*dz)
```

The last two functions **distance2D** and **distance3D** are used to compute distances between points v1 and v2, based on Pythagoras' theorem. The vectors v1 and v2 are arrays of three elements (v1[0], v1[1], v1[2]) corresponding to x, y and z values. All the geometric functions used hereinafter utilize this definition.

9.7   Boundary

As described above, a DTM is usually made up of a large number of points, including many points on the boundary of the domain. Not all the points on the boundary are necessary for a triangulation algorithm and therefore it is convenient to reduce the number of boundary points. Figure 9.2 shows the selected points on the boundary, obtained by applying the algorithm described here. Note how, with respect to the DTM of Fig. 9.1, the number of boundary points was reduced.

We implemented the Ramer–Douglas–Peucker algorithm (Ramer, 1972; Douglas and Peucker, 1973), which is a recursive procedure based on a search for the point that has the maximum distance from the line connecting the two extreme points of the set. The first function **getDtmBoundary** takes as input a DTM and a header structure (described above) and returns a selected set of boundary points.

The first operation that the function performs is to identify all the boundary points that are stored into a Boolean matrix called boundary. A point is on a boundary if it has a value and it meets either one of the following conditions: (1) it is on the border of the matrix (first or last row, first or last column); (2) one of the points in its neighbourhood

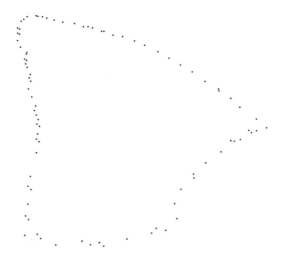

Figure 9.2 *Selected set of boundary points.*

is a NODATA point. These conditions are in the first nested `for` loop. After having identified all the boundary points, these points must be stored into an ordered list in an anticlockwise order. This operation is performed by searching the first point (in the upper left corner) by the function **getFirstPoint**. Then the next point (in an anticlockwise order) is searched for by the function **getNearestPoint**. This function is repeated until **getNearestPoint** finds a valid point. Each point included in the list is then removed from the boundary matrix. The code **getDtmBoundary** is as follows:

```
def getDtmBoundary(header, dtm, threshold, maxSide):
    pointList = []
    boundary = np.zeros((header.nrCols, header.nrRows), bool)
    for col in range(header.nrCols):
        for row in range(header.nrRows):
            boundary[col, row] = False
            if dtm[col, row] != header.flag:
                if ((row == 0) or (col == 0)
                or (row == (header.nrRows-1)) or (col ==
                                            (header.nrCols-1))):
                    boundary[col, row] = True
                if row > 0 and dtm[col, row-1] == header.flag:
                    boundary[col, row] = True
                elif col > 0 and dtm[col-1, row] == header.flag:
                    boundary[col, row] = True
                elif row < (header.nrRows-1) and dtm[col, row+1] ==
                                            header.flag:
                    boundary[col, row] = True
```

```
                    elif col < (header.nrCols-1) and dtm[col+1, row] ==
                                                          header.flag:
                        boundary[col, row] = True

            print ("reduce boundary points (Ramer-Douglas-Peucker
                                          algorithm)...")

            isFirstBoundary, col, row = getFirstPoint(header, boundary)
            while (isFirstBoundary):
                boundaryList = []
                isBoundary = True
                while (isBoundary):
                    myPoint = getPointFromColRow(header, dtm, col, row)
                    boundaryList.append(myPoint)
                    boundary[col, row] = False
                    isBoundary, col, row = getNearestPoint(header, boundary,
                                                          col, row)

                reducedList = RamerDouglasPeucker(boundaryList, threshold,
                                                  maxSide)

                for i in range(len(reducedList)):
                    p = reducedList[i]
                    pointList.append(reducedList[i])

                isFirstBoundary, col, row = getFirstPoint(header, boundary)

        return(pointList)
```

When the operation is finished, we obtain a closed object and the corresponding boundaryList is passed to the function **RamerDouglasPeucker**, described below, which will return a reduced boundary list. Each point of the reduced list is appended to the pointList, which is initially empty. All these operations can be repeated several times if the DTM includes separated areas (like islands), and this explains why it is included in a while loop. At the end of the loop, the function **getFirstPoint** is called again to check if there is at least one other boundary point. If this is true, the cycle starts again.

The Ramer–Douglas–Peucker algorithm (Ramer, 1972; Douglas and Peucker, 1973), is a recursive procedure, as depicted in Fig. 9.3.

The procedure, as implemented in the function **RamerDouglasPeucker**, begins with the selection of two initial points. The algorithm then performs four operations:

1. If the distance between the first and last points of the array $(0, n)$ is larger then maxSide, then the middle point of the array is identified and this point will be used to divide the array according to step 3.

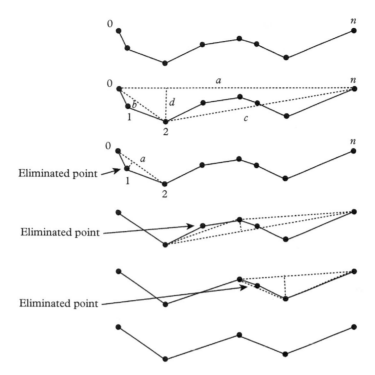

Figure 9.3 *Example of the Ramer–Douglas–Peucker algorithm. The dashed lines are the sides of the triangles used to find the perpendicular distances.*

2. Otherwise, all the internal points $(1, n-1)$ are scanned and the point with maximum perpendicular distance d with respect to the segment a is identified. In Fig. 9.3, this point is number 2.

3. The length d is compared with a threshold. If it is greater than this value, then the point is added to the key point list and the array is subdivided into two sub-arrays. In the example in Fig. 9.3, the first is from 0 to 2 and the second is from 2 to n. The function recursively calls itself on the two subarrays.

4. If neither of the conditions in steps 1 and 3 are true, then none of the internal points of the array (as for the case of point 1) are used, and the analysis for this subarray is completed.

At the end of the recursive procedure, only the key points of the boundary are stored into the array `resultList`, which the function returns. The module `PSP_RamerDouglasPeucker.py` is as follows:

```
#PSP_RamerDouglasPeucker.py
from __future__ import print_function, division
from PSP_gis import *

# return first boundary point, if exist
# scan grid from [0,0]
def getFirstPoint(header, boundary):
    for col in range(0, header.nrCols):
        for row in range(0, header.nrRows):
            if (boundary[col, row] == True):
                return(True, col, row)
    return(False, NODATA, NODATA)

# return first boundary point, if exist
# in the neighbours of [row,col], anti-clockwise order
def getNearestPoint(header, boundary, col, row):
    if isTrue(header, boundary, col, row+1):
        return(True, col, row+1)
    elif isTrue(header, boundary, col+1, row+1):
        return(True, col+1, row+1)
    elif isTrue(header, boundary, col+1, row):
        return(True, col+1, row)
    elif isTrue(header, boundary, col+1, row-1):
        return(True, col+1, row-1)
    elif isTrue(header, boundary, col, row-1):
        return(True, col, row-1)
    elif isTrue(header, boundary, col-1, row-1):
        return(True, col-1, row-1)
    elif isTrue(header, boundary, col-1, row):
        return(True, col-1, row)
    elif isTrue(header, boundary, col-1, row+1):
        return(True, col-1, row+1)
    return(False, NODATA, NODATA)

def triangleAreaErone(v1, v2, v3):
    a = distance3D(v2, v3)
    b = distance3D(v1, v3)
    c = distance3D(v1, v2)
    #semiperimeter
    sp = (a + b + c) * 0.5
    squaredArea = sp*(sp-a)*(sp-b)*(sp-c)
    if (squaredArea < 0.0):
        return(0.0)
    else:
        return sqrt(squaredArea)

def perpendicularDistance(p, v0, v1, base):
    area = triangleAreaErone(p,v0,v1)
    return (2.0 * area) / base
```

```
def firstBoundaryPoint(header, boundary):
    for col in range(0, header.nrCols):
        for row in range(0, header.nrRows):
            if (boundary[col, row] == True):
                return(True, col, row)
    return(False, NODATA, NODATA)

def isTrue(header, boolGrid, col, row):
    if isOutOfGridColRow(header, col, row):
        return(False)
    else:
        return boolGrid[col][row]

# reduce boundary points with Ramer-Douglas-Peucker algorithm
def RamerDouglasPeucker(pointList, threshold, maxSide):
    # Find the point with the maximum distance
    dmax = 0
    index = 0
    end = len(pointList)-1
    base = distance3D(pointList[0], pointList[end])
    if (base > maxSide):
        index = int(end / 2.)
        dmax = base
    else:
        for i in range (1, end):
            d = perpendicularDistance(pointList[i], pointList[0],
                                      pointList[end], base)
            if (d > dmax):
                index = i
                dmax = d
    # if distance > epsilon cut the array
    if (dmax > threshold):
        list1 = RamerDouglasPeucker(pointList[0:index+1], threshold,
                                    maxSide)
        list2 = RamerDouglasPeucker(pointList[index:len(pointList)],
                                    threshold, maxSide)

        resultList = []
        for j in range(len(list1)):
            resultList.append(list1[j])
        for j in range(1, len(list2)):
            resultList.append(list2[j])
    else:
        resultList = []
        resultList.append(pointList[0])
        resultList.append(pointList[end])

    return resultList
```

218 Triangulated Irregular Network

To compute the perpendicular distance between a line and a point, the algorithm employs the definition of an area of a triangle A, given by Heron of Alexandria (AD 10–70). Heron's formula was derived to find the area of a triangle by knowing the length of the three sides. Let a, b and c be the sides and S the semiperimeter, given by

$$S = \frac{a+b+c}{2} \tag{9.1}$$

Heron's formula for the area is

$$A = \sqrt{S(S-a)(S-b)(S-c)} \tag{9.2}$$

The perpendicular distance d is then

$$d = \frac{2A}{a} \tag{9.3}$$

where a is the base of the triangle. To find the sides of the triangle, the distance between the points must be computed, as described above. After having generated a list of reduced boundary points, it is useful to include a first approximation to be used as initial points for the triangulation. This is especially true for DTMs with large flat areas; otherwise the rules on subdivision based on elevation would not produce a correct subdivision. In the function **triangulate**, there is a flag used to select the possibility of using an initial partition.

9.8 Geometrical Properties of Triangles

Fundamental classes and functions to manage geometrical properties are implemented in the file `PSP_triangle.py`. Moreover, four additional functions are written to test geometrical properties that are linked to the triangulation algorithm. We define the classes `Crectangle` and `Ccircle`, which are used to represent dimensional geometrical objects in two dimensions. Then we define the classes `Cplane` and `Ctriangle`, which represent objects in three dimensions because the vertices of the triangle have coordinates x, y and z.

A series of functions is written to obtain specific instances of the classes. In particular, **getRectangle** returns the rectangle of minimum area containing the triangle projected on the horizontal plane. **getCircumCircle** returns the circle that passes through all the vertices of the triangle, projected onto the horizontal plane. **getPlane** returns the plane containing the three vertices of the triangle. **getZPlane** returns the coordinate z of a point for which x and y are known.

isInCircumCircle is used to check if a point (x, y) is inside the circumcircle of the triangle passed to the function. The function **isPointInside** if true if a point is inside a triangle (defined by its vertex v).

A point *P* belongs to a right or left semiplane (defined by a segment *AB*) depending on the cross-product between the vectors *A* – *P* and *A* – *B*. The cross-product on the plane (*x, y*) can be calculated as a determinant of the components of the two vectors $D = dx1 * dy2 - dy1 * dx2$. This formulation becomes the equation `crossProductXY` in the function **sign**, since the elements of *x* are on the index 0 and those of *y* on the index 1 of the points *P*, *A* and *B*. Finally, a point is inside a triangle *ABC* if the signs of the segments *AB, BC* and *CA* is consistent (they are all positive or all negative).

The function **getArea2D** returns the area of the projected triangle in two dimensions (*x, y*), while **getArea** returns the more general area in three dimensions using the function **magnitude** that returns the norm of a vector *v*. We use this function to obtain the norm of the cross-product from the vectors `v[1]-v[0]` and `v[2]-v[0]`, which represents the area of a parallelogram built by the two vectors. To obtain the area of the triangle, the area is divided by two. **getCosine** returns the cosine of the angle between two segments of the triangle, by employing Carnot's equation. This function is used by the function **getAngle** to obtain the angle itself, through the arc cosine. The function **getMinAngle** is given three vertices, obtains the three angles from the function **getAngle** and returns the smallest of the three. This function is used in the function written to refine the grid based on the angle properties.

Since each value of the domain of cosines [–1, 1] corresponds to two angles, the function **getAngle** cannot be used to order a set of vectors with respect to a reference vector. For instance, the value 0 is given by the cosine of two angles, $\pi/2$ and $3\pi/2$. Since in the Delaunay triangulation we need an angle sorting, and we are not interested in the real value of the angle but only in the order, we use an optimization procedure based on a definition of a pseudo-angle. In the function **getPseudoAngle**, we change the domain of the cosine from [–1, 1] to [–2, 2], where each angle has a separate value. The pseudo-angles are all calculated with reference to the first vector of the list. We obtain the cosine of this angle, add 1 to the cosine to obtain only positive values [0, 2], determine the sign of the new vector with respect to the reference (with a two-dimensional cross-product) and multiply by –1 all positive vectors. **isAdjacent** is a function to evaluate if two triangles are adjacent, i.e. if they have two shared vertices.

The function **isCircumCircleLeft** checks if the circumcircle of the triangle is entirely at the left of the input `point`; these conditions will be useful to accelerate the Delaunay triangulation, as described in Section 9.9.

The function **hasVertexInDomain** checks whether a triangle has one of its vertices in the external domain. The external domain is set by the function **firstTriangulation**, and the four points that belong to it are identified by having elevation equal to NODATA.

The function **isInsideDTM** determines if a triangle is inside the DTM. A triangle is not inside the DTM if its circumcentre does not correspond to a valid point in the DTM and if its three vertices are on the external boundary.

The module `PSP_triangle.py` with the definition of the geometrical functions is as follows:

```
#PSP_triangle.py
from __future__ import division
from PSP_gis import *
MINIMUMANGLE = pi/200.

class Crectangle:
    x0 = x1 = NODATA
    y0 = y1 = NODATA

class Ccircle:
    def __init__(self, xc, yc, radiusSquared, isCorrect):
        if (isCorrect):
            self.x = xc
            self.y = yc
            self.radiusSquared = radiusSquared
            self.radius = sqrt(radiusSquared)
        self.isCorrect = isCorrect

class Cplane:
    # plane equation: ax + by + cz + d = 0
    a = b = c = d = NODATA

class Ctriangle:
    def __init__(self, v = np.zeros((3, 3), float)):
        self.v = v
        if (not np.all(v == 0.0)):
            #centroid
            self.x = sum(v[:,0]) / 3.0
            self.y = sum(v[:,1]) / 3.0
            self.circle = getCircumCircle(v)
            self.isRefinedZ = False
            self.isRefinedAngle = False

def getRectangle(v):
    rect = Crectangle()
    x = v[:,0]
    y = v[:,1]
    rect.x0 = min(x[0], x[1], x[2])
    rect.y0 = min(y[0], y[1], y[2])
    rect.x1 = max(x[0], x[1], x[2])
    rect.y1 = max(y[0], y[1], y[2])
    return(rect)

def getCircumCircle(v):
    if (getMinAngle(v) < MINIMUMANGLE):
        return Ccircle(NODATA, NODATA, NODATA, False)
    x = v[:,0]
    y = v[:,1]
```

```
    a = x[0]*x[0] + y[0]*y[0]
    b = x[1]*x[1] + y[1]*y[1]
    c = x[2]*x[2] + y[2]*y[2]
    d = 2*(x[0]*(y[1]-y[2]) + x[1]*(y[2]-y[0]) + x[2]*(y[0]-y[1]))
    xc = (a*(y[1] - y[2]) + b*(y[2] - y[0]) + c*(y[0] - y[1])) / d
    yc = (a*(x[2] - x[1]) + b*(x[0] - x[2]) + c*(x[1] - x[0])) / d
    dx = fabs(xc - x[0])
    dy = fabs(yc - y[0])
    radiusSquared = dx * dx + dy * dy
    return Ccircle(xc, yc, radiusSquared, True)

def getPlane(v):
    plane = Cplane()
    x = v[:,0]
    y = v[:,1]
    z = v[:,2]
    plane.a = (y[1]-y[0])*(z[2]-z[0]) - (y[2]-y[0])*(z[1]-z[0])
    plane.b = (x[2]-x[0])*(z[1]-z[0]) - (x[1]-x[0])*(z[2]-z[0])
    plane.c = (x[1]-x[0])*(y[2]-y[0]) - (x[2]-x[0])*(y[1]-y[0])
    plane.d = -(plane.a*x[0] + plane.b*y[0] + plane.c*z[0])
    return(plane)

def getZplane(plane, x, y):
    z = -(plane.a*x + plane.b*y + plane.d) / plane.c
    return(z)

# checks if a point is in circumcircle of the triangle.
def isInCircumCircle(point, triangle):
    dx = point[0] - triangle.circle.x
    dy = point[1] - triangle.circle.y
    if (((dx * dx) + (dy * dy)) <= triangle.circle.radiusSquared):
                                    return(True)
    else: return(False)

def sign(P, A, B):
    crossProductXY = (P[0]-A[0])*(B[1]-A[1]) - (P[1]-A[1])*(B[0]-A[0])
    if crossProductXY > 0.0:
        return 1.0
    else:
        return -1.0

def isPointInside(point, v):
    s1 = sign(point, v[0], v[1])
    s2 = sign(point, v[1], v[2])
    s3 = sign(point, v[2], v[0])
    if (s1 == s2) and (s2 == s3):
        return True
    else:
        return False
```

```
# Area = 1/2 |x1(y2-y3) - x2(y1-y3) + x3(y1 - y2)|
def getArea2D(v):
    x = v[:,0]
    y = v[:,1]
    return 0.5 * fabs(x[0]*(y[1]-y[2]) - x[1]*(y[0]-y[2]) + x[2]*
        (y[0]-y[1]))

def magnitude(v):
    return(np.sqrt(v.dot(v)))

def getArea(v):
    return 0.5 * magnitude(np.cross(v[1] - v[0], v[2] - v[0]))

# Carnot equation
def getCosine(v0, v1, v2):
    a = distance2D(v0, v2)
    b = distance2D(v0, v1)
    c = distance2D(v2, v1)
    denom = 2.0 * b * c
    denom = max(denom, 0.0001)
    return (b*b + c*c - a*a) / denom

def getAngle(v0, v1, v2):
    cosine = getCosine(v0, v1, v2)
    cosine = min(cosine, 1.0)
    cosine = max(cosine, -1.0)
    return acos(cosine)

def getPseudoAngle(v0, v1, v2):
    cosine = getCosine(v0, v1, v2)
    pseudoAngle = (cosine + 1.0)                    #[ 0, 2]
    if sign(v0, v1, v2) > 0:
        pseudoAngle *= -1                           #[-2, 2]
    return pseudoAngle

def getMinAngle(v):
    angle0 = getAngle(v[0],v[1],v[2])
    angle1 = getAngle(v[1],v[2],v[0])
    angle2 = getAngle(v[2],v[0],v[1])
    return min(angle0, angle1, angle2)

#adjacent: 2 shared vertices with polygon v
def isAdjacent(u, v):
    nrSharedVertices = 0
    for i in range(3):
        for j in range(len(v)):
            if np.all(u[i] == v[j]):
                nrSharedVertices += 1
```

```
                if (nrSharedVertices == 2):
                    return True
        return False

def isCircumCircleLeft(triangle, point):
    xMax = triangle.circle.x + triangle.circle.radius
    currentX = point[0]
    return(currentX > xMax)

def hasVertexInDomain(triangle):
    for i in range(3):
        if (triangle.v[i][2] == NODATA): return(True)
    else: return(False)

def isInsideGrid(triangle, internalPointList, header, grid):
    zCircumCenter = getValueFromXY(header, grid, triangle.circle.x,
                                   triangle.circle.y)
    if (zCircumCenter != header.flag):
        return True
    else:
        for i in range (len(internalPointList)):
            for j in range(3):
                if np.all(triangle.v[j] == internalPointList[i]):
                    return True
    return False
```

9.9 Delaunay Triangulation

As we have described in Section 9.5, the DTM triangulation starts by selecting the key points of the boundary (by employing the algorithms detailed in Section 9.7) and storing them in the set `pointList`. The program then defines an initial internal partition (if the option `isInitialPartition` is set to `True`) by using the function **addPartitionPoints** contained in the module `PSP_utility` and adding these points to the set (Fig 9.4). To speed up the following algorithms, all the points of the `pointList` are sorted for incremental values of *x* by the function **sortPointList** of the `PSP_utility` module.

Then the triangulation algorithm starts in the function **firstTriangulation** of the module `PSP_Delaunay`. The inputs to the function are the DTM structures `header` and `dtm` and the two lists `pointList` and `internalPointList`, which contain only the points of the internal partition. It will be empty if `isInitialPartition` is `False`.

firstTriangulation begins by generating a rectangular domain, slightly larger than the DTM, and its vertices are identified as [A, B, C, D]. This domain is then subdivided into two triangles [A,B,C] and [C,D,A], which make the initial list, called `triangleList`. The first triangulation proceed by iteratively updating this list for each point of the `pointList` by using the Delaunay triangulation algorithm. Delaunay triangulation is

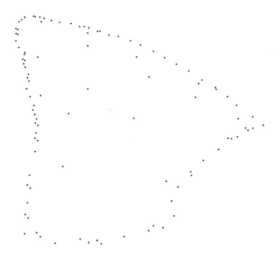

Figure 9.4 *Selected set of boundary points, with points used for initial partition inside the DTM.*

an algorithm to create a set of triangles for a set of points in a plane, such that no point is inside the circumcircle of any triangle in the set. Moreover, this algorithm maximizes the minimum angle of all the triangles in the triangulation to avoid generating triangles having very small angles. Details about Delaunay triangulation can be found in Cignoni *et al.* (1998).

The algorithm is implemented in the function **Delaunay**, which takes as input the current `triangleList` and the `newPoint` that has to be included in the triangulation.

The function **Delaunay** initializes the lists necessary to the algorithm, then it scans all the triangles of the current `triangleList`, checking if the `newPoint` is inside the circumcircle of one of these. If this is true, the triangle is appended in the `deleteList` and its vertices are inserted into the `vertexList`, which will be used to generate the new triangles. The vertices are ordered by the function **insertVertexClockwise**. This function checks if the new vertex is already present in the list; otherwise it inserts it in clockwise order with respect to the angle between `newVertex` and `vertexList[0]`, centred on the `newPoint`. To speed up the sorting procedure of the angles, the function **insertVertexClockwise** uses the pseudo-angle computed in the range $[-2, 2]$ by the function **getPseudoAngle**, described in Section 9.8.

When the `vertexList` is complete and clockwise-ordered, for each vertex of the list, a new triangle is generated. The vertices of this new triangle are `newPoint`, the current vertex `vertexList[i]` and the next vertex in the ordered list `vertexList[(i+1) nrVertices]`. The use of the function **module** allows for the next vertex following the last to be the first of the sorted list. Each new triangle generated by the constructor of the class **Ctriangle** (as described in Section 9.8) is appended to the `newTriangles` list. The domain to be triangulated can be initially subdivided in sub-domain or not, as shown in Fig. 9.5. If the original DTM has many flat areas an initial subdivision is

(a)   (b)

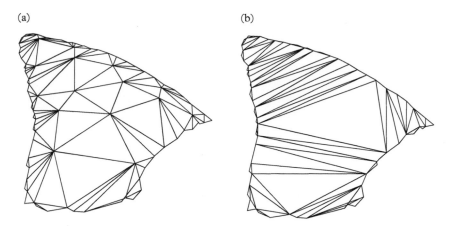

Figure 9.5 *First triangulation, with initial subdivision (a) and without (b).*

necessary to obtain a better triangulation, otherwise very large and irregular triangles would be generated in the flat areas where the differences in elevation are small.

Finally, the function **Delaunay** deletes all the triangles in the `deleteList` and appends all the triangles in the `newTriangles`, so the `triangleList` is updated. If, during the generation of new triangles, one having angles close to zero is generated, the new generation is not accepted and `newPoint` is excluded from the `pointList`. Evidently, it is a point that lies on the segment that connects to neighbouring points. This condition is generated by the property `myTriangle.circle.isCorrect`, computed by the constructor of the **Ctriangle** class.

The iterative triangulation scheme of the function **firstTriangulation** has a computational complexity of order 2 with respect to the number of points of the `pointList`; therefore, to speed up the procedure, the sorted property on the x axis described above is exploited. At each iteration, before calling the function **Delaunay**, a check is made of which triangles can be excluded from the next controls, through the function **isCircumCircleLeft**. If the circumcircle of one triangle is entirely on the left of the current vertex, it does not need to be checked again in the algorithm (since all the subsequent points are going to be on the right of the latter).

```python
#PSP_Delaunay.py
from __future__ import print_function, division
from PSP_triangle import getPseudoAngle
from PSP_utility import *
import PSP_triangle as triangle

def insertVertexClockwise(newVertex, center, vertexList, angleList):
    #is first point
    v = copy(newVertex)
    if (len(vertexList) == 0):
        vertexList.append(v)
        angleList.append(-2)
        return()
```

```
    #check duplicate
    for i in range(len(vertexList)):
        if ((v[0] == vertexList[i][0])
        and (v[1] == vertexList[i][1])):
            return()
    #compute false angle (no trigonometric functions) [-2,2]
    angle = getPseudoAngle(v, center, vertexList[0])
    #check position
    i = 1
    while ((i < len(angleList)) and (angle > angleList[i])):
        i += 1
    vertexList.insert(i, v)
    angleList.insert(i, angle)

def Delaunay(triangleList, newPoint, header, dtm):
    deleteList = []
    vertexList = []
    angleList = []
    newTriangles = []
    # erase a triangle if the point is inside his circumcircle.
    for i in range(len(triangleList)):
        myTriangle = triangleList[i]
        if triangle.isInCircumCircle(newPoint, myTriangle):
            deleteList.append(i)
            for j in range(3):
                insertVertexClockwise(myTriangle.v[j],
                            newPoint, vertexList, angleList)
    # create new triangles
    nrVertices = len(vertexList)
    for i in range(nrVertices):
        v = np.zeros((3, 3), float)
        v[0] = newPoint
        v[1] = vertexList[i]
        v[2] = vertexList[(i+1) % nrVertices]
        myTriangle = triangle.Ctriangle(v)
        if myTriangle.circle.isCorrect:
            newTriangles.append(myTriangle)

    if len(newTriangles) == 0: return False

    for i in range(len(deleteList)-1, -1, -1):
        triangleList.pop(deleteList[i])
    for i in range(len(newTriangles)):
        triangleList.append(newTriangles[i])
    return True

def firstTriangulation(pointList, header, dtm):
    print("Delaunay...")
```

```
step = header.cellSize * (header.nrCols + header.nrRows) / 50
xMin = header.xllCorner - step
xMax = header.xllCorner + header.nrCols * header.cellSize + step
yMin = header.yllCorner - step
yMax = header.yllCorner + header.nrRows * header.cellSize + step

A = np.array([xMin, yMin, NODATA])
B = np.array([xMin, yMax, NODATA])
C = np.array([xMax, yMax, NODATA])
D = np.array([xMax, yMin, NODATA])

# boundary: domain divided in two triangles
triangleList = []
triangleList.append(triangle.Ctriangle(np.array([A,B,C])))
triangleList.append(triangle.Ctriangle(np.array([C,D,A])))

triangleListOutput = []

pointIndex = 0
while pointIndex < len(pointList):
    currentPoint = pointList[pointIndex].copy()
    # move 'completed' triangles to the output
    i = 0
    while (i < len(triangleList)):
        myTriangle = triangleList[i]
        if (triangle.isCircumCircleLeft(myTriangle, currentPoint)
        and not triangle.hasVertexInDomain(myTriangle)):
            if triangle.isInsideDTM(myTriangle, internalPointList,
                                    header, dtm):
                triangleListOutput.append(myTriangle)
            triangleList.pop(i)
        else: i += 1

    if Delaunay(triangleList, currentPoint):
        pointIndex += 1
    else:
        pointList.pop(pointIndex)

# remove all the triangles belonging to the boundary
# and move the remaining triangles to the output
while (len(triangleList) > 0):
    if not triangle.isBoundary(triangleList[0]):
        if triangle.isInsideGrid(triangleList[0], header, dtm):
            triangleListOutput.append(triangleList[0])
    triangleList.pop(0)

pointList = sortPointList(pointList)
return(pointList, triangleListOutput)
```

9.10   Refinement

An important aspect of triangulation for three-dimensional surfaces is the refinement of the grid based on elevation. This is a key component of the computational grid, since the simulation of processes such as surface runoff is strongly dependent on the refinement and shape of the network. For instance, in areas having a strong variation in slope, the triangles have to be smaller than in areas having uniform slope. The PSP_refinement.py module contains the refinement functions for the triangulation both on the elevation and on the angles. As an example, Fig. 9.6 depicts the TIN during the refinement procedure.

The function **requireRefinement** evaluates if a triangle must be refined or not, based on the elevation data. This function determines the difference between the elevation z of the original DTM data and the elevation resulting from the triangulation procedure, for all points of the triangle. The function saves the point having maximum elevation difference between the DTM value and the elevation of the horizontal plane of the triangle dzMax. If dzMax is larger than a tolerance value called zThreshold, the triangle must be refined by adding a new point to the set employed for the triangulation procedure.

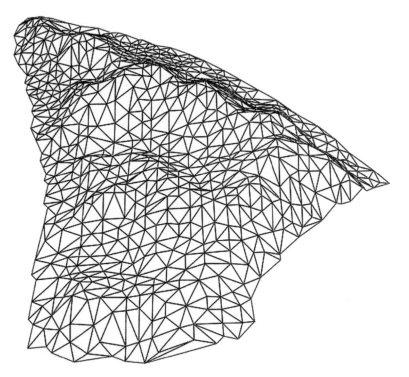

Figure 9.6 *Triangulated irregular network showing the triangles during the refinement procedure.*

The function **refinementZ** computes the refinement based on elevation. For every triangle of the current `triangleList`, the function checks if its area is larger than `areaMin` and if it has to be refined on elevation, using **requireRefinement**. If one of these conditions is met, the property `isRefinedZ` of the triangle will be set to `True` and the triangle will not be considered again for this procedure. If both conditions are `True`, then the function **refinement** is passed the new vertex to be included, specifying that it is a refinement based on elevation. If the refinement is successful, the function **refinement** returns the index of the `triangleList` updated until that moment, and then procedure starts again from that index value. Each new point is added to the `pointList`, which will be sorted at the end of the function.

The function **refinementAngle** computes the refinement based on angle properties. For every triangle of the current `triangleList`, the function evaluates if its minimum angle is less than `angleThreshold` and if its area is larger than `areaMin`. If one of these conditions is met, the triangle's property `isRefinedAngle` will be set to `True` and the triangle will not be evaluated again in this procedure. If both conditions are `True`, then a point corresponding to the circumcentre of the triangle is added, since the insertion of this point ensures the definition of triangles having larger angles in the subsequent refinement. If this point belongs to the grid, it is passed to the **refinement** function, specifying that the refinement is on the angles. Analogously to the function **refinementZ**, the procedure is performed starting from the new index `newTriangleIndex` returned by the function.

Finally, the function **refinement** refines the current triangulation, adding a new vertex and applying the Delauney algorithm on the subset of selected triangles. First, it identifies the triangles that must be deleted from the current triangulation and then inserts these triangles into the `deleteList`. In particular, the triangles to be eliminated are the current triangle (on which the refinement is applied) and all the triangles that had an angle that was eliminated. To speed up the computation, first a list of triangles to be evaluated is created, called `checkList`. This list is populated by potential triangles created by new points that are geometrically inside the circumcentre. Then an iterative procedure eliminates the triangles that are adjacent to those that were already eliminated, by using the function **isAdjacent**, until no triangles are left to be eliminated. These triangles are added to the list `deleteList`. During the elimination procedure, their vertices are added to the `vertexList` in clockwise order, as was done in the function **Delaunay**. A special additional control is performed in the case in which the refinement is performed on the angles, by checking that the new vertex is at least in one of the eliminated triangles.

As soon as the `vertexList` has been populated, the triangulation procedure continues as described in the function **Delaunay**, by updating the visualization of the triangles and inserting the new triangles, starting from the index of the first eliminated triangle.

```python
#PSP_refinement.py
from __future__ import print_function, division
from PSP_Delaunay import insertVertexClockwise
import PSP_triangle as triangle
import PSP_visual3D as visual3D
from PSP_utility import *


REFINEMENTZ = 1
REFINEMENTANGLE = 2

def requireRefinement(header, dtm, flag,
currentTriangle, zThreshold, step):
    plane = triangle.getPlane(currentTriangle.v)
    rect = triangle.getRectangle(currentTriangle.v)
    colMin, rowMin = getColRowFromXY(header, rect.x0, rect.y1)
    colMax, rowMax = getColRowFromXY(header, rect.x1, rect.y0)
    dzMax = 0.0
    for col in np.arange(colMin, colMax+1, step):
        for row in np.arange(rowMin, rowMax+1, step):
            if isTrue(header, flag, col, row):
                point = getPointFromColRow(header, dtm, col, row)
                if triangle.isPointInside(point, currentTriangle.v):
                    x, y, z = point
                    z0 = triangle.getZplane(plane, x, y)
                    dz = fabs(z - z0)
                    if (dz > dzMax):
                        dzMax = dz
                        newRow = row
                        newCol = col
    if (dzMax > zThreshold):
        return True, newCol, newRow
    else:
        return False, NODATA, NODATA


def refinementZ(pointList, triangleList, header, dtm,
zThreshold, areaMin, flagMatrix, step):
    print ("Z refinement...")
    triangleIndex = 0
    while triangleIndex < len(triangleList):
        currentTriangle = triangleList[triangleIndex]
        refinementDone = False
        if not currentTriangle.isRefinedZ:
            if triangle.getArea2D(currentTriangle.v) > areaMin:
                required, newCol, newRow = requireRefinement(header,
                                     dtm, flagMatrix, currentTriangle,
                                     zThreshold, step)
                if required:
                    newPoint = getPointFromColRow(header,
```

```
                    dtm, newCol, newRow)
                newTriangleIndex = refinement(triangleList,
                triangleIndex, newPoint, header,
                                        dtm, REFINEMENTZ)
                if newTriangleIndex != NODATA:
                    pointList.append(newPoint)
                    flagMatrix = updateFlagMatrix(flagMatrix,
                    header, newCol, newRow, step)
                    triangleIndex = newTriangleIndex
                    refinementDone = True

        triangleList[triangleIndex].isRefinedZ = True
        if (not refinementDone): triangleIndex += 1

    pointList = sortPointList(pointList)
    return pointList, triangleList, flagMatrix

def refinementAngle(pointList, triangleList, header,
dtm, angleThreshold, areaMin, flagMatrix, step):
    print ("angle refinement...")
    triangleIndex = 0
    while triangleIndex < len(triangleList):
        currentTriangle = triangleList[triangleIndex]
        refinementDone = False
        if not currentTriangle.isRefinedAngle:
            if triangle.getMinAngle(currentTriangle.v) <
                                                angleThreshold:
                if triangle.getArea2D(currentTriangle.v) > areaMin:
                    newCol, newRow = getColRowFromXY(header,
                            currentTriangle.circle.x,
                            currentTriangle.circle.y)
                    if isTrue(header, flagMatrix, newCol, newRow):
                        newPoint = getPointFromColRow(header, dtm,
                                            newCol, newRow)
                        newTriangleIndex = refinement(triangleList,
                         triangleIndex, newPoint,
                         header, dtm, REFINEMENTANGLE)
                        if newTriangleIndex != NODATA:
                            pointList.append(newPoint)
                            flagMatrix = updateFlagMatrix(flagMatrix,
                            header, newCol, newRow, step)
                            triangleIndex = newTriangleIndex
                            refinementDone = True

        triangleList[triangleIndex].isRefinedAngle = True
        if (not refinementDone): triangleIndex += 1

    pointList = sortPointList(pointList)
    return pointList, triangleList, flagMatrix
```

```
def refinement(triangleList, triangleIndex, newPoint, header, dtm,
               refinementType):
    deleteList = []
    vertexList = []
    angleList = []
    newTriangles = []
    checkList = []

    currentTriangle = triangleList[triangleIndex]
    if refinementType == REFINEMENTANGLE:
        isInside = triangle.isPointInside(newPoint, currentTriangle.v)

    deleteList.append(triangleIndex)
    for i in range(3):
        insertVertexClockwise(currentTriangle.v[i],
                newPoint, vertexList, angleList)

    for i in range(len(triangleList)):
        if (i != triangleIndex):
            myTriangle = triangleList[i]
            dx = newPoint[0] - myTriangle.circle.x
            if (dx <= myTriangle.circle.radius):
                dy = newPoint[1] - myTriangle.circle.y
                if (dy <= myTriangle.circle.radius):
                    if (((dx * dx) + (dy * dy)) <=
                                myTriangle.circle.radiusSquared):
                        checkList.append(i)
    i = 0
    while (i < len(checkList)):
        index = checkList[i]
        myTriangle = triangleList[index]
        if  triangle.isAdjacent(myTriangle.v, vertexList):
            for j in range(3):
                insertVertexClockwise(myTriangle.v[j],
                        newPoint, vertexList, angleList)
            if refinementType == REFINEMENTANGLE and (not isInside):
                isInside = triangle.isPointInside(newPoint,
                                            myTriangle.v)
            orderedInsert(index, deleteList)
            checkList.pop(i)
            i = 0
        else: i+=1

    if refinementType == REFINEMENTANGLE and (not isInside):
                    return NODATA

    # create new triangles
    nrVertices = len(vertexList)
```

```
    for i in range(nrVertices):
        v = np.zeros((3, 3), float)
        v[0] = newPoint
        v[1] = vertexList[i]
        v[2] = vertexList[(i+1) % nrVertices]
        myTriangle = triangle.Ctriangle(v)
        if myTriangle.circle.isCorrect:
            newTriangles.append(myTriangle)
        else:
            return NODATA

    if len(newTriangles) <= len(deleteList): return NODATA

    firstIndex = deleteList[0]
    for i in range(len(deleteList)-1, -1, -1):
        triangleList.pop(deleteList[i])
        visual3D.delTriangle(deleteList[i])

    for i in range(len(newTriangles)):
        myTriangle = newTriangles[i]
        triangleList.insert(firstIndex+i, myTriangle)
        visual3D.addTriangle(firstIndex+i, myTriangle, header)
    return firstIndex
```

9.11 Utilities

The PSP_utility.py module contains utilities for computation and file management.
After a list of reduced boundary points has been generated, it is useful to include a first
approximation to be used as initial points for the triangulation. This is especially true for
DTMs with large flat areas; otherwise the rules on subdivision based on elevation would
not produce a correct subdivision.

```
#PSP_triangulation.py
from __future__ import print_function, division
from PSP_gis import *

def initializeFlagMatrix(header, dtm, pointList, step):
    flagMatrix = np.zeros((header.nrCols, header.nrRows), bool)

    for col in range(header.nrCols):
        for row in range(header.nrRows):
            if dtm[col][row] == header.flag:
                flagMatrix[col, row] = False
            else:
                flagMatrix[col, row] = True
```

```python
    for i in range(len(pointList)):
        p = pointList[i]
        col, row = getColRowFromXY(header, p[0], p[1])
        updateFlagMatrix(flagMatrix, header, col, row, step)
    return flagMatrix

def updateFlagMatrix(flagMatrix, header, col, row, step):
    for dCol in np.arange(1-step, step):
        for dRow in np.arange(1-step, step):
            if not isOutOfGridColRow(header, col+dCol, row+dRow):
                flagMatrix[col+dCol, row+dRow] = False
    return flagMatrix

def searchPosition(x, sortPointList, first, last):
    if x <= sortPointList[first][0]:
        return first
    elif x > sortPointList[last][0]:
        return (last+1)
    elif (last - first) < 2:
        return last
    else:
        m = int((first+last)/2)
        if (x <= sortPointList[m][0]):
            return(searchPosition(x, sortPointList, first, m))
        else:
            return(searchPosition(x, sortPointList, m, last))

def sortPointList(pointList):
    sortList = [pointList[0]]
    for i in range(1, len(pointList)):
        x = pointList[i][0]
        index = searchPosition(x, sortList, 0, len(sortList) - 1)
        sortList.insert(index, pointList[i])
    return(sortList)

def addPartitionPoints(pointList, header, dtm, intervalStep,
                       randomFactor, flagMatrix, flagStep):
    np.random.seed()
    col = 0
    isLastCol = False
    partitionPointList = []
    while not isLastCol:
        row = 0
        isLastRow = False
        while not isLastRow:
            delta = intervalStep * randomFactor * \
                    (2*np.random.random() - 1)
            c = col
```

```
                r = row
                if col != 0 and col < (header.nrCols -1): c += int(delta)
                if row != 0 and row < (header.nrRows -1): r += int(delta)
                if isTrue(header, flagMatrix, c, r):
                    point = getPointFromColRow(header, dtm, c, r)
                    pointList.append(point)
                    partitionPointList.append(point)
                    flagMatrix = updateFlagMatrix(flagMatrix, header, c,
                                              r, flagStep)
                if (row == header.nrRows -1):
                    isLastRow = True
                else:
                    row += intervalStep
                    if row >= (header.nrRows-intervalStep/3.): row =
                              header.nrRows -1
            if (col == header.nrCols -1):
                isLastCol = True
            else:
                col += intervalStep
                if col >= (header.nrCols-intervalStep/3.): col =
                          header.nrCols -1
    return partitionPointList

def orderedInsert(index, indexList):
    #is first
    if (len(indexList) == 0):
        indexList.append(index)
        return()
    #check duplicate
    for i in range(len(indexList)):
        if (index == indexList[i]):
            return()
    #check position
    i = 0
    while ((i < len(indexList)) and (index > indexList[i])):
        i += 1
    indexList.insert(i, index)

def isAdjacentIndex(t1, t2):
    shareVertices = 0
    for i in range(3):
        for j in range(3):
            if (t1[i] == t2[j]):
                shareVertices += 1
                if shareVertices == 2:
                    return True
    return False
```

```python
def getNeighbours(triangleList):
    nrTriangles = len(triangleList)
    neighbourList = np.zeros((nrTriangles, 3), int)
    nrNeighbours = np.zeros(nrTriangles, int)
    for i in range(nrTriangles):
        index = 0
        neighbourList[i] = [NOLINK, NOLINK, NOLINK]
        j = 0
        while (j < nrTriangles) and (index < 3):
            if (nrNeighbours[j]<3):
                if isAdjacentIndex(triangleList[i], triangleList[j]):
                    if (j != i):
                        neighbourList[i, index] = j
                        nrNeighbours[j] += 1
                        index += 1
            j += 1
    return(neighbourList)

def writeTIN(triangleList, pointList, header, dtm,
            fnVertices, fnTriangles, fnNeighbours):
    print("Save TIN...")
    #save vertices
    file = open(fnVertices, "w")
    for i in range(len(pointList)):
        p = pointList[i]
        file.write(str(p[0]) +","+ str(p[1]) +","+
                    format(p[2],".1f") +"\n")

    #save triangle vertices
    triangleVertexList = np.zeros((len(triangleList),3), int)
    for i in range(len(triangleList)):
        for j in range(3):
            x = triangleList[i].v[j][0]
            pointIndex = searchPosition(x, pointList, 0,
                                        len(pointList)-1)
            while not(np.all(triangleList[i].v[j] ==
                            pointList[pointIndex])):
                pointIndex += 1
            triangleVertexList[i,j] = pointIndex

    file = open(fnTriangles, "w")
    for i in range(len(triangleVertexList)):
        t = triangleVertexList[i]
        file.write(str(t[0]) +","+ str(t[1]) +","+ str(t[2]) +"\n")

    neighbourList =  getNeighbours(triangleVertexList)
    file = open(fnNeighbours, "w")
    for i in range(len(neighbourList)):
        n = neighbourList[i]
        file.write(str(n[0]) +","+ str(n[1]) +","+ str(n[2]) +"\n")
```

9.12   Visualization

The visualization part of the program is written to visualize the different steps of the triangulation procedure. In the program there is a stop between the executions of the different algorithms to select the boundary points, the triangulation and refinement. This stop has been introduced to visualize the results of the different steps. Therefore the user must hit any keyboard key to proceed with the next operations. The module PSP_color.py defines the colours used by the plotting algorithms:

```python
#PSP_colour.py
from __future__ import print_function, division
import numpy as np

colorScaleDtm = np.array([[],[]], float)

def setColorScale(nrLevels, keyColors):
    for i in range (len(keyColors)):
        for j in range (3):
            keyColors[i,j] /= 256.

    nrIntervals = len(keyColors)-1
    step = int(max(nrLevels / nrIntervals, 1))
    myScale = np.zeros((nrLevels, 3), float)

    for i in range (nrIntervals):
        dRed = (keyColors[i+1,0] - keyColors[i,0]) / step
        dGreen = (keyColors[i+1,1] - keyColors[i,1]) / step
        dBlue = (keyColors[i+1,2] - keyColors[i,2]) / step

        for j in range (step):
            index = step * i + j
            myScale[index, 0] = keyColors[i,0] + (dRed * j)
            myScale[index, 1] = keyColors[i,1] + (dGreen * j)
            myScale[index, 2] = keyColors[i,2] + (dBlue * j)

    lastIndex = index
    if (lastIndex < (nrLevels-1)):
        for i in range(lastIndex, nrLevels):
            myScale[i] = myScale[lastIndex]
    return (myScale)

def setColorScaleDtm():
    global colorScaleDtm
    keyColors = np.zeros((4,3),float)
    keyColors[0] = (32, 128, 16)        #green
    keyColors[1] = (255, 196, 18)       #yellow
    keyColors[2] = (118, 64, 18)        #brown
    keyColors[3] = (160, 160, 160)      #grey
    colorScaleDtm = setColorScale(256, keyColors)
```

```
def getDTMColor(z, header):
    zRelative = (z - header.zMin) / header.dz
    index = int(zRelative * (len(colorScaleDtm)-1))
    index = min(len(colorScaleDtm)-1, max(index, 0))
    return(colorScaleDtm[index])
```

The module `PSP_visual3D.py` contains the function to plot DTM and TIN in three dimensions.

```
#PSP_visual3D.py
from __future__ import print_function, division
import visual
import sys
from copy import copy
from PSP_color import *


NODATA = -9999

pointList3D = []
pointListDtm3D = []
triangleList3D = []
surfaceList3D = []

global TINScene, DTMScene

def initialize(header):
    global TINScene, DTMScene
    cX = header.xllCorner + header.nrCols * header.cellSize * 0.5
    cY = header.yllCorner + header.nrRows * header.cellSize * 0.45
    cZ = header.zMin + header.dz * 0.5
    DTMScene = visual.display(x = 0, y = 0, width = 600, height = 600)
    DTMScene.title = "DTM"
    DTMScene.background = visual.color.white
    DTMScene.ambient = 0.33
    DTMScene.center = (cX, cY, cZ*header.magnify)
    DTMScene.forward = (0, 1, -1)

    TINScene = visual.display(x = 600, y = 0, width = 600,
                              height = 600)
    TINScene.title = "TIN"
    TINScene.background = visual.color.white
    TINScene.ambient = 0.33
    TINScene.center = (cX, cY, cZ*header.magnify)
    TINScene.up = (0,0,1)
    TINScene.forward = (0, 1, -1)
    setColorScaleDtm()

def drawDTM(pointList, header):
    for i in range(len(pointList)):
```

```
        p = pointList[i]
        myColor = getDTMColor(p[2], header)
        myPos = ([p[0], p[1], p[2] * header.magnify])
        myPoint = visual.points(display = DTMScene, pos = myPos,
                                    size = 3.0, color=myColor)
        pointListDtm3D.append(myPoint)
    key = DTMScene.kb.getkey()

def cleanAllPoints():
    while (len(pointList3D) > 0):
        pointList3D[0].visible = False
        myPoint = pointList3D.pop(0)
        del myPoint

def delTriangle(i):
    triangleList3D[i].visible = False
    myTriangle = triangleList3D.pop(i)
    del myTriangle

def cleanAllTriangles():
    lastIndex = len(triangleList3D) - 1
    for i in np.arange(lastIndex, -1, -1):
        delTriangle(i)

def cleanAll():
    cleanAllPoints()
    cleanAllTriangles()

def drawSurface(myColor, v):
    mySurface = visual.faces(display = TINScene, color=myColor,
                                pos = v)
    mySurface.make_twosided()
    mySurface.make_normals()
    mySurface.smooth()
    surfaceList3D.append(mySurface)

def drawAllPoints(pointList, header):
    cleanAll()
    myPosition = []
    for i in range(len(pointList)):
        p = pointList[i]
        myPosition.append([p[0], p[1], p[2] * header.magnify])
    myPoints = visual.points(display = TINScene, pos = myPosition,
                                size = 3.0, color=visual.color.black)
    pointList3D.append(myPoints)
    key = TINScene.kb.getkey()

def addTriangle(index, triangle, header):
    v = copy(triangle.v)
    v[:,2] *= header.magnify
```

```
    myTriangle = visual.curve(pos=[v[0],v[1],v[2],v[0]],
                                  color=visual.color.black)
    triangleList3D.insert(index, myTriangle)
    if sys.version_info < (3, 0): visual.wait()

def drawAllTriangles(triangleList, header):
    cleanAll()
    for i in range(len(triangleList)):
        addTriangle(i, triangleList[i], header)
    key = TINScene.kb.getkey()

def drawAllSurfaces(triangleList, header):
    cleanAll()
    nrTriangles = len(triangleList)
    for i in range(nrTriangles):
        v = copy(triangleList[i].v)
        z = sum(v[:,2]) / 3.0
        myColor = getDTMColor(z, header)
        v[:,2] *= header.magnify
        drawSurface(myColor, v)
    if sys.version_info < (3, 0): visual.wait()
```

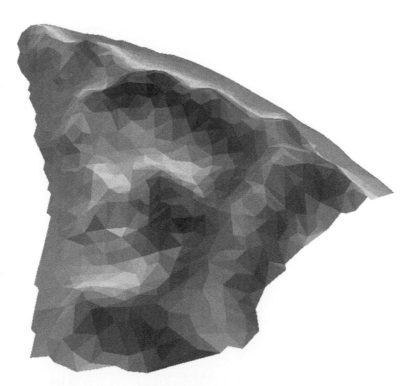

Figure 9.7 *Triangulated irregular network (maximum quality) of dimension 2603 vertices, 5134 triangles.*

Figure 9.7 shows the TIN obtained from the algorithm presented in this chapter. In this example, the TIN was selected to be of the highest quality (maximum number of possible triangles) and there was a reduction by a factor of 8 between the number of pixels in the original DTM file and the number of triangles in the TIN.

. . . .

9.13 EXERCISE

9.1. Change the quality parameter and observe the generated TIN.

10

Water Flow in Three Dimensions

In Chapter 8, one-dimensional numerical solutions of transient water flow were presented, while in Chapter 9, we presented a program to generate a triangulated irregular grid from a regular grid obtained from a DTM. In practical applications, it is often necessary to compute surface runoff, infiltration and redistribution in a two- or three-dimensional domain, for instance when water balance studies are necessary in hilly or mountaineous areas. In this chapter, we present a three-dimensional solution of water flow, solved by employing the cell-centred finite volume scheme presented in Chapter 8. As described in Chapter 9, irregular networks are employed since they better represent the often irregular geometries of natural landscapes. Therefore, in this chapter, the three-dimensional solution is implemented into a triangle mesh, obtained from the program of Chapter 9. Moreover, in this chapter, we also couple the three-dimensional solution for infiltration and redistribution with an algorithm for surface runoff.

10.1 Governing Equations

In Chapter 8, we presented the solution for infiltration and redistribution. In this chapter, we also include surface runoff by coupling the infiltration and runoff processes. The coupling algorithm was developed by Bittelli *et al.* (2010) and is based on the following rationale. A volumetric term V is used to account for the change in volume of the surface computational element. Specifically, for subsurface flow, the element volume is constant (we assume a rigid soil that does not undergo swelling and shrinking), while the volumetric water content within the element volume changes according to Richards' equation. On the other hand, for surface flow, the surface element volume may change as function of the surface hydraulic height (depending on precipitation and on fluxes in and out from other surface nodes), while the water content is always unity ($\theta = 1$), since the surface element volume represent the water running over the soil surface.

The scheme and the surface flow variables are depicted in Fig. 10.1, where h_s represents the surface water, which is designated as immobile or pond water h_{pond}, and mobile water h'_s, defined as $h'_s = \max(0, (h_s - h_{pond}))$. The height of the pond water h_{pond} [m], is defined by the user as input to the model. The surface water h_s [m] is computed within the time step, depending on the boundary conditions (i.e. precipitation input).

Soil Physics with Python. First Edition. Marco Bittelli, Gaylon S. Campbell and Fausto Tomei.
© Marco Bittelli, Gaylon S. Campbell and Fausto Tomei 2015. Published in 2015 by Oxford University Press.

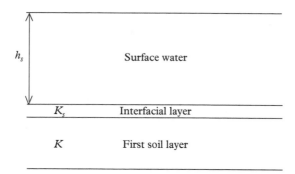

Figure 10.1 *Schematic of the surface flow variables for coupling infiltration and surface runoff.*

The interfacial layer has hydraulic conductivity equal to the saturated hydraulic conductivity. The differential continuity equation for water flow is

$$\frac{\partial(V\theta)}{\partial t} = -\text{div}(f_w) + u \tag{10.1}$$

where f_w is the water flux, V is the total available volume at a node, θ is the volumetric water content and u is the water input or output (sink–source term expressed as a volume per unit time). As introduced above, this general equation is solved adopting two different laws to describe fluxes within the soil matrix (infiltration and redistribution) and on the soil surface (runoff). In the first case, we employ Richards' equation

$$V\frac{d\theta}{dH}\frac{\partial H}{\partial t} = \text{div}[K(h) \cdot \text{grad}(H)] + u \tag{10.2}$$

where $K(h)$ is the hydraulic conductivity and H is the total hydraulic head ($H = h + z$). The total hydraulic head H [m] is given by the sum of the elevation z (or gravitational potential) and the matric potential h. Note that here we use the symbol h for the matric potential. In this chapter, the total water potential will be referred as hydraulic head with units of metres, since it is easier to formulate St Venant's and Manning's equations for surface runoff by expressing the total water potential as hydraulic head with units of metres.

For surface flow, the volume upon which the mass balance is computed varies according to the surface hydraulic depth h_s, which results in a surface storage term $V = Ah_s$, where A is the planimetric surface area [m^2]. This assumption allows us to rewrite the first term in eqn (10.1) as

$$\frac{\partial(V\theta)}{\partial t} = A\frac{\partial h_s}{\partial t} = A\frac{\partial H}{\partial t} \tag{10.3}$$

Assuming that the depth h_s is much smaller than the flow width, the velocity along the coordinate x is given by Manning's equation, with h_s adopted as the hydraulic radius:

$$v_x = -\left(\frac{h_s^{0.67}}{M}\right)\sqrt{\frac{\partial H}{\partial x}} \tag{10.4}$$

where M is the Manning roughness parameter. In two dimensions, Manning's equation can be written as (Di Giammarco *et al.*, 1996)

$$v_m = -K^*(H, h_s)\ \text{grad}\ H \tag{10.5}$$

where K^* is a conveyance function depending on H and h_s according to

$$K^* = \frac{\left(\dfrac{h_s^{0.67}}{M}\right)}{\sqrt[4]{\left(\dfrac{\partial H}{\partial x}\right)^2 + \left(\dfrac{\partial H}{\partial y}\right)^2}} \tag{10.6}$$

Substitution of eqns (10.3) and (10.5) into eqn (10.1) yields

$$A\frac{\partial H}{\partial t} = \text{div}[K^*(H, h_s) \cdot \text{grad}\ (H)] + u \tag{10.7}$$

which is the well-known parabolic (or 'diffusion wave') approximation of St Venant's equation.

Richards' equation for flow in variably saturated soil and the parabolic St Venant's equation for surface flow have equivalent mathematical structure that can be expressed in the unified form

$$C\frac{\partial H}{\partial t} = \text{div}[K' \cdot \text{grad}(H)] + u \tag{10.8}$$

where

$$C = \begin{cases} A & \text{for surface flow} \\ V\dfrac{d\theta}{dH} & \text{for subsurface flow} \end{cases} \tag{10.9}$$

and

$$K' = \begin{cases} K^* & \text{for surface flow} \\ K & \text{for subsurface flow} \end{cases} \tag{10.10}$$

The surface flow height is considered as the total height of water on the surface minus the inactive water height that is retained in microtopography (tillage, different cropping systems). This mathematical formulation is convenient, since it allows us to couple the numerical solution by implementing the equations into a single matrix system.

10.2 Numerical Formulation

The solution of the governing equations described above is based on the cell-centred finite volume (also called integrated finite difference) method introduced in Chapter 8. This method consists in the integration of the differential continuity equation over a finite domain D, leading to the integral equation

$$\iiint \frac{\partial(V\theta)}{\partial t} \, dx \, dy \, dz = \iiint f_w \, dx \, dy \, dz + \iiint u \, dx \, dy \, dz \qquad (10.11)$$

Gauss' theorem states that the integral of f_w over the total volume of the grid cell can be replaced by the fluxes over the boundaries of the grid cell, leading to

$$\iiint \frac{\partial(V_t\theta)}{\partial t} \, dx \, dy \, dz = \int_A f_w \cdot \vec{e} \, dA + \iiint u \, dx \, dy \, dz \qquad (10.12)$$

where A is the surface of the computational domain, and $\vec{e}$ its unit normal vector. Equation (10.12) can be applied over a simulation volume within which the material properties are constant.

As described in Chapter 8, when the simulation domain is approximated by a grid of nodes, eqn (10.12) is equivalent to the mass balance equation for the volume surrounding each node:

$$V\frac{\partial\theta}{\partial t} = \sum_{j=1}^{n} F_{ij} + u_i \qquad\qquad \forall i \neq j \qquad (10.13)$$

where θ is the amount of stored water in the volume surrounding the node, F_{ij} is the flux between the ith and jth nodes, and u_i is the sink–source term. We can write a system of equations for all nodes, with the unknowns being the hydraulic heads H.

The flux F_{ij} is described by Darcy's law in finite difference form:

$$F_{ij} = -K_{ij}A_{ij}\frac{H_i - H_j}{L_{ij}} \qquad (10.14)$$

where A_{ij} is the interfacial area between nodes i and j, L_{ij} is the distance between the two nodes, H_i is the total hydraulic head at node i, and K_{ij} is the internode hydraulic conductivity.

For subsurface nodes, the nodal mass balance of eqn (10.13) can be written as

$$V_i \left(\frac{d\theta}{dH} \right)_i \frac{\partial H_i}{\partial t} = \sum_{j=1}^{n} \zeta_{ij} K(\theta)_{ij} (H_j - H_i) + u_i \tag{10.15}$$

where V_i is the soil bulk volume, θ_i is the soil water content, K is the hydraulic conductivity, H_i is the hydraulic head (total water potential) of node i, H_j is the hydraulic head of node j and ζ_{ij} is the ratio of bulk exchange area between nodes i and j over the internode distance. The term V_i, as introduced above, is used to account for the change in volume of the computational element. For subsurface flow, the element volumes are constant and the volumetric water content changes (Richards' equation). In surface flow, the element volumes may change as functions of the surface hydraulic height, while the water content of the node is always saturated ($\theta = 1$).

The dependence on saturation of the hydraulic conductivity is given by Campbell (1974). The formulation proposed by Mualem (1976) and van Genuchten (1980) can also be used; however, this formulation provides unreliable computation of hydraulic conductivity. If van Genuchten's (1980) formulation for the water retention curve is employed, it is advisable to use the modified model proposed by Ippisch *et al.* (2006), as described in Chapter 5. The term $(d\theta/dH)_i$ is evaluated through the soil water function previously described.

For surface nodes, K becomes conveyance, dependent upon water depth, ζ_{ij} is the ratio of exchange area between nodes i and j, and the term $W_i(d\theta/dH)_i$ is replaced by the extension of the topographic surface assigned to each surface node. The exchange area between nodes i and j is the product of the average water depth in nodes i and j with the internodal exchange length.

Equation (10.15) can be written for each of the n nodes (here n is used for number of nodes) of the discretization domain, leading to a system of equations in the form

$$\mathbf{C} \frac{\partial H}{\partial t} = \mathbf{A} H + q \tag{10.16}$$

where $\mathbf{C}$ is the diagonal mass matrix with elements

$$\mathbf{C}_{ii} = W_i \left(\frac{d\theta}{dH} \right)_i \tag{10.17}$$

for subsurface nodes and $\mathbf{C}_{ii} = \mathbf{S}_{ii}$ for surface nodes, while $\mathbf{A}$ is the symmetric stiffness matrix, the elements of which are

$$\mathbf{A}_{ij} = -\sum_{j=1}^{n} \zeta_{ij} K(H)_{i,j} \quad \text{for } i = j \tag{10.18}$$

$$\mathbf{A}_{ij} = \zeta_{ij} K(H)_{i,j} \quad \text{for } i \neq j \tag{10.19}$$

Both **C** and **A** are strongly dependent on the unknown hydraulic heads, with a nonlinear relationship. The time derivative $\partial H/\partial t$ is replaced by its incremental ratio

$$\frac{\partial H}{\partial t} = \frac{H^{t+1} - H^t}{\Delta t} \tag{10.20}$$

where the superscripts denote generic times t and $t+1$, with Δt being the time step. By approximating the right-hand-side of eqn (10.16), we obtain

$$H = \eta H^{t+1} + (1-\eta)H^t \tag{10.21}$$

where η is a weighting factor that ranges from 0 to 1. For the three-dimensional water flow algorithm, we used an implicit scheme ($\eta = 1$).

Equation (10.16) is then written as

$$\mathbf{C}\frac{H^{t+1} - H^t}{\Delta t} = \mathbf{A}\left[\eta H^{t+1} - (1-\eta)H^t\right] + q \tag{10.22}$$

or, rearranged to highlight the unknowns, as

$$\left(\frac{\mathbf{C}}{\Delta t} - \eta\mathbf{A}\right)H^{t+1} = \left((1-\eta)\mathbf{A} + \frac{\mathbf{C}}{\Delta t}\right)H^t + q \tag{10.23}$$

With the definitions

$$\mathbf{M} = \frac{\mathbf{C}}{\Delta t} - \eta\mathbf{A}, \qquad \mathbf{N} = (1-\eta)\mathbf{A} + \frac{\mathbf{C}}{\Delta t} \tag{10.24}$$

eqn (10.23) reads

$$\mathbf{M}H^{t+1} = \mathbf{N}H^t + q \tag{10.25}$$

Both **C** and **A** depend on H, which is approximated by H^t, allowing solution of the linearized system of equations. A first-approximation solution is $H(0)^{t+1}$. With this solution, **A** and **C** can be recalculated and the procedure iterated until a certain tolerance in the variation of the subsequent H^{t+1} is met. The tolerance is then checked against the mass balance error. The linear and nonlinear defect, or the infinite-norm methods, can be chosen as convergence criteria.

10.3   Coupling Surface and Subsurface Flow

The surface flow variables are described in Fig. 10.1, where h_s represents the surface water, designated as immobile or pond water h_{pond}, and mobile water h_s', defined as $h_s' = \max(0, (h_s - h_{\mathrm{pond}}))$. The depth of the ponded water, h_{pond}, is defined by the user as input to the model.

Moreover, consideration of the mass exchanges in the surface nodes requires distinguishing the following cases depending on the domain to which each node pair belongs. If nodes i and j are both on the surface, then Manning's equation applies in the form

$$Q_{ij} = \frac{(h'_{s_i})^{5/3} B_{ij} \left(\dfrac{H_i - H_j}{L_{ij}} \right)^{0.5}}{M_{ij}} \tag{10.26}$$

where the node i is the one with the highest hydraulic head, B_{ij} is the width of the interface between nodes i and j, $M_{ij} = (M_i + M_j)/2$ is the average internode roughness, L_{ij} is the horizontal distance between the nodes, and $(H_i - H_j)/L_{ij}$ is the slope of the hydraulic heads.

For surface flow, the matrix $\mathbf{A}$ is modified as follows:

$$\mathbf{A}_{ij} = \eta_{ij} K_{i,j} \qquad\qquad \text{for } i \neq j \tag{10.27}$$

with η_{ij} given by

$$\eta_{ij} = \frac{B_{ij}}{M_{ij} (L_{ij})^{0.5}} \tag{10.28}$$

and

$$K_{ij} = \frac{(h'_{s_i})^{5/3}}{|H_i - H_j|^{0.5}} \qquad\qquad \text{for } h'_s > 0 \tag{10.29}$$

otherwise

$$K_{ij} = 0 \tag{10.30}$$

If node i is on the surface and node j is below the surface, a coupled system is determined between surface and subsurface, where the exchange of water is analogous to a one-dimensional variably-saturated form of Darcy's law.

Therefore, for node i on the surface and node j in the subsurface, the matrix $\mathbf{A}$ is modified as follows:

$$\mathbf{A}_{ij} = \zeta_{ij} K'_j \qquad\qquad \text{for } i \neq j \tag{10.31}$$

where

$$K'_j = \text{mean}(K(H)_j, Ks_j) \tag{10.32}$$

The main variable is the interface conductivity K'_j, which is computed, analogously to the subsurface conductivities, as a geometric, harmonic or logarithmic mean between the conductivity of the first soil node and the saturated conductivity of the interface layer. With this formulation, semi-impermeable surfaces (such as roads or sealed soils) can also be considered, by setting very low values of Ks_j.

On the other hand, excessive infiltration rates may occur in the case of high gradients such as rainfall over a very dry soil. For this case, a limit on the value of K'_j is imposed by

$$K''_j = \min(K'_j, K_{\max_{ij}}) \qquad (10.33)$$

where $K_{\max_{ij}}$ is the value of K in the case in which the flux F_{ij} exchanges all the available water present in the surface h_s within a time step Δt. This value can be computed by using the flux equation (10.14):

$$K_{\max_{ij}} = \frac{h_s}{\Delta t} \frac{L_{ij}}{|H_i - H_j|} \qquad (10.34)$$

This formulation allows a single continuity equation to be used for both surface and subsurface flow to allow Richards' and St Venant's equations to be solved simultaneously, without decoupling the two processes.

Computation of the derivative $d\theta/dH$, which is utilized in the mass matrix $\mathbf{C}$, can be performed at the beginning of the time step, but this value will not be correct for the whole time step if there are variations of the two variables θ and H. In fact, preliminary numerical tests showed that significant over- or under-estimation of water fluxes were experienced, depending on the increasing or decreasing behaviour of the derivative during the time step. To address this problem in the code, the analytical derivative of Campbell's equation or of the Ippisch–van Genuchten equation is computed first at the beginning of the time step, while at the end of each approximation the derivative is recomputed by the incremental ratio of the new values of θ and H obtained at each iteration. The final value that is used is an intermediate value between the first and last values. By using this approach, it is possible to obtain a value of the ratio also in transition between unsaturated and saturated conditions.

The numerical implementation of the model allows the user to use different numerical methods for solving the system of equations, such as the Gauss–Seidel (Morton and Mayers, 1994), over-relaxation or conjugate gradient methods (Burden and Faires, 1997). In this program, the system of equations was solved with the Gauss–Seidel method, with an adaptive time-step algorithm, since in the performed numerical tests for this problem, the Gauss–Seidel method provided the most accurate solution (Tomei, 2005).

10.4　Numerical Implementation

The model described above was implemented in the C programming language by Pistocchi and Tomei (2003) and then further developed and corroborated by Bittelli *et al.* (2010), including simulation of topography-dependent solar radiation, snow

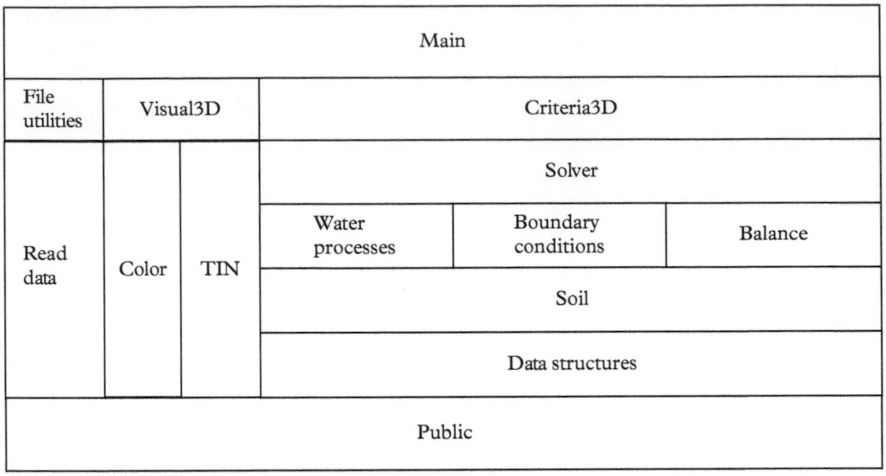

Main					
File utilities	Visual3D	Criteria3D			
Read data	Color	TIN	Solver		
			Water processes	Boundary conditions	Balance
			Soil		
			Data structures		
Public					

Figure 10.2 *Diagram with the software architecture of the modules for the three-dimensional water flow program.*

melting, evaporation and plant transpiration. This version of the Criteria3D model is developed in the Python programming language and includes all the main water processes. Figure 10.2 depicts the software architecture for the program. This consists of 14 modules, as shown in Fig. 10.2, and it is called PSP_Criteria3D.

The following is a list of the modules depicted in Fig. 10.2. The PSP_readDataFile is the same module presented in previous chapters.

1. Read data (PSP_readDataFile.py)
2. Public (PSP_public.py)
3. Data structures (PSP_dataStructures.py)
4. TIN (PSP_tin.py)
5. Soil (PSP_soil.py)
6. Water processes (PSP_waterProcesses.py)
7. Boundary conditions (PSP_boundaryConditions.py)
8. Solver (PSP_solver.py)
9. Balance (PSP_balance.py)
10. Criteria3D (PSP_criteria3D.py)
11. Main (main.py)
12. Visual3D (PSP_visual3D.py)
13. Color (PSP_color.py)
14. File utilities (PSP_fileUtilities.py)

10.4.1 Public

The module `PSP_Public` defines critical variables and constants used by other modules. In particular, the class `C3DParameters` contains the main modelling parameters that can be modified by the user:

- the water retention curve parameters
- the mean used for computing the hydraulic conductivity at the interface between two elements
- the initial value of water potential in the soil profile
- the name of the file containing the observed precipitation and its time step in minutes (default 15 minutes)
- `computeOnlySurface`: the flag to activate the simulation for surface runoff only
- `isFreeDrainage`: the flag for lower boundary conditions (free drainage or impermeable)
- minimum and maximum thickness of soil layers and the factor for generating the geometrical grid in the vertical direction
- roughness parameter (Manning's equation) and pond (minimum height for surface hydraulic head to start runoff)
- initial, minimum and maximum value for time step
- maximum number of approximations and iterations for each approximation
- tolerance threshold for the Gauss–Seidel method
- tolerance threshold for mass balance error (as a percentage)
- ratio for horizontal to vertical hydraulic conductivity

The program `PSP_public.py` is shown below with these options. The user can use his or her own precipitation data to simulate the conditions of a specific site by changing the data in the file `precipitation.txt`. To investigate the hydrological response of a catchment, the observed precipitation time step can also be changed.

```
#PSP_public.py
from math import sqrt, fabs, log

CAMPBELL = 1
IPPISCH_VG = 2

LOGARITHMIC = 0
HARMONIC = 1
GEOMETRIC = 2

# user choices
class C3DParameters:
```

```
      waterRetentionCurve = IPPISCH_VG
      meanType = LOGARITHMIC
      initialWaterPotential = -2.0
      precFileName = "data\\precipitation.txt"
      obsPrecTimeLength = 15
      computeOnlySurface = False
      isFreeDrainage = True
      minThickness = 0.01
      maxThickness = 0.1
      geometricFactor = 1.2
      roughness = 0.24
      pond = 0.002
      currentDeltaT = 60.0
      deltaT_min = 6.0
      deltaT_max = obsPrecTimeLength * 60.0
      maxIterationsNr = 100
      maxApproximationsNr = 10
      residualTolerance = 1E-12
      MBRThreshold = 1E-2
      conductivityHVRatio = 10.0

EPSILON_METER = 1E-5
UP = 1
DOWN = 2
LATERAL = 3
NODATA = -9999.
NOLINK = -1
BOUNDARY_RUNOFF = 1
BOUNDARY_FREEDRAINAGE = 2
BOUNDARY_FREELATERALDRAINAGE = 3
BOUNDARY_NONE = 99
OK = 1
INDEX_ERROR = -1111
LINK_ERROR = -5555
```

10.4.2   Data structures

The PSP_dataStructures module defines the data structures required by the model
to store all geometrical and hydraulic properties of cells during the computation. The
class C3DStructure includes general information on the dimensions of the system: the
number of layers characterizing the soil, the numbers of triangles generated by the triangulation algorithm, the total number of cells (equal to the number of surface triangles
multiplied by the number of layers) and the total area of the domain. These variables are
dynamically computed when the TIN is loaded and the numbers of layers are selected
by the user. On the other hand, the parameters nrLateralLinks and nrMaxLinks
are fixed and depend on the geometrical structure. Each element has three lateral links,
one upper link and one lower link.

The class `Clink` defines the properties of each of the links between the ith cell and another cell. The class `Cboundary` defines the properties of a boundary cell: the type of boundary conditions, the area of water exchange, its slope (only for lateral boundary) and the computed flow due to the boundary in the time step. The class `Ccell` includes geometrical and hydraulic properties of a cell. The geometrical properties are its location (x, y, z), the area of the triangle at the base of the prism, its volume and a flag to define if the prism is on the surface (`isSurface`). Once these properties have been defined, they will not change during the computation.

The hydraulic properties are the degree of saturation S_e, the current and previous total water potentials H_0 and H, the hydraulic conductivity, the water sink–source (i.e. precipitation or evapotranspiration, and the total flow for the time step (due to sink–source and boundary flow). Obviously, this last property will change during the computation. Finally, the class `Ccell` includes an instance of the class `Cboundary` for boundary properties and five instances of the class `Clink` (three for the possibly lateral links plus an upLink and a downLink).

```
#PSP_dataStructures.py
from PSP_public import  *

class C3DStructure:
    nrLayers = 0
    nrTriangles = 0
    nrCells = 0
    nrLateralLinks = 3
    nrMaxLinks = 5
    totalArea = 0

class Clink:
    def __init__(self):
        self.index = NOLINK
        self.area = NODATA
        self.distance = NODATA

class Cboundary:
    def __init__(self):
        self.type = BOUNDARY_NONE
        self.area = NODATA
        self.slope = NODATA
        self.flow = NODATA

class Ccell:
    def __init__(self):
        self.x = NODATA
        self.y = NODATA
        self.z = NODATA
        self.area = NODATA
```

```
        self.volume = NODATA
        self.isSurface = False
        self.Se = NODATA
        self.H = NODATA
        self.H0 = NODATA
        self.k = NODATA
        self.sinkSource = NODATA
        self.flow = NODATA
        self.boundary = Cboundary()
        self.upLink = Clink()
        self.downLink = Clink()
        self.lateralLink = [Clink(), Clink(), Clink()]
#global
C3DCells = []
```

10.4.3   TIN

Criteria3D describes the surface topography using a TIN. The triangulated network is produced by the code described in Chapter 9, which produces three output files: `triangles.csv`, `vertices.csv` and `neighbours.csv`. Therefore the code implements functions to manage the topographical information contained in these three files. For this reason, the module `PSP_tin` uses and combines the functions described in the modules `GIS` and `triangle`, described in Chapter 9.

The new function **getHeader** defines the minimum and maximum topographical values of the factor `magnify`, similar to what was performed by the function **computeMinMaxGrid** of the module **GIS**. With respect to the original version of the class **Ctriangle**, there are three additional properties: `isBoundary`, `boundarySlope` and `boundarySide`. These properties are necessary to memorize if a triangle is on the boundary, the side and the slope of the boundary. They are computed in the functions **getBoundaryProperties** and **getSlope**. The module `PSP_tin` is as follows:

```
#PSP_tin.py
from __future__ import print_function, division
from PSP_public import *
import numpy as np
from copy import copy

class CheaderTin():
    xMin = NODATA
    xMax = NODATA
    yMin = NODATA
    yMax = NODATA
    zMin = NODATA
    zMax = NODATA
    dz = NODATA
    magnify = NODATA
```

```
class Ctriangle:
    def __init__(self, v = np.zeros((3, 3), float)):
        self.isBoundary = False
        self.boundarySlope = NODATA
        self.boundarySide = NODATA
        self.v = copy(v)
        if (not np.all(v == 0.0)):
            self.centroid = (v[0]+v[1]+v[2])/3.0
            self.area = getArea2D(self.v)

#global structures
header = CheaderTin()
C3DTIN = []

def magnitude(v):
    return(np.sqrt(v.dot(v)))

def getArea(v):
    return 0.5 * magnitude(np.cross(v[1] - v[0], v[2] - v[0]))

# Area = 1/2 |x1(y2-y3) - x2(y1-y3) + x3(y1 - y2)|
def getArea2D(v):
    x = v[:,0]
    y = v[:,1]
    return 0.5 * fabs(x[0]*(y[1]-y[2]) - x[1]*(y[0]-y[2]) + x[2]
                *(y[0]-y[1]))

def getHeader(triangleList):
    header = CheaderTin()
    header.xMin = triangleList[0].centroid[0]
    header.yMin = triangleList[0].centroid[1]
    header.zMin = triangleList[0].centroid[2]
    header.xMax = header.xMin
    header.yMax = header.yMin
    header.zMax = header.zMin

    for i in range(1, len(triangleList)):
        x = triangleList[i].centroid[0]
        y = triangleList[i].centroid[1]
        z = triangleList[i].centroid[2]
        header.xMin = min(header.xMin, x)
        header.yMin = min(header.yMin, y)
        header.zMin = min(header.zMin, z)
        header.xMax = max(header.xMax, x)
        header.yMax = max(header.yMax, y)
        header.zMax = max(header.zMax, z)

    dx = header.xMax - header.xMin
    dy = header.yMax - header.yMin
```

```
        header.dz = header.zMax - header.zMin
        dtmRatio = (sqrt(dx*dy) / header.dz) * 0.1
        header.magnify = max(2.0, min(6.0, dtmRatio))
        return(header)

def distance2D(v1, v2):
    dx = fabs(v1[0] - v2[0])
    dy = fabs(v1[1] - v2[1])
    return sqrt(dx*dx + dy*dy)

def distance3D(v1, v2):
    dx = fabs(v1[0] - v2[0])
    dy = fabs(v1[1] - v2[1])
    dz = fabs(v1[2] - v2[2])
    return sqrt(dx*dx + dy*dy + dz*dz)

def getSlope(TIN, i, j):
    dz = TIN[i].centroid[2] - TIN[j].centroid[2]
    dxy = distance2D(TIN[i].centroid, TIN[j].centroid)
    return dz/dxy

def getBoundaryProperties(TIN, index, neighbours):
    slope = getSlope(TIN, int(neighbours[0]), index)
    if  (int(neighbours[1]) != NOLINK):
        slope1 = getSlope(TIN, int(neighbours[1]), index)
        slope = max(slope, slope1)
    TIN[index].boundarySlope = slope
    TIN[index].boundarySide = distance3D(TIN[index].v[0],
                               TIN[index].v[1])

def getAdjacentVertices(t1, t2):
    isFirst = True
    for i in range(3):
        for j in range(3):
            if (t1[i] == t2[j]):
                if isFirst:
                    index1 = t1[i]
                    isFirst = False
                else:
                    index2= t1[i]
                    return (index1, index2)
    return NOLINK, NOLINK

def getAdjacentSide(i, j, vertexList, triangleList):
    triangle1 = triangleList[i]
    triangle2 = triangleList[j]
    index1, index2 = getAdjacentVertices(triangle1, triangle2)
    v1 = vertexList[index1]
    v2 = vertexList[index2]
    return distance2D(v1, v2)
```

10.4.4 Soil

The `PSP_soil` module is very similar to that presented in Chapter 8, where the functions for the hydraulic properties (water retention curve, hydraulic conductivity and water capacity) are implemented. The two retention curves used in `Criteria3D` are those of Campbell and Ippisch–van Genuchten.

The new functions, with respect to the other version of `Soil`, are **setLayers** that allow one to set `depth` and `thickess` of the different layers, on the basis of the parameters set in the classes `C3DParameters` of the module `public`, and **readHorizon**, which reads a single line (row *i*) of the `soilFileName` file and stores its parameters for the Campbell and Ippisch–van Genuchten curves in the **CsoilHorizon** class. The new functions are shown below.

It is possible to modify the code to load and manage soils having different horizons, as is commonly the case in natural soils, using the same methods as described in Chapter 8, by using the horizon list and by assigning to each node the corresponding horizon. To write this modification, it is necessary to load and store the soil properties of each horizon in an array of the class **CsoilHorizon**, instead of loading a single element. The user must then add to the class **Ccell** an index of the horizon. The horizon index can be easily determined by the property `depth` of the soil layer and by the properties `upperDepth` and `lowerDepth` of each horizon. Finally, the hydraulic functions of the module `Soil` must be modified such that they use the properties of the different horizons.

The module reads a soil file having just one soil horizon and it assigns the same hydraulic properties parameters to all the nodes of the domain.

```
# set depth and thickness of layers
def setLayers(totalDepth, minThickness, maxThickness, factor):
    nrLayers = 1
    prevThickness = minThickness
    depth = minThickness * 0.5
    while (depth < totalDepth):
        nextThickness = min(maxThickness, prevThickness * factor)
        depth = depth + (prevThickness + nextThickness) * 0.5
        prevThickness = nextThickness
        nrLayers += 1

    z = np.zeros(nrLayers, np.float64)
    thickness = np.zeros(nrLayers, np.float64)
    z[0] = 0.0
    thickness[0] = 0.0
    for i in range(1, nrLayers):
        if (i == 1):
            thickness[i] = minThickness
        else:
            thickness[i] = min(maxThickness, thickness[i-1] * factor)
        z[i] = z[i-1] + (thickness[i-1] + thickness[i]) * 0.5
    return(nrLayers, z, thickness)
```

10.4.5   Water processes

The module `PSP_waterProcesses` defines the functions for the computation of the water transfer processes, as described above. The function **redistribution** computes the soil water redistribution and returns the elements A_{ij} of the **A** matrix according to eqn (10.19). The function **infiltration** computes the movement of water from a surface node into the first underlying soil node according to eqn (10.31). If the hydraulic head of the subsurface node is larger than the topographic elevation of the surface node, the control functions on the maximum value of hydraulic conductivity are deactivated and the computation uses the saturated hydraulic conductivity. In this case, the soil profile is in a condition of full saturation (i.e. saturated soil column or capillary rise).

The function **runoff** computes the surface flow according to eqns (10.26)–(10.30) by using Manning's equation to compute the surface water velocity. Since the solution of the system is fully implicit, both functions that involve the surface layer may oscillate during the approximations following the time step. This condition of numerical instability may occur when the flux is very high with respect to the available water volume in the surface layer; therefore the estimated values of h_s may oscillate. This condition may occur in the case of intense precipitation over a dry soil surface To solve this problem, both functions that compute flux (infiltration and surface runoff) do so at an intermediate value within the time step and use an arithmetic average value of the hydraulic heads H and H_0.

During the first approximation, where only the value of H_0 is known and the node has not yet experienced the effect of a possible precipitation event, a value of half of the precipitation occurring during the whole time step is added as a source to the node. This procedure is written to make the value of h_s closer to the actual value that it will experience during the time step. Finally, the Courant number C_{ij} between two surface nodes is computed using the flux velocity v_{ij}, the value of the time step Δt and the topographic distance L_{ij} between the two nodes:

$$C_{ij} = v_{ij}\frac{\Delta t}{L_{ij}} \tag{10.35}$$

The condition of Courant–Friedrichs–Levy prevents the time step from being larger than the time necessary for the wave to cover the distance between two nodes and is commonly used in explicit numerical schemes. Numerical tests show that its use also significantly improves the numerical stability of this implicit scheme, at the price of using small time steps when surface fluxes are high.

As we will see in Section 10.4.7, Courant's condition is computed by storing the maximum Courant number in the variable `maxCourant`. When this value is larger than 1, the numerical solution is interrupted and restarts with a smaller time step. The code for the `PSP_waterProcesses` module is as follows:

```
#PSP_waterProcesses.py
from __future__ import division
from PSP_dataStructures import *
import PSP_soil as soil
import PSP_balance as balance
```

```
def redistribution(i, link, isLateral, deltaT):
    j = link.index
    k = soil.meanK(C3DParameters.meanType, C3DCells[i].k,
                   C3DCells[j].k)
    if (isLateral):
        k *= C3DParameters.conductivityHVRatio
    return (k * link.area) / link.distance

def infiltration(surf, sub, link, deltaT, isFirstApprox):
    if (C3DCells[surf].z > C3DCells[sub].H):
        #unsaturated
        Havg = (C3DCells[surf].H + C3DCells[surf].H0) * 0.5
        Hs = Havg - C3DCells[surf].z
        if isFirstApprox:
            rain =  (C3DCells[surf].sinkSource
                       / C3DCells[surf].area) * (deltaT / 2.0)
            Hs += rain
        if (Hs < 1E-12): return 0.0
        interfaceK = soil.meanK(C3DParameters.meanType,
                                C3DCells[sub].k, soil.C3DSoil.Ks)
        dH = C3DCells[surf].H - C3DCells[sub].H
        maxK = (Hs / deltaT) * (link.distance / dH)
        k = min(interfaceK , maxK)
    else:
        #saturated
        k = soil.C3DSoil.Ks
    return (k  * link.area) / link.distance

def runoff(i, link, deltaT, isFirstApprox):
    j = link.index
    zmax = max(C3DCells[i].z, C3DCells[j].z)
    Hmax = max((C3DCells[i].H + C3DCells[i].H0)/ 2.0,
               (C3DCells[j].H + C3DCells[j].H0)/ 2.0)
    Hs = Hmax - (zmax + C3DParameters.pond)
    if isFirstApprox:
        rain = (C3DCells[i].sinkSource / C3DCells[i].area) *
               (deltaT / 2.0)
        Hs += rain
    if (Hs <= EPSILON_METER): return(0.0)

    dH = fabs(C3DCells[i].H - C3DCells[j].H)
    if (dH < EPSILON_METER): return (0.0)

    # [m/s] Manning equation
    v = (pow(Hs, 2.0 / 3.0) * sqrt(dH/link.distance)) /
        C3DParameters.roughness
    Courant = v * deltaT / link.distance
    balance.maxCourant = max(balance.maxCourant, Courant)
```

```
# on surface: link.area = side length [m]
area = link.area * Hs
return (v / dH) * area
```

10.4.6   Boundary Conditions

The module PSP_boundaryConditions computes the fluxes due to the boundary of the computational domain and adds them to the sink–source terms eventually present in the nodes, thereby obtaining the total value for the sink–source term for each node, corresponding to the values of u_i of eqn (10.13). This computation is performed by the function **updateBoundary**.

The three types of flow at the boundary of the domain are the surface runoff (BOUNDARY_RUNOFF), the vertical free drainage (BOUNDARY_FREEDRAINAGE) and the lateral free flow (BOUNDARY_FREELATERALDRAINAGE). The atmospheric fluxes (precipitation or evaporation) or the sink–source by plant transpiration are passed to the node through the property sinkSource of each node. Clearly, computation of evapotranspiration fluxes requires the implementation of a specific module that is not provided here. However, it is possible to include the modules presented in Chapters 4, 11, 14 and 15 to include heat transfer, soil evaporation, plant transpiration and atmospheric conditions. In this module, only precipitation flux is used as an upper boundary condition.

Different conditions must be associated with the boundary nodes of the system, depending upon the system topography. This task is performed at a higher level of the program, by the module main. The nodes on the lateral boundaries of the computational domain will be associated to the runoff if they are on the surface or to the lateral subsurface flow if they are soil nodes. The nodes at the lower boundary will be associated to the vertical free drainage. For the lower boundary nodes, it is possible to set a condition of free drainage through the flag isFreeDrainage, already described in the module public. When the flag is set to True, a condition of free drainage is assigned to the last layer. If the condition is set to False, the last layer is set as impermeable, allowing for simulation of a perched water table. All the domain boundaries can be described as Neumann boundary conditions, which are computed with the same rules presented in the module waterProcesses. For the lateral boundary nodes on the surface (lateral runoff), the same computation is applied in the function **runoff**, using Manning's equation. The calculation imposes the conditions that the hydraulic head of the external node has the same value as the boundary node and that the slope of the external node is the same as that of the neighbouring nodes, computed by the function **getSlope** in the module tin. On the soil boundary nodes, a similar condition of free drainage is applied, using Darcy's equation and assuming that there is no matric potential gradient, and therefore the flux is determined only by differences in the hydraulic head given by differences in elevation.

For the surface and subsurface lateral fluxes, only the nodes having slope greater than zero are considered in the computation; therefore water flows from the external nodes are not considered. However, this condition can be removed by deactivating the control over the slope. No Dirichlet boundary conditions are implemented in the model. In cases where a Dirichlet boundary condition is needed, it is advisable to formulate it as an imposed sink–source term and compute it in this module. The reason for this choice will be clearer when the modules regarding the computation of the water balance are described and the numerical condition for the choice of the time step is discussed. Every Dirichlet boundary condition determines an input or output flux, which would be included in the final mass balance computation. Moreover, in some cases, Dirichlet boundary conditions require a rearrangement of the matrix. It is therefore easier to compute the sink–source determined by the Dirichlet condition and assign it to the boundary.flow term of the node. In this way, the mass balance algorithm needs no changes. The module PSP_boundaryConditions is as follows:

```python
#PSP_boundaryConditions.py
from __future__ import print_function, division
from PSP_dataStructures import *
import PSP_soil as soil


def updateBoundary(deltaT):
    retentionCurve = C3DParameters.waterRetentionCurve
    for i in range(C3DStructure.nrCells):
        #sink/source: precipitation, evapotranspiration
        if (C3DCells[i].sinkSource != NODATA):
            #[m3 s-1]
            C3DCells[i].flow = C3DCells[i].sinkSource
        else:
            C3DCells[i].flow = 0.0

        if (C3DCells[i].boundary.type != BOUNDARY_NONE):
            C3DCells[i].boundary.flow = 0.0
            slope = C3DCells[i].boundary.slope
            meanH = (C3DCells[i].H + C3DCells[i].H0) * 0.5;

            if (C3DCells[i].boundary.type == BOUNDARY_RUNOFF):
                if (slope > 0.0):
                    Hs = meanH - (C3DCells[i].z + C3DParameters.pond)
                    if (Hs > EPSILON_METER):
                        boundaryArea = C3DCells[i].boundary.area * Hs
                        maxFlow = (Hs * C3DCells[i].area) / deltaT
                        # Manning equation [m3 s-1]
                        flow = ((boundaryArea /
                                    C3DParameters.roughness) *
                                    (Hs**(2./3.)) * sqrt(slope))
                        C3DCells[i].boundary.flow = -min(flow,
                                                            maxFlow)

            elif (C3DCells[i].boundary.type ==
                                    BOUNDARY_FREELATERALDRAINAGE):
```

```
    if (slope > 0.0):
        signPsi = meanH - C3DCells[i].z
        Se = soil.degreeOfSaturation(retentionCurve,
                                     signPsi)
        k = soil.hydraulicConductivity(retentionCurve, Se)
        k *= C3DParameters.conductivityHVRatio
        C3DCells[i].boundary.flow = - k *
                    C3DCells[i].boundary.area * slope

elif (C3DCells[i].boundary.type == BOUNDARY_FREEDRAINAGE):
    signPsi = meanH - C3DCells[i].z
    Se = soil.degreeOfSaturation(retentionCurve, signPsi)
    k = soil.hydraulicConductivity(retentionCurve, Se)
    C3DCells[i].boundary.flow = -k *
                    C3DCells[i].upLink.area

C3DCells[i].flow += C3DCells[i].boundary.flow
```

10.4.7   Solver

The module `PSP_solver` implements the numerical solution. At the beginning, the arrays for the solution are defined: C is the diagonal mass matrix of eqn (10.17), x is the vector of unknowns H^{t+1} and A is the symmetric stiffness matrix of eqns (10.19) and (10.27). The matrix A will be rearranged during the computation to define the matrix M of the linearized system of eqn (10.25) and the second term of the equation will be stored in the array b.

The two-dimensional vector `indices` is used to compress the dimension of the original matrix A. A is a symmetric sparse matrix (typical of multidimensional domains) where each row contains a maximum number of non-zero values defined by the variable `nrMaxLinks`, plus a value on the diagonal. In this version of the model (triangular grid), the maximum number of possible links is five (three laterals and two vertical). It is therefore possible to greatly reduce the memory allocation and the computational cost by storing only the non-zero values. To perform this task in the function **setCriteria3DArrays**, the matrix A is initialized with a number of columns equal only to `nrLinks`. To identify the address of each non-zero element, we define another matrix called `indices`, having also a number of columns equal to `nrLinks`, in which the column number j of each non-zero value of the original matrix is stored. Finally, the diagonal of the matrix will be dynamically computed during the rearrangment of the matrix.

The function **computeStep** builds the linear system and solves it for each time step. The process begins by initializing the values that will not change during the time step: the initial hydraulic head H_0 and the value of the diagonal matrix C for the surface nodes, since the latter is always equal to the surface area of the node, according to eqn (10.17). Moreover, the value x is initialized equal to H_0.

The next `while` loop manages all the possible approximations related to the solution of the system. As we will see, it is possible to leave this loop if the prescribed conditions are not met and the time step must be reduced. At each approximation, the degree of saturation `Se`, the hydraulic conductivity `K` and the new water capacity $(d\theta/dH)_i$ are computed only for the subsurface nodes. At this point, all the domain is updated and it is possible to compute the boundary fluxes by calling the function **updateBoundary** described in Section 10.4.6.

The inner `for` loop builds the ith row of the matrix $\mathbf{A}$ for each node of the domain and then rearranges it into the corresponding row of the matrix $\mathbf{M}$ of the linearized system of eqn (10.25). The flows on lateral and vertical links of the ith node are computed and the corresponding elements $\mathbf{A}_{ij}$ are assigned by the function **newMatrixElement** using the functions of the module `PSP_waterProcesses`.

If the nodes involved in the computation are both surface nodes, the function **runoff** is called; if both nodes are subsurface nodes, the function **redistribution** is called; finally, if one is a surface node and the other is a subsurface node, the function **infiltration** is called. If the flux resulting from the ith node and one of its links is equal to zero, the corresponding value is not saved, either in the matrix $\mathbf{A}$ or in the matrix `indices`, to further reduce the memory occupied and the computational cost. When all the non-zero values of the matrix $\mathbf{A}$ have been assigned, the matrix is rearranged by the function **arrangeMatrix**.

For each row of the matrix $\mathbf{A}$, the function **arrangeMatrix** computes the value on the diagonal $\mathbf{A}_{ii}$ (stored in the variable `D`) as a summation of the elements $\mathbf{A}_{ij}$ (according to eqn (10.18)) and builds the corresponding row of the rearranged matrix $\mathbf{M}$ (according to eqn (10.24)). The function then computes the second term of eqn (10.25), which is stored in the vector `b`. At this point, the whole linear system of equations (10.25) is available in the form $Ax = b$. To improve the conditioning of the system, $\mathbf{A}$ and `b` are divided by the value of the diagonal `D`. This procedure avoids the storage of the diagonal (which becomes a unit diagonal) and improves the numerical solution of the system.

After the linear system has been built, a numerical control based on the Courant number is performed. The Courant numbers of the flows on each surface node are computed by the function **runoff** of the module `PSP_waterProcesses`. If the maximum Courant number is greater than 1, there could be instabilities in the numerical solution for the surface flow. Therefore the time step is halved by the function **halveTimeStep** for a number of time sufficient to bring the Courant number to less than 1 (`balance.maxCourant < 1.0`) and the procedure exits from the loop by returning a flag `False`. This procedure forces a restart of the computation with the new, smaller time step.

If Courant's condition is met, the linear system is solved by the function **solveMatrix** using the Gauss–Seidel iterative method. The new values of x are computed, corresponding to the new H^{t+1}. The convergence of the solution is guaranteed, since the matrix is symmetric and positive-definite. Moreover, only the storage vector x is required by the solver, since in the Gauss–Seidel method the elements can be overwritten while they are computed. The method ends when a sufficiently small residual is reached or a number of iterations equal to `maxIterationsNr` have been performed.

An additional control is performed over the computed x values for the surface nodes. These values cannot be less than the value of the topographic elevation, otherwise this would correspond to a negative value of surface water, which is physically nonsense. Finally, the new x values are assigned to the hydraulic head H and the new values of degree of saturation are computed.

Finally, the function **waterBalance** is called, which checks if the solution of the current step is accepted, depending on the mass balance algorithm. We will see this function in Section 10.4.8. The function **waterBalance** can reject the solution, forcing restart of the computation with a smaller time step (when the flag balance.forceExit is true), or it can decide that the solution is improving but has not yet reached the prescribed tolerance for the mass balance error and therefore another approximation is required. Otherwise, if the mass balance is respected, the function will return True and computations for the new time step are executed. The code for the PSP_solver.py is as follows:

```
#PSP_solver.py
from __future__ import print_function, division
from PSP_waterProcesses import *
import PSP_boundaryConditions as boundary
import PSP_visual3D as visual3D
import numpy as np

C = np.array([], np.float64)
A = np.array([[],[]], np.float64)
x = np.array([], np.float64)
b = np.array([], np.float64)
indices = np.array([[],[]], int)

def setCriteria3DArrays(nrCells, nrLinks):
    global x, b, C, D, A, indices
    x.resize(nrCells)
    b.resize(nrCells)
    C.resize(nrCells)
    A.resize((nrCells, nrLinks))
    indices.resize((nrCells, nrLinks))

def computeStep(deltaT):
    global x, C, indices
    #initialize
    approximation = 1
    isValidStep = False
    for i in range(C3DStructure.nrCells):
        C3DCells[i].H0 = C3DCells[i].H
        x[i] = C3DCells[i].H0
        if (C3DCells[i].isSurface):
            C[i] = C3DCells[i].area
```

```
while ((not isValidStep)
and (approximation <= C3DParameters.maxApproximationsNr)):
    isFirstApprox = (approximation == 1)
    balance.maxCourant = 0.0
    for i in range(C3DStructure.nrCells):
        if (not C3DCells[i].isSurface):
            C3DCells[i].Se = soil.getDegreeOfSaturation(i)
            C3DCells[i].k = soil.getHydraulicConductivity(i)
            C[i] = C3DCells[i].volume * soil.getdTheta_dH(i)
    boundary.updateBoundary(deltaT)

    print ("approximation nr:", approximation)
    print ("Sum flows (abs) [m^3]:", balance.sumWaterFlow(deltaT,
                                                        True))
    visual3D.redraw(False)

    for i in range(C3DStructure.nrCells):
        k = 0
        if (newMatrixElement(i, C3DCells[i].upLink, k,
                        False, deltaT, isFirstApprox)):
            k += 1
        for l in range(C3DStructure.nrLateralLinks):
            if (newMatrixElement(i, C3DCells[i].lateralLink[l], k,
                        True, deltaT, isFirstApprox)):
                k += 1
        if (newMatrixElement(i, C3DCells[i].downLink, k,
                        False, deltaT, isFirstApprox)):
            k += 1
        if (k < C3DStructure.nrMaxLinks):
            indices[i][k] = NOLINK

        arrangeMatrix(i, deltaT)

    if ((balance.maxCourant > 1.0)
    and (deltaT > C3DParameters.deltaT_min)):
        print ("Courant is too high:", balance.maxCourant)
        while (balance.maxCourant > 1.0):
            balance.halveTimeStep()
            balance.maxCourant *= 0.5
        return(False)

    if not solveMatrix(approximation):
        balance.halveTimeStep()
        print("System is not convergent.")
        return(False)
    # check surface error
    for i in range(C3DStructure.nrTriangles):
        if (C3DCells[i].isSurface):
```

```
                if (x[i] < C3DCells[i].z):
                    x[i] = C3DCells[i].z
        # new hydraulic head
        for i in range(0, C3DStructure.nrCells):
            C3DCells[i].H = x[i]
            C3DCells[i].Se = soil.getDegreeOfSaturation(i)
        # balance
        isValidStep = balance.waterBalance(deltaT, approximation)
        if (balance.forceExit): return(False)
        approximation += 1
    return (isValidStep)


def  newMatrixElement(i, link, k, isLateral, deltaT, isFirstApprox):
    global A, indices
    j = link.index
    if (j == NOLINK): return (False)

    value = 0.0
    if C3DCells[i].isSurface:
        if C3DCells[j].isSurface:
            value = runoff(i, link, deltaT, isFirstApprox)
        else:
            value = infiltration(i, j, link, deltaT, isFirstApprox)
    else:
        if C3DCells[j].isSurface:
            value = infiltration(j, i, link, deltaT, isFirstApprox)
        else:
            value = redistribution(i, link, isLateral, deltaT)
    if (value == 0.0): return(False)

    indices[i][k] = j
    A[i][k] = value
    return (True)


def arrangeMatrix(i, deltaT):
    global b, C, A
    mySum = 0.0
    for j in range(C3DStructure.nrMaxLinks):
        if (indices[i][j] == NOLINK): break
        mySum += A[i][j]
        A[i][j] *= -1.0

    # diagonal and vector b
    D = (C[i] / deltaT) + mySum
    b[i] = (C[i] / deltaT) * C3DCells[i].H0
    if (C3DCells[i].flow != NODATA):
        b[i] += C3DCells[i].flow
```

```
    # matrix conditioning
    b[i] /= D
    for j in range(C3DStructure.nrMaxLinks):
        if (indices[i][j] == NOLINK): break
        A[i][j] /= D

def solveMatrix(approximation):
    ratio = (C3DParameters.maxIterationsNr
            / C3DParameters.maxApproximationsNr)
    maxIterationsNr = max(10, ratio * approximation)

    iteration = 0
    norm = 1000.
    bestNorm = norm
    while ((norm > C3DParameters.residualTolerance)
    and (iteration < maxIterationsNr)):
        norm = GaussSeidel()
        if norm > (bestNorm * 10.0): return(False)
        bestNorm = min(norm, bestNorm)
        iteration += 1
    return(True)

def GaussSeidel():
    global x
    norm = 0.0
    for i in range(C3DStructure.nrCells):
        new_x = b[i]
        for j in range(C3DStructure.nrMaxLinks):
            n = indices[i][j]
            if (n == NOLINK): break
            new_x -= (A[i][j] * x[n])

        dx = fabs(new_x - x[i])
        # infinite norm
        if (dx > norm): norm = dx
        x[i] = new_x
    return(norm)
```

10.4.8 Balance

The module `PSP_Balance` contains the structures and functions needed by the mass balance algorithm and the time step management. The new class **C3DBalance** is defined, which includes variables for the total water storage, the total water flow, the mass balance error (MBE) and the mass balance ratio (MBR). Three instances of this class are defined: one for the current time step, one for the previous time step and the final one for the whole simulation time (`allSimulation`). The next global variables are used by the mass balance algorithm. Among these variables, `maxCourant` has already been described. The first two functions are written to double (**doubleTimeStep**) or halve (**halveTimeStep**) the current time step. The two following functions **incMBRThreshold** and **decMBRThreshold** double or halve the parameter related to the tolerance threshold (`MBRThreshold`). In this case, the global variable `MBRMultiply` is defined and used to store how many times the threshold has been doubled with respect to the default value, indicated in `C3DParameters`. The threshold will be doubled only in the case of extreme conditions (when the system is very unstable), and it is brought back to the default value as soon as the system stabilizes.

The functions **initializeBalance** and **updateBalance** allows initialization of the class instances in **C3DBalance**. The function **getWaterStorage** returns the water storage for the whole domain, while the following functions return the sum (over the whole domain) of all the fluxes due to the boundary. This information is provided to the sink–source (**sumSinkSource**) and to the sum of all the components (**sumWaterFlow**). In the function **sumWaterFlow**, it is possible to store and print the values for all the fluxes in and out of the domain, to be used to compute the percentage of the mass balance error.

The function **computeBalanceError** computes the current mass balance error. The function `deltaStorage` calculates the variation of water storage with respect to the previous time step; it then computes the sum of the water flows in the current time step (`currentStep.waterFlow`) and defines the MBE as a difference between the two. Similarly to other numerical softwares (Simunek *et al.*, 2005), the percentage error is computed by dividing the MBE by the absolute value of the sum of the fluxes (`sumAbslouteFlow`). This ratio is defined as the mass balance ratio (MBR). To avoid values close to zero in the denominator, a threshold for a minimum flux is defined against which the flux can be compared. If the flux is smaller than this value, the minimum value is used. This value is proportional to the total area and the time step. The function **waterBalance** contains the mass balance algorithm. The function calls **computeBalanceError** and performs the test based on the MBR value.

The best scenario is when the error is less than the tolerance. In this case, the solution is accepted and the flag `True` is returned to the solver. If conditions are very stable (fewer than three approximations are needed for convergence, the Courant number is less than 0.5 and the error is less than the minimum threshold), the time step is doubled and the default value for the tolerance is used.

In intermediate conditions, the error has decreased with respect to the previous time step, but the mass balance is still above the tolerance. In this case, the new MBR is stored into the variable `bestMBR`, but the flag `False` is returned, indicating that the solution has not yet been accepted and another approximation is necessary. Finally, the worst condition is that the error is higher than that computed in the previous time step. This is the typical numerical instability condition where the system does not converge but oscillates back and forth between two states. In this case, the time step is reduced to the lowest possible value. If the reduction of the time step does not improve the convergence, then the tolerance is increased to overcome this condition of instability. When the system returns to stable conditions, the tolerance can be brought back to the default value and the time step can be increased.

The code for the `PSP_balance.py` is as follows:

```
#PSP_balance.py
from __future__ import print_function, division
from PSP_soil import *
import sys

class C3DBalance:
    waterStorage = NODATA
    waterFlow = NODATA
    MBE = NODATA
    MBR = NODATA

currentStep = C3DBalance()
previousStep = C3DBalance()
allSimulation = C3DBalance()
maxCourant = 0.0
totalTime = 0.0
MBRMultiply = 1.0
bestMBR = NODATA
forceExit = False

def doubleTimeStep():
    C3DParameters.currentDeltaT = min(C3DParameters.currentDeltaT *
                                2.0, C3DParameters.deltaT_max)
def halveTimeStep():
    C3DParameters.currentDeltaT = max(C3DParameters.currentDeltaT *
                                0.5, C3DParameters.deltaT_min)
def incMBRThreshold():
    global MBRMultiply
    MBRMultiply *= 2.0
    C3DParameters.MBRThreshold *= 2.0

def decMBRThreshold():
    global MBRMultiply
    if (MBRMultiply > 1.0):
        MBRMultiply *= 0.5
        C3DParameters.MBRThreshold *= 0.5

def initializeBalance():
    global totalTime
    totalTime = 0.0
    storage = getWaterStorage()
    currentStep.waterStorage = storage
    previousStep.waterStorage = storage
    allSimulation.waterStorage = storage
    previousStep.waterFlow = 0.0
    currentStep.waterFlow = 0.0
```

```
    allSimulation.waterFlow = 0.0
    currentStep.MBR = 0.0
    currentStep.MBE = 0.0
    allSimulation.MBE = 0

def updateBalance(deltaT):
    global totalTime
    totalTime += deltaT
    previousStep.waterStorage = currentStep.waterStorage
    previousStep.waterFlow = currentStep.waterFlow
    allSimulation.waterFlow += currentStep.waterFlow
    allSimulation.MBE += currentStep.MBE

def getWaterStorage():
    waterStorage = 0.0
    for i in range(C3DStructure.nrCells):
        if (C3DCells[i].isSurface):
            if (C3DCells[i].H > C3DCells[i].z):
                waterStorage += (C3DCells[i].H - C3DCells[i].z) *
                                C3DCells[i].area
        else:
            waterStorage += (getVolumetricWaterContent(i) *
                            C3DCells[i].volume)
    return waterStorage

def sumBoundaryFlow(deltaT):
    mySum = 0.0
    for i in range(C3DStructure.nrCells):
        if (C3DCells[i].boundary.type != BOUNDARY_NONE):
            if (C3DCells[i].boundary.flow != NODATA):
                mySum += C3DCells[i].boundary.flow * deltaT
    return (mySum)

def sumSinkSource(deltaT):
    mySum = 0.0
    for i in range(C3DStructure.nrCells):
        if (C3DCells[i].sinkSource != NODATA):
            mySum += C3DCells[i].sinkSource * deltaT
    return (mySum)

def sumWaterFlow(deltaT, isAbsoluteValue):
    mySum = 0.0
    for i in range(C3DStructure.nrCells):
        if (C3DCells[i].flow != NODATA):
            if isAbsoluteValue:
                mySum += fabs(C3DCells[i].flow * deltaT)
            else:
                mySum += C3DCells[i].flow * deltaT
    return (mySum)
```

```
def computeBalanceError(deltaT):
    currentStep.waterStorage = getWaterStorage()
    currentStep.waterFlow = sumWaterFlow(deltaT, False)
    deltaStorage = currentStep.waterStorage - previousStep.waterStorage
    currentStep.MBE = deltaStorage - currentStep.waterFlow

    sumFlow = sumWaterFlow(deltaT, True)
    minimumFlow = C3DStructure.totalArea * EPSILON_METER *
                  (deltaT / 3600.0)
    if (sumFlow < minimumFlow):
        currentStep.MBR = fabs(currentStep.MBE) / minimumFlow
    else:
        currentStep.MBR = fabs(currentStep.MBE) / sumFlow
    print ("Mass Balance Error [m^3]:", format(currentStep.MBE,".5f"))
    print ("Mass Balance Ratio:", format(currentStep.MBR,".5f"))

def waterBalance(deltaT, approximation):
    global bestMBR, forceExit
    if (approximation == 1): bestMBR = 100.0
    computeBalanceError(deltaT)
    isLastApprox = (approximation == C3DParameters.maxApproximationsNr)

    forceExit = False
    # case 1: error < tolerance
    if currentStep.MBR <= C3DParameters.MBRThreshold:
        updateBalance(deltaT)
        if ((approximation < 3) and (maxCourant < 0.5)
        and (currentStep.MBR < (C3DParameters.MBRThreshold * 0.5))):
                print("Good MBR!")
                if (deltaT >= (C3DParameters.deltaT_min * 10)):
                    decMBRThreshold()
                doubleTimeStep()
        return True
    # case 2: error improves
    if (currentStep.MBR < bestMBR):
        bestMBR = currentStep.MBR
        if isLastApprox:
            if (deltaT == C3DParameters.deltaT_min):
                updateBalance(deltaT)
                return True
            else:
                halveTimeStep()
                forceExit = True
                return False
        return False
    # case 3: error worsens
    if (deltaT > C3DParameters.deltaT_min):
        print("Solution is not convergent: time step decreased")
```

```
        halveTimeStep()
        forceExit = True
        return False
    else:
        print("Solution is not convergent: error tolerance increased")
        incMBRThreshold()
        return False
```

10.4.9   Criteria3D

The PSP_criteria3D module implements the interface between the model and the software (i.e. at a higher level). The implementation of this intermediate module allows the creation of different computational domains and boundaries, without changing the code that performs the numerical solution, but simply by calling the appropriate functions. In its original form (written in C), it is a library and the module PSP_criteria3D contains the functions that the library exports. In this case, the caller is in the **main** function and for simplicity the setting functions are written directly into the class **C3DParameters**. Also, the number of possible lateral links of each element was directly defined in this class, while in the complete version it is possible to manage elements with more links.

In PSP_criteria3D, there are functions to build the computational domain and to run the model. The function **memoryAllocation** computes the total number of cells, determined by the number of surface triangles (TIN) and the number of vertical layers. The function builds the list C3DCells made by an instance of the class **Ccell**, for each node of the domain. The other arrays and matrices needed for the compution are allocated by the function **setCriteria3DArrays**, already described in the solver.

The instances of the class **Ccell** at the beginning have values set to NODATA; therefore it is necessary that the functions define the geometrical and hydraulic properties to assign values to the cells. The first function is **setCellGeometry**, which assigns the geometrical properties for each cell i: the position (x,y,z) of the centre of the cell, the area of the triangle and the volume. The function **setCellProperties** assigns a flag needed to indicate if the cell is on the surface, in the subsurface or on the boundary. If it is not on the boundary, it will be called BOUNDARYNONE. Only for the cell on the boundary does the function setBoundaryProperties specify the properties of the corresponding class: the interfacial area and its slope. The function getCellDistance computes the distance between the two cells and is used by SetCellLink, which assigns the properties of the links on the ith node. The links can be of three types (UP, DOWN and LATERAL) and, as shown for the class **Clink**, their properties are the index of the neighbouring cell, the interfacial area and the distance between the centres of the two nodes.

The function **setMatricPotential** specifies the initial condition for the soil matric potential of each node i. The matric potential is summed with the gravitational potential to compute the total water potential (H), then the degree of saturation and the hydraulic conductivity are computed. If the cell is on the surface, the degree of saturation is always set equal to 1 and the hydraulic conductivity equal to the saturated hydraulic conductivity.

The **setRainfall** function assigns the sinkSource property [m³ s⁻¹] to all the surface cells, starting from the precipitation value [mm] for a given amount of time [s]. In the original model (Bittelli *et al.*, 2010), there are other possible sink–source terms (namely evaporation and transpiration) that are not implemented here. However, the user can include these terms by using the algorithm presented in the other chapters. The function **compute** runs the model for a certain time (timeLength) in seconds, typically equal to the time step defined in the precipitation input data. This value can be chosen by changing precFileName and obsPrecTimeLength in the class **C3DParameters**. The function **compute** calls the solver until the simulation timeLength is completed. If the solution for a time step is not accepted, then the function **restoreWater** is called and the time step brought back to the initial value. The code for the PSP_criteria3D.py is as follows:

```
#PSP_criteria3D.py
from __future__ import print_function, division
from PSP_dataStructures import *
from PSP_tin import distance3D
import PSP_solver as solver
import PSP_balance as balance
import PSP_soil as soil

def memoryAllocation(nrLayers, nrTriangles):
    C3DStructure.nrTriangles = nrTriangles
    C3DStructure.nrLayers = nrLayers
    nrCells = nrLayers * nrTriangles
    C3DStructure.nrCells = nrCells
    solver.setCriteria3DArrays(nrCells, C3DStructure.nrMaxLinks)
    for i in range(nrCells):
        C3DCells.append(Ccell())

def setCellGeometry(i, x, y, z, volume, area):
    C3DCells[i].x = x;
    C3DCells[i].y = y;
    C3DCells[i].z = z;
    C3DCells[i].volume = volume;
    C3DCells[i].area = area;

def setCellProperties(i, isSurface, boundaryType):
    C3DCells[i].isSurface = isSurface
    C3DCells[i].boundary.type = boundaryType

def setBoundaryProperties(i, area, slope):
    C3DCells[i].boundary.area = area
    C3DCells[i].boundary.slope = slope

def getCellDistance(i, j):
    v1 = [C3DCells[i].x, C3DCells[i].y, C3DCells[i].z]
```

```
    v2 = [C3DCells[j].x, C3DCells[j].y, C3DCells[j].z]
    return distance3D(v1, v2)


#-------------------------------------------------------------
# direction:        UP, DOWN, LATERAL
# interfaceArea     [m^2]
#-------------------------------------------------------------
def SetCellLink(i, linkIndex, direction, interfaceArea):
    if (direction == UP):
        C3DCells[i].upLink.index = linkIndex
        C3DCells[i].upLink.area = interfaceArea
        C3DCells[i].upLink.distance = fabs(C3DCells[i].z -
                                           C3DCells[linkIndex].z)
        return(OK)
    elif(direction == DOWN):
        C3DCells[i].downLink.index = linkIndex
        C3DCells[i].downLink.area = interfaceArea
        C3DCells[i].downLink.distance = fabs(C3DCells[i].z -
                                           C3DCells[linkIndex].z)
        return(OK)
    elif (direction == LATERAL):
        for j in range(C3DStructure.nrLateralLinks):
            if (C3DCells[i].lateralLink[j].index == NOLINK):
                C3DCells[i].lateralLink[j].index = linkIndex
                C3DCells[i].lateralLink[j].area = interfaceArea
                C3DCells[i].lateralLink[j].distance = getCellDistance
                                                    (i, linkIndex)

                return(OK)
    else:
        return(LINK_ERROR)

def setMatricPotential (i, signPsi):
    if (C3DCells[i].isSurface):
        C3DCells[i].H = C3DCells[i].z + max(signPsi, 0.0)
        C3DCells[i].Se = 1.
        C3DCells[i].k = soil.C3DSoil.Ks
    else:
        C3DCells[i].H = C3DCells[i].z + signPsi
        C3DCells[i].Se = soil.getDegreeOfSaturation(i)
        C3DCells[i].k = soil.getHydraulicConductivity(i)
    C3DCells[i].H0 = C3DCells[i].H
    return(OK)


#-------------------------------------------------------------
# rain              [mm]
# duration          [s]
#-------------------------------------------------------------
def setRainfall(rain, duration):
    rate = (rain * 0.001) / duration
```

```
    for i in range(C3DStructure.nrTriangles):
        if (C3DCells[i].isSurface):
            area = C3DCells[i].area
            C3DCells[i].sinkSource = rate * area

def restoreWater():
    for i in range(C3DStructure.nrCells):
        C3DCells[i].H = C3DCells[i].H0

def compute(timeLength):
    currentTime = 0
    while (currentTime < timeLength):
        residualTime = timeLength - currentTime
        acceptedStep = False
        while (not acceptedStep):
            deltaT = min(C3DParameters.currentDeltaT, residualTime)
            print ("\ntime step [s]: ", deltaT)
            print ("sink/source [m^3]:", balance.sumSinkSource(deltaT))

            acceptedStep = solver.computeStep(deltaT)
            if not acceptedStep: restoreWater()
        currentTime += deltaT
```

10.4.10   Main

The **main** function is the main body of the program. The code imports the essential modules for the model computation, the TIN management and the interface with three-dimensional visualization.

The function begins by building the TIN using the output files produced by the triangulation algorithm presented in Chapter 9. Specifically, the program uses the three files `vertices.csv`, `triangles.csv` and `neighbours.csv`, containing information about the triangle vertices, properties and neighbours.

The file `vertices.csv` is made of three columns, corresponding to the coordinates (x, y, z) of each vertex of the triangles. The index of each vertex will be equal to the row number (with first row equal to zero). The file `triangles.csv` is made of three columns, corresponding to the three vertex indices of each triangle. As for the vertices, the index of each triangle will be equal to the row number. The file `neighbours.csv` is made of three columns, corresponding to the three indices of the triangles neighbouring each triangle. If the triangle has minus three neighbours, the NOLINK value (−1) will be used.

The following instructions read the soil data from the `soil.txt` file. This file contains the hydraulic properties parameters. The layer properties (depth and thickness) are then returned by the function **setLayers**. It uses the maximum depth of the soil profile, the minimum and maximum thicknesses of the computational layers, and the geometric factor. Note that, also for this code, the vertical grid can be linear or geometric.

The geometric factor sets up the geometric grid. The code then calls the function **memoryAllocation** of the module `PSP_Criteria3D.py` to allocate the memory space for all the nodes and elements of the system.

The nested loops following the instruction `print("set initial conditions...")` loop over all the triangles and the soil layers and they comprise the builder used to call all the core functions of the program. At the beginning for each node, it computes or assigns geometrical properties, such as the position of the triangle centroid, the elevation and the volume. It then passes these properties to the function **setCellGeometry**. The function then sets up the boundary of the simulated domain, the links and the initial conditions. To provide atmospheric boundary conditions of prescribed flux, the **readDataFile** reads a `precipitation.txt` file with hourly precipitation data. Finally, it calls the function **computeCriteria3D**, which runs the program for the weather data, and it visualizes the output during simulation.

```
from __future__ import print_function, division
from PSP_criteria3D import *
from PSP_readDataFile import readDataFile
import PSP_visual3D as visual3D
import PSP_tin as tin
import numpy as np

def main():
    print ("Load TIN...")
    vertexList, isFileOk = readDataFile("data\\vertices.csv", 0, ",",
                                        False)
    if (not isFileOk): return
    triangleList, isFileOk = readDataFile("data\\triangles.csv", 0,
                                          ",", False)
    if (not isFileOk): return
    nrTriangles = len(triangleList)
    neighbourList, isFileOk = readDataFile("data\\neighbours.csv", 0,
                                           ",", False)
    if (not isFileOk): return
    print ("Nr. of triangles:", nrTriangles)

    v = np.zeros((3, 3), float)
    for i in range(nrTriangles):
        for j in range(3):
            v[j] = vertexList[int(triangleList[i,j])]
        tin.C3DTIN.append(tin.Ctriangle(v))
        C3DStructure.totalArea += tin.C3DTIN[i].area
    print ("Total area [m^2]:", C3DStructure.totalArea)

    tin.header = tin.getHeader(tin.C3DTIN)
```

```
print ("Set boundary...")
for i in range(nrTriangles):
    tin.C3DTIN[i].isBoundary = False
    if (neighbourList[i,2] == NOLINK):
        tin.C3DTIN[i].isBoundary = True
        tin.getBoundaryProperties(tin.C3DTIN, i, neighbourList[i])

print ("Load soil...")
soil.C3DSoil = soil.readHorizon("data\\soil.txt", 1)
if (C3DParameters.computeOnlySurface):
    totalDepth = 0
else:
    totalDepth = soil.C3DSoil.lowerDepth
print("Soil depth [m]:", totalDepth)

nrLayers, soil.depth, soil.thickness = soil.setLayers(totalDepth,
        C3DParameters.minThickness, C3DParameters.maxThickness,
        C3DParameters.geometricFactor)
print("Nr. of layers:", nrLayers)
print("Depth:", soil.depth)

# Initialize memory
memoryAllocation(nrLayers, nrTriangles)
print("Nr. of cells: ", C3DStructure.nrCells)

print("Set cell properties...")
for i in range(nrTriangles):
    for layer in range(nrLayers):
        [x, y, z] = tin.C3DTIN[i].centroid
        index = i + nrTriangles * layer
        elevation = z - soil.depth[layer]
        volume = float(tin.C3DTIN[i].area * soil.thickness[layer])
        setCellGeometry(index, x, y,
                            elevation, volume, tin.C3DTIN[i].area)
        if (layer == 0):
            # surface
            if tin.C3DTIN[i].isBoundary:
                setCellProperties(index, True, BOUNDARY_RUNOFF)
                setBoundaryProperties(index,
                            tin.C3DTIN[i].boundarySide,
                            tin.C3DTIN[i].boundarySlope)
            else:
                setCellProperties(index, True, BOUNDARY_NONE)
            setMatricPotential(index, 0.0)

        elif (layer == (nrLayers-1)):
            # last layer
            if C3DParameters.isFreeDrainage:
```

```
                            setCellProperties(index, False,
                                        BOUNDARY_FREEDRAINAGE)
                    else:
                        setCellProperties(index, False, BOUNDARY_NONE)
                    setMatricPotential(index,
                            C3DParameters.initialWaterPotential)

            else:
                if tin.C3DTIN[i].isBoundary:
                    setCellProperties(index, False,
                                BOUNDARY_FREELATERALDRAINAGE)
                    setBoundaryProperties(index,
                                    tin.C3DTIN[i].boundarySide
                                    * soil.thickness[layer],
                                    tin.C3DTIN[i].boundarySlope)
                else:
                    setCellProperties(index, False, BOUNDARY_NONE)
                setMatricPotential(index,
                            C3DParameters.initialWaterPotential)

    print("Set links...")
    for i in range(nrTriangles):
        # UP
        for layer in range(1, nrLayers):
            exchangeArea = tin.C3DTIN[i].area
            index = nrTriangles * layer + i
            linkIndex = index - nrTriangles
            SetCellLink(index, linkIndex, UP, exchangeArea)
        # LATERAL
        for j in range(len(neighbourList[i])):
            neighbour = int(neighbourList[i,j])
            if (neighbour != NOLINK):
                linkSide = tin.getAdjacentSide(i, neighbour,
                                        vertexList, triangleList)
                for layer in range(nrLayers):
                    if (layer == 0):
                        #surface: boundary length [m]
                        exchangeArea = linkSide
                    else:
                        #sub-surface: boundary area [m2]
                        exchangeArea = soil.thickness[layer] * linkSide
                    index = nrTriangles * layer + i
                    linkIndex = nrTriangles * layer + neighbour
                    SetCellLink(index, linkIndex, LATERAL, exchangeArea)
        # DOWN
        for layer in range(nrLayers-1):
            exchangeArea = tin.C3DTIN[i].area
            index = nrTriangles * layer + i
```

```
            linkIndex = index + nrTriangles
            SetCellLink(index, linkIndex, DOWN, exchangeArea)
    print("Initial water potential [m]:",
            C3DParameters.initialWaterPotential)
    balance.initializeBalance()
    print("Initial water storage [m^3]:",
            balance.currentStep.waterStorage)

    print("Read precipitation data...")
    data, isFileOk = readDataFile(C3DParameters.precFileName, 1,
                                  "\t", False)
    if (not isFileOk):
        print("Error! Wrong precipitation file.")
        return
    prec = data[:,1]
    nrObsPrec = len(prec)
    timeLength = C3DParameters.obsPrecTimeLength * 60   # [s]
    print("Time lenght [s]:", timeLength)
    print("Total simulation time [s]:", nrObsPrec * timeLength)

    visual3D.initialize(1280)
    visual3D.redraw(True)

    visual3D.isPause = True
    print("\nPress key (on a window):")
    print("'l' to load a saved state")
    print("'r' to start with initial conditions")
    while visual3D.isPause:
        visual3D.visual.wait()

    # main cycle
    for i in range(nrObsPrec):
        visual3D.currentPrec = prec[i] / timeLength * 3600
        setRainfall(prec[i], timeLength)
        compute(timeLength)

    visual3D.redraw(False)
    print ("\nEnd simulation.")
main()
```

10.5 Simulation

Two simulations were performed for two different catchments: the Fontanafredda catchment described in Chapter 9 as an example for the triangulation program and the Troy catchment used by Bittelli *et al.* (2010) for experimental corroboration of the model. The latter is a smaller catchment, so the computational time is shorter. Python is a bytecode-interpreted language and therefore the computational speed is slower than for machine-compiled languages like C. For this reason, we also used a Python extension called Cython that allows compilation of Python modules using variables and functions of the C language. We have selected the most computationally expensive functions of the solver module and translated these functions into the Cython language in the module `PSP_solverC.pyx` (the extension `pyx` is specific for Cython). C-arrays are also used instead of the arrays of numpy. The code was compiled using the module `PSP_setup.py`, which specifies which modules have to be 'cythonized', thus obtaining the compiled file `PSP_solverC.pyd`.

On Windows machines, since the `.pyd` file is already compiled, it is sufficient to run the module `main.py` of the `PSP_C_criteria3D` project, instead of `PSP_criteria3D`. On Linux or Macintosh machines, after downloading and installing Cython, it is necessary to compile the new module. This requires a C compiler (`gcc` or Xcode). This can be done by selecting the directory where the code is located in the terminal and typing the instruction

```
python PSP_setup.py build_ext --inplace
```

The compiler creates a `.c` and a `.pyd` in the same directory. Finally, the module can be executed by running the module `main.py`. This version has a runtime that is around one-tenth of the runtime of the original Python code.

10.6   Visualization and Results

10.6.1   Troy Catchment

Figure 10.3 shows the results of the simulation for the Troy catchment. The plot at the top depicts the surface water, while the plots below show the degree of saturation for three soil layers (–2, –5 and –20 cm). The legend at the right shows the different shades for the degree of saturation, as well as current states of the numerical solution. From top to bottom the following information is printed. First, the degree of saturation of the soil layer and the depth of the soil layer are shown. Below this box is a box that allows the user to visualize different soil layers by clicking the letter 'u' (up) or 'd' (down) on the keyboard. On doing this, VisualPython will display the corresponding soil layer and its degree of saturation. The third box is used to pause the simulation by clicking the letter 'p' (pause) and change the colour scale 'c' (colourscale). If the option 'c' is selected, the program prompts the choice of maximum and minimum values of the degree of saturation. The prompt is

```
Set min. value (Degree of saturation): 0.5
Set max. value (Degree of saturation): 1
```

For system analysis purposes, it is often convenient to start the simulation from specific initial conditions of the system. We wrote an option in the program that saves the

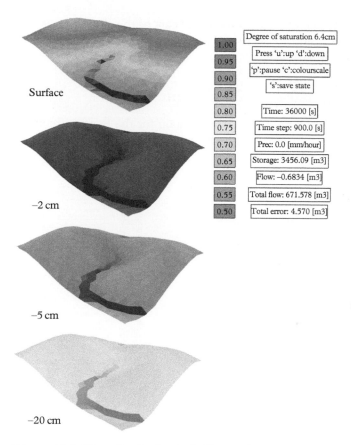

Surface

−2 cm

−5 cm

−20 cm

1.00	Degree of saturation 6.4cm
0.95	Press 'u':up 'd':down
0.90	'p':pause 'c':colourscale
0.85	's':save state
0.80	Time: 36000 [s]
0.75	Time step: 900.0 [s]
0.70	Prec: 0.0 [mm/hour]
0.65	Storage: 3456.09 [m3]
0.60	Flow: −0.6834 [m3]
0.55	Total flow: 671.578 [m3]
0.50	Total error: 4.570 [m3]

Figure 10.3 *Water distribution in the Troy catchment. The top plot depicts the distribution of surface water, while the other plots show the soil water content distribution at three different depths. The water content is shown as degree of saturation.*

condition of saturation of the system at any particular time, decided by the user. The box with the option 's' (save) allows saving of the conditions. The saved conditions are then reloaded when a new simulation is performed by typing the letter 'l' (load) at the beginning of the simulation. The saved files are comma-separated files .csv, where the water potential values H are saved for each triangle (rows) and for each layer (columns). The first row contains the layer depths. A state can be loaded only if it was generated by the same TIN, since the numbers of triangles and layers are specific for a given simulation domain. The remaining boxes indicate simulation parameter such as the elapsed time, the time step, the amount of precipitation, the total water storage in the catchment, the surface flow, the total flow out of the boundary domains and the total mass balance error.

The accumulation of water in the lower part of the catchment is clearly visible, both as accumulation of surface water and as a higher degree of saturation. Redistribution of water in the headwaters is also visible.

10.6.2 Fontanafredda Catchment

Figure 10.4 shows the results of the simulation for the Fontanafredda catchment. Since the Fontanafredda catchment is much larger than the Troy catchment, the Cython version of the program was used. In general, for catchments larger than a few hectares,

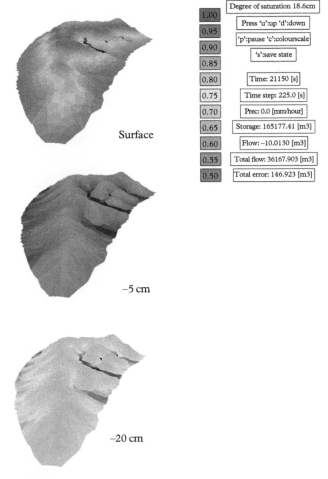

Figure 10.4 *Water distribution in the Fontanafredda catchment. The top plot depicts the distribution of surface water, while the other plots show the soil water content distribution at three different depths. The water content is shown as degree of saturation.*

the user should use the Cython version of the program to speed up the computation. The surface water distribution (top plot) exhibited higher accumulation in an area at higher slopes. The other channels determined the formation of a smaller amount of run-off water. The two plots below this depict the soil water content at –5 and –20 cm. The distribution again represents the topographical distribution of soil water in the area at higher slopes.

10.7 EXERCISES

10.1. Collect (or create) a precipitation data file on your own and run the simulation to investigate the hydrological response of the Troy catchment.

10.2. Run a simulation using a DTM that you collected. DTMs can be downloaded for free from websites of governmental agencies. Use a freeware GIS software to select a small catchment from the DTM and save the resulting grid in the ESRI binary format `.flt`. Load the `.flt` file with the `PSP_triangulation` program to create the corresponding TIN as input for Criteria3D.

10.3. Change the parameter values for the hydraulic properties in the file `soil.txt`. Run the simulation for a sandy and a clay soil. Compare the results.

10.4. Compile the program using Cython and run it for the Fontanafredda catchment. Run the program with both Python and Cython and record the computational time for the example simulation.

11

Evaporation

As water is being redistributed in the soil, it is also evaporating from the soil surface. The amount of water that evaporates depends on soil properties and environmental conditions. Under some conditions, most of the precipitation received at a soil surface may be lost by evaporation. Even when the soil is covered by vegetation, evaporation is typically at least 10 % of evapotranspiration. These observations indicate that evaporation at the soil surface is always an important component of the water balance. In dry-land farming and desert soils, evaporation can be the largest loss component in the water balance.

11.1   General Concepts

If the surface of a soil column were wet, and water loss were measured as a function of time, the evaporation rate would stay nearly constant for some time and then suddenly decrease, as shown in Fig. 11.1. Two stages of drying are often identified in classical drying theory.[1] During first stage, the soil surface is wet, and the evaporation rate is determined by the vapour concentration difference between the surface and the air and by the boundary-layer resistance of the air above the soil surface. When the soil dries sufficiently that water cannot be supplied to the surface fast enough to meet the evaporative demand, the soil surface dries and the evaporation rate is reduced. This reduction is caused by the increased diffusion resistance of the dry soil at the surface. As the depth of the dry layer increases, the evaporation rate decreases.

Following first-stage drying, evaporation can be analysed either as a vapour diffusion problem or as a liquid flux problem. The depth of the dry layer that forms is determined by the rate at which liquid water can be supplied to the drying front. If the liquid water supply is too slow, the dry layer deepens, causing an increase in vapour diffusion resistance. This reduces the evaporation rate until supply and demand are brought back into balance.

---

[1] According to Or *et al.* (2013), the first stage is not constant and, under high atmospheric demand ($> 5\,\mathrm{mm\,d^{-1}}$), a declining evaporation rate may be observed. The authors reported that the soil water retention curve contains information for predicting the drying front depth and mass loss at the end of stage one. In their analysis, they maintain that the diffusion process is dominant over advection in generating evaporation fluxes from soil surfaces.

Soil Physics with Python. First Edition. Marco Bittelli, Gaylon S. Campbell and Fausto Tomei.
© Marco Bittelli, Gaylon S. Campbell and Fausto Tomei 2015. Published in 2015 by Oxford University Press.

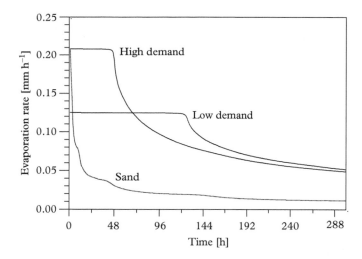

Figure 11.1 *Evaporation rate for loam at high and low evaporative demand and for sand at high evaporative demand.*

Because of these self-adjusting features of the system, some aspects of the evaporation process are relatively easy to analyse. A complete analysis, however, must take into account both liquid and vapour flux and must consider effects of temperature gradients on water movement. Isothermal vapour flux will be considered first, after which thermally induced flux will be included. In Chapter 12, an implementation of coupled heat, liquid and vapour water flux will be presented to provide a complete analysis of soil evaporation, including the energy balance computation to quantify the contribution of latent heat fluxes.

11.2   Simultaneous Transport of Liquid and Vapour in Isothermal Soil

The flux density of water vapour is described by Fick's law:

$$f_v = -D_v \frac{dc_v}{dz} \tag{11.1}$$

where D_v is the vapour diffusivity in soil [m^2 s^{-1}] and c_v is the soil vapour concentration [g m^{-3}]. The diffusivity in soil is computed from eqns (3.12)–(3.14). We will assume $\zeta = 0.66\phi_g$. The vapour concentration can be written as

$$c_v = h_r c_v' \tag{11.2}$$

where h_r is the relative humidity and c_v' is the saturation vapour concentration at soil temperature. If the soil is isothermal, then

$$\frac{dc_v}{dz} = c_v' \frac{dh_r}{dz} \tag{11.3}$$

The relative humidity, in turn, is a function of the soil water potential, eqn (5.44):

$$h_r = \exp\left(\frac{M_w \psi}{RT_K}\right) \tag{11.4}$$

where ψ is the soil water potential [J kg^{-1}], R is the gas constant (8.3143 J mol^{-1} K^{-1}), T_K is the temperature [K] and M_w is the molecular mass of water (0.018 kg mol^{-1}). Using the chain rule,

$$\frac{dh_r}{dz} = \frac{dh_r}{d\psi}\frac{d\psi}{dz} \tag{11.5}$$

Using eqn (11.4),

$$\frac{dh_r}{d\psi} = \frac{h_r M_w}{RT_K} \tag{11.6}$$

Using eqns (11.1)–(11.6), we can write the flux of vapour driven by a water potential gradient:

$$f_v = -K_v \frac{d\psi}{dz} \tag{11.7}$$

where the combination of terms leads to the definition of vapour conductivity as

$$K_v = \frac{D_v c_v' h_r M_w}{RT_K} \tag{11.8}$$

The vapour conductivity can be added to the liquid conductivity to calculate the total water flux (liquid + vapour). The vapour conductivities for sand, loam and clay soils calculated using eqn (11.8), are shown in Fig. 11.2.

The initial increase in K_v with decreased potential results from the increase in D_v as pores empty and ϕ_g increases. The subsequent decrease is due to the decrease in h_r with water potential. It is interesting to note that vapour transport is negligible compared with liquid transport until soils become quite dry. Equation (11.8) indicates that K_v has about the same temperature dependence as c_v', which is around 5 % per degree at 20 °C. The total flux in a soil is the sum of the liquid and vapour fluxes, and both are functions of the matric potential gradient. Osmotic potential gradients are somewhat more difficult to treat, since they are as effective as matric potential gradients in driving vapour flux but have little effect on liquid flux. This fact must be taken into account if one is modelling vapour transport in soils that have high salt concentrations. The effect of the gravitational potential gradient on vapour flux is negligibly small.

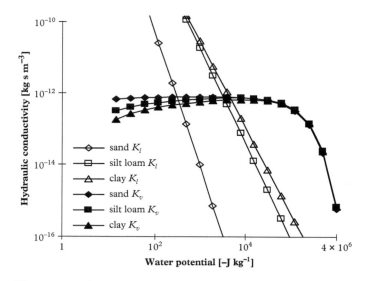

Figure 11.2 *Comparison of liquid and vapour conductivities for typical sand, silt loam and clay soils. Note that both scales are logarithmic.*

11.3 Modelling evaporation

Incorporation of the ideas discussed so far in this chapter into the numerical models for water flux is now the next step for obtaining a coupled numerical model. Here we present a code for solution of water and vapour flux. For simplicity, the code is not yet coupled with a heat flux model.

It is therefore useful to consider the information needed from the model before deciding how far to go in incorporating vapour flux. The simplest option, and one that is often used in modelling soil water flow, is to neglect vapour flux except at the soil surface. The evaporation during second-stage drying is then controlled by the relative humidity at the evaporating surface and liquid flux in the soil. If we assume that the potential evaporation rate is known, then the actual evaporation rate at any time is

$$E = E_p \frac{h_s - h_a}{1 - h_a} \tag{11.9}$$

where E_p is the potential evaporation rate, h_s is the relative humidity at the soil surface (eqn (11.4)) and h_a is the atmospheric relative humidity. The value given for E_p is only important during first-stage drying (when $E = E_p$). The value chosen for h_a is relatively unimportant, since, after the first stage, E is determined by liquid flux to the surface and not by the values chosen for eqn (11.9).

This approach can give quite reasonable estimates of the evaporation rate during the first stages of drying, when most of the water is lost. However, it will not correctly predict the water content or water potential profile in the soil. For isothermal conditions, vapour fluxes in each element can be computed from

$$f_{vi} = -D_v \bar{c}_v' \frac{h_{i+1} - h_i}{z_{i+1} - z_i} \tag{11.10}$$

11.4   Numerical Implementation

The following is an example program for simulating vapour flux in soil. The project is called PSP_evaporation and comprises seven files:

1. main.py
2. PSP_grid.py
3. PSP_readDataFile.py
4. PSP_soil.py
5. PSP_vapor1D.py
6. PSP_ThomasAlgorithm.py
7. soil.txt

11.4.1   Main

The program PSP_evaporation_1D.py is a modification of the program PSP_infiltration_1D to allow for isothermal vapour flow within the profile and evaporation at the soil surface. The main file main.py contains the call to the other file PSP_vapor1D.py, where the solvers for the numerical solution are implemented. This file (described below) calls the file used to build the grid (PSP_grid.py) and the files where the soil data structure is implemented (PSP_soil.py). Similarly to the code for solution of water flux, the soil hydraulic parameters are saved in the file soil.txt; there are the parameters for the Campbell, van Genuchten and Ippisch–van Genuchten equations. In this solution, only the Campbell equation is used; however, the code can be modified to utilize the other parameterizations.

The Campbell formulation is selected by choosing funcType = 1, while the numerical scheme (in this case Newton–Raphson) is selected by choosing solver = 2. The initial water content of the soil profile is set by thetaIni = 0.22 [m^3 m^{-3}] and the lower boundary condition is set to free drainage by setting the flag isFreeDrainage = True. The total simulation time is set by the variable simulationLenght = 24 [h].

The upper boundary condition is a constant water potential in the atmosphere. The model also allows other upper boundary conditions to be set up, such as atmospheric boundaries, which are simulated by applying a prescribed flux. The implementation of atmospheric boundary conditions will be presented in Chapter 12. The lower boundary condition can be chosen as free drainage or constant water potential. The solution is based on the Newton–Raphson algorithm, with the matric potential solution.

The code is as follows:

```
#PSP_evaporation
from __future__ import print_function, division
import matplotlib.pyplot as plt
import PSP_vapor1D as vap

def main():
    isSuccess, soil = vap.readSoil("soil.txt")
    if not isSuccess:
        print("warning: wrong soil file.")
        return

    print ('Water Retention Curve: Campbell')
    funcType = 1

    print ('Solver: Matric Potential with Newton-Raphson')
    solver = 2

    #Initial conditions
    thetaIni = 0.22
    isFreeDrainage = True

    simulationLenght = 24
    vap.initializeWater(funcType, soil, thetaIni, solver)
    endTime = simulationLenght * 3600
    maxTimeStep = 3600
    dt = 600
    time = 0
    sumEvaporation = 0

    plt.ion()
    f, myPlot = plt.subplots(2, figsize=(8, 8), dpi=80)
    myPlot[0].set_xlim(0, soil.thetaS)

    myPlot[1].set_xlim(0, simulationLenght)
    myPlot[1].set_ylim(0, 0.25)
    myPlot[1].set_ylabel("Evaporation Rate [mm h$^{-1}$]")
    myPlot[1].set_xlabel("Time [h]")

    while (time < endTime):
        dt = min(dt, endTime - time)
        (success, iterations, evaporationFlux =
         vap.NewtonRapsonMP(funcType, soil, dt, isFreeDrainage))
```

```
        if success:
            for i in range(vap.n+1):
                vap.oldTheta[i] = vap.theta[i]
                vap.oldvapor[i] = vap.vapor[i]
            sumEvaporation += evaporationFlux * dt
            time += dt

            print("time =", int(time), "\tdt =", int(dt), "\tIter. =",
                iterations, "\tsumEvap:", format(sumEvaporation, '.3f'))

            myPlot[0].clear()
            myPlot[0].set_xlim(0, soil.thetaS)
            myPlot[0].set_xlabel("Water content [m$^3$ m$^{-3}$]")
            myPlot[0].set_ylabel("Depth [m]")
            myPlot[0].plot(vap.theta, -vap.z, 'yo')
            myPlot[1].plot(time/3600., evaporationFlux*3600., 'ro')
            plt.draw()

            if (float(iterations / vap.maxNrIterations) < 0.1):
                    dt = min(dt*2.0, maxTimeStep)

        else:
            print("time =", int(time), "\tdt =", int(dt),
                    "\tIter. =", iterations, "No convergence")

            for i in range(vap.n+1):
                vap.theta[i] = vap.oldTheta[i]
                vap.vapor[i] = vap.oldvapor[i]
                if (solver == vap.NEWTON_RAPHSON_MFP):
                    vap.psi[i] = vap.MFPFromTheta(soil, vap.theta[i])
                else:
                    vap.psi[i] = vap.waterPotential(funcType, soil,
                                                    vap.theta[i])
            dt = max(dt / 2.0, 1)

    plt.ioff()
    plt.show()
main()
```

11.4.2 Soil

The module `PSP_soil.py` is similar to that presented in Chapter 8. In this module, the equations necessary to compute the vapour conductivity, the vapour concentration gradients from the matric potential gradients, Kelvin's equation (to obtain relative humidity from water potential) and the equation for the evaporation flux are implemented. The program is as follows:

```
#PSP_soil
from __future__ import division
from PSP_readDataFile import *
from math import sqrt, log, exp

g = 9.8065
waterDensity = 1000.
area = 1
mw = 0.018
R = 8.31
T = 293
dv = 0.000024
vp = 0.017
E_p = 0.21/3600.
h_a = 0.5

NODATA = -9999.

CAMPBELL = 1
RESTRICTED_VG = 2
IPPISCH_VG = 3
VAN_GENUCHTEN = 4

CELL_CENT_FIN_VOL = 1
NEWTON_RAPHSON_MP = 2
NEWTON_RAPHSON_MFP = 3

LOGARITHMIC = 0
HARMONIC = 1
GEOMETRIC = 2

class Csoil:
    upperDepth = NODATA
    lowerDepth = NODATA
    Campbell_he = NODATA
    Campbell_b = NODATA
    CampbellMFP_he = NODATA
    VG_alpha = NODATA
    VG_n = NODATA
    VG_m = NODATA
    VG_he = NODATA
    VG_alpha_mod = NODATA
    VG_n_mod = NODATA
    VG_m_mod = NODATA
    VG_Sc = NODATA
    VG_thetaR = NODATA
    Mualem_L = NODATA
```

```
        thetaS = NODATA
        Ks = NODATA


def readSoil(soilFileName):
    soil = Csoil()
    A, isFileOk = readDataFile(soilFileName, 1, ',', False)
    if ((not isFileOk) or (len(A[0]) < 12)):
        return False, soil

    soil.upperDepth = A[0,0]
    soil.lowerDepth = A[0,1]
    soil.Campbell_he = A[0,2]
    soil.Campbell_b = A[0,3]
    soil.Campbell_n = 2.0 + (3.0 / soil.Campbell_b)
    soil.VG_he = A[0,4]
    soil.VG_alpha = A[0,5]
    soil.VG_n = A[0,6]
    soil.VG_m =  1. - (1. / soil.VG_n)
    soil.VG_alpha_mod = A[0,7]
    soil.VG_n_mod = A[0,8]
    soil.VG_m_mod =  1. - (1. / soil.VG_n_mod)
    soil.VG_Sc = ((1. + (soil.VG_alpha_mod * abs(soil.VG_he))**
                           soil.VG_n_mod)**(-soil.VG_m_mod))
    soil.VG_thetaR = A[0,9]
    soil.thetaS = A[0,10]
    soil.Ks = A[0,11]
    soil.Mualem_L = 0.5
    soil.CampbellMFP_he = soil.Ks * soil.Campbell_he /
                           (1.0 - soil.Campbell_n)
    return True, soil


def airEntryPotential(funcType, soil):
    if (funcType == CAMPBELL):
        return(soil.Campbell_he)
    elif (funcType == IPPISCH_VG):
        return(soil.VG_he)
    elif (funcType == RESTRICTED_VG):
        return(0)
    else:
        return(NODATA)


def waterPotential(funcType, soil, theta):
    psi = NODATA
    Se = SeFromTheta(funcType, soil, theta)
    if (funcType == RESTRICTED_VG):
        psi = -(1./soil.VG_alpha)*((1./Se)**(1./soil.VG_m) - 1.)**
                                       (1./soil.VG_n)
```

```
    elif (funcType == IPPISCH_VG):
        psi = -((1./soil.VG_alpha_mod)*((1./(Se*soil.VG_Sc))**
                    (1./soil.VG_m_mod)-1.)**(1./soil.VG_n_mod))
    elif (funcType == CAMPBELL):
        psi = soil.Campbell_he * Se**(-soil.Campbell_b)
    return(psi)


def SeFromTheta(funcType, soil, theta):
    if (theta >= soil.thetaS): return(1.)
    if (funcType == CAMPBELL):
        Se = theta / soil.thetaS
    else:
        Se = (theta - soil.VG_thetaR) / (soil.thetaS - soil.VG_thetaR)
    return (Se)


def thetaFromSe(funcType, soil, Se):
    if (funcType == RESTRICTED_VG) or (funcType == IPPISCH_VG):
        theta = (Se * (soil.thetaS - soil.VG_thetaR) + soil.VG_thetaR)
    elif (funcType == CAMPBELL):
        return(Se * soil.thetaS)
    return(theta)


def degreeOfSaturation(funcType, soil, psi):
    if (psi >= 0.): return(1.)
    Se = NODATA
    if (funcType == IPPISCH_VG):
        if (psi >= soil.VG_he): Se = 1.
        else:
            Se = (1./soil.VG_Sc) * pow(1.+pow(soil.VG_alpha_mod
                            * abs(psi), soil.VG_n_mod), -soil.VG_m_mod)
    elif (funcType == RESTRICTED_VG):
        Se = 1 / pow(1 + pow(soil.VG_alpha * abs(psi), soil.VG_n),
                            soil.VG_m)
    elif (funcType == CAMPBELL):
        if psi >= soil.Campbell_he: Se = 1.
        else: Se = pow(psi / soil.Campbell_he, -1. / soil.Campbell_b)
    return(Se)


def thetaFromPsi(funcType, soil, psi):
    Se = degreeOfSaturation(funcType, soil, psi)
    theta = thetaFromSe(funcType, soil, Se)
    return(theta)


def hydraulicConductivityFromTheta(funcType, soil, theta):
    k = NODATA
    if (funcType == RESTRICTED_VG):
        Se = SeFromTheta(funcType, soil, theta)
        k = (soil.Ks * pow(Se, soil.Mualem_L) *
```

```
            (1. -pow(1. -pow(Se, 1./soil.VG_m), soil.VG_m))**2)
    elif (funcType == IPPISCH_VG):
        Se = SeFromTheta(funcType, soil, theta)
        num   = 1. - pow(1. - pow(Se * soil.VG_Sc, 1./ soil.VG_m_mod),
                                soil.VG_m_mod);
        denom = 1. - pow(1. - pow(soil.VG_Sc, 1./ soil.VG_m_mod),
                        soil.VG_m_mod);
        k = soil.Ks * pow(Se, soil.Mualem_L) * pow((num / denom), 2.)
    elif (funcType == CAMPBELL):
        psi = waterPotential(funcType, soil, theta)
        k = soil.Ks * (soil.Campbell_he / psi)**soil.Campbell_n
    return(k)

def dTheta_dPsi(funcType, soil, psi):
    airEntry = airEntryPotential(funcType, soil)
    if (psi > airEntry): return 0.0

    if (funcType == RESTRICTED_VG):
        dSe_dpsi = soil.VG_alpha * soil.VG_n * (soil.VG_m
                    * pow(1. + pow(soil.VG_alpha * abs(psi), soil.VG_n),
                    -(soil.VG_m + 1.)) *
                    pow(soil.VG_alpha * abs(psi), soil.VG_n - 1.))
        return dSe_dpsi * (soil.thetaS - soil.VG_thetaR)
    elif (funcType == IPPISCH_VG):
        dSe_dpsi = soil.VG_alpha_mod * soil.VG_n_mod * (soil.VG_m_mod
                    * pow(1. + pow(soil.VG_alpha_mod * abs(psi),
                    soil.VG_n_mod), -(soil.VG_m_mod + 1.)) *
                    pow(soil.VG_alpha_mod * abs(psi), soil.VG_n_mod - 1.))
        dSe_dpsi *= (1. / soil.VG_Sc)
        return dSe_dpsi * (soil.thetaS - soil.VG_thetaR)
    elif (funcType == CAMPBELL):
        theta = soil.thetaS * degreeOfSaturation(funcType, soil, psi)
        return -theta / (soil.Campbell_b * psi)

def MFPFromTheta(soil, theta):
    return (soil.CampbellMFP_he * (theta / soil.thetaS)**
            (soil.Campbell_b + 3.0))

def MFPFromPsi(soil, psi):
    return (soil.CampbellMFP_he * (psi / soil.Campbell_he)**
            (1.0 - soil.Campbell_n))

def thetaFromMFP(soil, MFP):
    if (MFP > soil.CampbellMFP_he):
        return(soil.thetaS)
    else:
        return(soil.thetaS * (MFP / soil.CampbellMFP_he)**(1.0/
            (soil.Campbell_b + 3.0)))
```

```
def hydraulicConductivityFromMFP(soil, MFP):
    b3 = (2.0 * soil.Campbell_b + 3.0) / (soil.Campbell_b + 3.0)
    k = soil.Ks * (MFP / soil.CampbellMFP_he)**b3
    return(k)

def dTheta_dH(funcType, soil, H0, H1, z):
    psi0 = H0 + g*z
    psi1 = H1 + g*z
    if (abs(psi1-psi0) < 1E-5):
        return dTheta_dPsi(funcType, soil, psi0)
    else:
        theta0 = thetaFromPsi(funcType, soil, psi0)
        theta1 = thetaFromPsi(funcType, soil, psi1)
        return (theta1 - theta0) / (psi1 - psi0)

 def meanK(meanType, k1, k2):
    if (meanType == LOGARITHMIC):
        if (k1 != k2):
            k = (k1-k2) / log(k1/k2)
        else:
            k = k1
    elif (meanType == HARMONIC):
        k = 2.0 / (1.0 / k1 + 1.0 / k2)
    elif (meanType == GEOMETRIC):
        k = sqrt(k1 * k2)
    return k

def hydraulicConductivityFromPsi(funcType, soil, psi):
    if (funcType == RESTRICTED_VG):
        psi = abs(psi)
        num = (1. - pow(soil.VG_alpha * psi, soil.VG_m * soil.VG_n)
               *pow(1. + pow(soil.VG_alpha*psi, soil.VG_n),
                    -soil.VG_m))**2
        denom = (pow(1. + pow(soil.VG_alpha*psi, soil.VG_n),
                     soil.VG_m * soil.Mualem_L))
        k = soil.Ks * (num / denom)
    elif (funcType == IPPISCH_VG):
        k = NODATA
    elif (funcType == CAMPBELL):
        k = soil.Ks * (soil.Campbell_he / psi)**soil.Campbell_n
    return(k)

def vaporConductivityFromPsiTheta(soil, psi, theta):
    humidity = exp(mw*psi/(R*T))
    k = 0.66*(soil.thetaS-theta)*dv*vp*humidity*mw/(R*T)
    return(k)

def dvapor_dPsi(funcType, soil, psi, theta):
```

```
      humidity = exp(mw*psi/(R*T))
      capacity_vapor = ((soil.thetaS-theta)*vp*humidity*(mw/(R*T))-
                             dTheta_dPsi(funcType,soil,psi)*vp*
                             humidity)
      return(capacity_vapor)

def vaporFromPsi(soil, psi, theta):
      humidity = exp(mw*psi/(R*T))
      vapor = (soil.thetaS - theta)*vp*humidity
      return(vapor)

def evaporation_flux(psi):
      h_s = exp(mw*psi/(R*T))
      return(E_p * (h_s-h_a)/(1.-h_a))
```

11.4.3   Solvers

The solution for vapour flow is implemented in the file `PSP_vapor1D.py` shown below.
The file imports the files `PSP_grid` and `PSP_ThomasAlgorithm` to create the grid
and the algorithm to invert the matrix as described in the previous chapters and detailed
in Appendix B. The variables for the numerical solution are then defined. Note that here
there are two variables `vapor` and `oldvapor` that are defined to compute the vapour
concentration at the end and beginning of the time step.

Vapour and liquid conductivities should be added to give the total conductivity. Using
the Newton–Raphson procedure, vapour fluxes can be explicitly calculated and added
directly to the mass balance equation. The derivative of the vapour flux must also be
added to the diagonal elements of the Jacobian to ensure rapid convergence to the correct
potentials. The relative humidity at each node is needed for the calculation.

```
#PSP_evaporation_solver
from __future__ import print_function, division

import PSP_grid as grid
from PSP_ThomasAlgorithm import ThomasBoundaryCondition
from PSP_soil import *

waterDensity = 1000.
area = 1
maxNrIterations = 100
tolerance = 1e-6

n = 40
z = np.zeros(n+2, float)
vol = np.zeros(n+2, float)
a = np.zeros(n+2, float)
```

```
b = np.zeros(n+2, float)
c = np.zeros(n+2, float)
d = np.zeros(n+2, float)

dz = np.zeros(n+2, float)
psi = np.zeros(n+2, float)
dpsi = np.zeros(n+2, float)
theta = np.zeros(n+2, float)
oldTheta = np.zeros(n+2, float)
vapor = np.zeros(n+2, float)
oldvapor = np.zeros(n+2, float)
C = np.zeros(n+2, float)
k = np.zeros(n+2, float)
u = np.zeros(n+2, float)
du = np.zeros(n+2, float)
f = np.zeros(n+2, float)
H = np.zeros(n+2, float)
H0 = np.zeros(n+2, float)

def initializeWater(funcType, soil, theta_0, solver):
    global z
    z = grid.geometric(n, soil.lowerDepth)
    vol[0] = 0
    for i in range(n+1):
        dz[i] = z[i+1]-z[i]
        if (i > 0): vol[i] = area * (z[i+1] - z[i-1]) / 2.0

    psi_0 = waterPotential(funcType, soil, theta_0)
    k_0 = (hydraulicConductivityFromTheta(funcType, soil, theta_0)
            + vaporConductivityFromPsiTheta(soil, psi_0, theta_0))

    psi[0] = 0
    for i in range(1, n+2):
        oldTheta[i] = theta_0
        theta[i] = theta_0
        oldvapor[i] = vaporFromPsi(soil, psi[i], theta[i])
        vapor[i] = vaporFromPsi(soil, psi[i], theta[i])
        psi[i] = psi_0
        H[i] = psi[i] - z[i]*g
        k[i] = k_0

def NewtonRapsonMP(funcType, soil, dt, isFreeDrainage):
    airEntry = airEntryPotential(funcType, soil)

    if (isFreeDrainage):
        psi[n+1] = psi[n]
        theta[n+1] = theta[n]
        k[n+1] = k[n]
```

```
nrIterations = 0
massBalance = 1
while ((massBalance > tolerance) and (nrIterations <
        maxNrIterations)):
    massBalance = 0
    for i in range(1, n+1):
        k[i] = (hydraulicConductivityFromTheta(funcType, soil,
        theta[i]) + vaporConductivityFromPsiTheta(soil,
        psi[i], theta[i]))
        u[i] = g * k[i]
        du[i] = -u[i] * soil.Campbell_n / psi[i]
        capacity = dTheta_dPsi(funcType, soil, psi[i])
        capacity_vapor = dvapor_dPsi(funcType, soil, psi[i],
                                     theta[i])
        C[i] = (vol[i] * (waterDensity*capacity + capacity_vapor))
               / dt

    for i in range (1, n+1):
        f[i] = ((psi[i+1] * k[i+1] - psi[i] * k[i])
               / (dz[i] * (1 - soil.Campbell_n))) - u[i]
        if (i == 1):
            a[i] = 0.
            c[i] = -k[i+1] / dz[i]
            b[i] =  k[i] / dz[i] + C[i] + du[i]
            d[i] = evaporation_flux(psi[i]) - f[i] + vol[i]*
                        (waterDensity * (theta[i] - oldTheta[i]) +
                        (vapor[i]-oldvapor[i])) /dt
        else:
            a[i] = -k[i-1] / dz[i-1] - du[i-1]
            c[i] = -k[i+1] / dz[i]
            b[i] = k[i] / dz[i-1] + k[i] / dz[i] + C[i] + du[i]
            d[i] = f[i-1] - f[i] + (waterDensity * vol[i]
                                * (theta[i] - oldTheta[i])) /dt
        massBalance += abs(d[i])

    ThomasBoundaryCondition(a, b, c, d, dpsi, 1, n)

    for i in range(1, n+1):
        psi[i] -= dpsi[i]
        psi[i] = min(airEntry, psi[i])
        theta[i] = thetaFromPsi(funcType, soil, psi[i])
        vapor[i] = vaporFromPsi(soil, psi[i], theta[i])
    nrIterations += 1

    if (isFreeDrainage):
        psi[n+1] = psi[n]
        theta[n+1] = theta[n]
        k[n+1] = k[n]
```

```
if (massBalance < tolerance):
    flux = evaporation_flux(psi[1])
    return True, nrIterations, flux
else:
    return False, nrIterations, 0
```

The results of the simulation, after a total time of 83 400 s, are shown below. The program prints on screen the time [s], the time step [s], the number of iterations for each time step needed to reach convergence [number s^{-1}] and the cumulative evaporation [kg m^{-2}].

```
Water Retention Curve: Campbell
Solver: Matric Potential with Newton-Raphson
Water Retention Curve: Campbell
Solver: Matric Potential with Newton-Raphson
time = 600    dt = 600    Iter. = 3   sumEvap: 0.035
time = 1800   dt = 1200   Iter. = 3   sumEvap: 0.105
time = 4200   dt = 2400   Iter. = 3   sumEvap: 0.245
time = 7800   dt = 3600   Iter. = 3   sumEvap: 0.455
time = 11400  dt = 3600   Iter. = 3   sumEvap: 0.664
time = 15000  dt = 3600   Iter. = 2   sumEvap: 0.874
time = 18600  dt = 3600   Iter. = 2   sumEvap: 1.084
time = 22200  dt = 3600   Iter. = 2   sumEvap: 1.294
time = 25800  dt = 3600   Iter. = 2   sumEvap: 1.504
time = 29400  dt = 3600   Iter. = 2   sumEvap: 1.713
time = 33000  dt = 3600   Iter. = 2   sumEvap: 1.923
time = 36600  dt = 3600   Iter. = 2   sumEvap: 2.133
time = 40200  dt = 3600   Iter. = 2   sumEvap: 2.343
time = 43800  dt = 3600   Iter. = 2   sumEvap: 2.552
time = 47400  dt = 3600   Iter. = 2   sumEvap: 2.762
time = 51000  dt = 3600   Iter. = 2   sumEvap: 2.972
time = 54600  dt = 3600   Iter. = 2   sumEvap: 3.182
time = 58200  dt = 3600   Iter. = 2   sumEvap: 3.391
time = 61800  dt = 3600   Iter. = 2   sumEvap: 3.601
time = 65400  dt = 3600   Iter. = 2   sumEvap: 3.811
time = 69000  dt = 3600   Iter. = 2   sumEvap: 4.021
time = 72600  dt = 3600   Iter. = 2   sumEvap: 4.230
time = 76200  dt = 3600   Iter. = 2   sumEvap: 4.440
time = 79800  dt = 3600   Iter. = 2   sumEvap: 4.650
time = 83400  dt = 3600   Iter. = 2   sumEvap: 4.859
```

Here we show a simulation example where the model was run for 240 h, for an initial water content of 0.2 m^3 m^{-3}. The soil profile is 1 m, the soil water retention curve parameters are for a silt loam soil (the parameters are listed in Table 5.3 and 5.4). Figure 11.3 shows the results of the simulations.

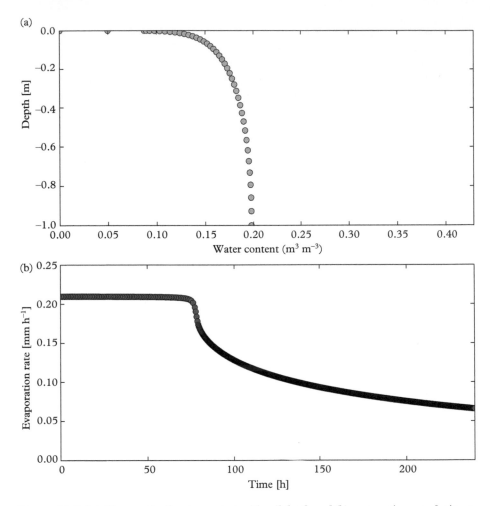

Figure 11.3 *(a) Change of soil water content with soil depth and (b) evaporation rate during a 240 h period for a silt loam soil.*

Specifically, soil water content as function of soil depth and evaporation rate as a function of time are shown. Clearly the evaporation rate exhibits a higher rate at the beginning of the simulation and decreases with time. The distinction between first- and second-stage evaporation is clearly visible. The drying front deepens with increasing simulation time, revealing the evolution of the drying front given by progressive loss of water from the surface and transfer of water from lower soil layers towards the surface. As shown in the printout above, the total evaporation after 23.16 h was 4.8 mm. As described above, this code does not account for vapour flux determined by thermal gradients and it does not consider heat flow. In Chapter 12, a fully coupled model for heat, liquid water and vapour flux will be presented.

....

11.5 EXERCISES

11.1. Simulate first- and second-stage evaporation with varying potential evaporation rates and soil texture. Plot the results. What effect does each of these have on the length of first-stage evaporation?

11.2. Plot cumulative evaporation for varying potential evaporation rates. What effect does rate have on cumulative evaporation?

11.3. Compare moisture profiles for simulations with and without vapour flux in soil. Use a loam soil, starting at around $0.15\,m^3\,m^{-3}$ initial water content.

11.4. Using the files `clay.txt` and `sand.txt`, compare the evaporative fluxes obtained for the two soils having different textures. The user may need to increase the simulation time to two or three days to observe the change from first-stage to second-stage evaporation.

12

Modelling Coupled Transport

In the previous we have described gas, heat, liquid water and vapour flow as separate processes. However, in soils, these processes are coupled. In this chapter, we describe a coupled model for heat, liquid water and vapour flow. The most important process determining the coupling between water and heat is the transport of latent heat by vapour flow within the soil and at the interface between soil and atmosphere. Therefore, the programs for heat, liquid water and water vapour must be linked to describe coupled transport. As described in previous chapters, accounting for the different transport processes is very important for a variety of applications. For instance, soil water content and soil temperature affect microbial transformations in soil, as well as the nitrogen and carbon cycle. Therefore heat and water transport are fundamental processes that must be correctly quantified.

There are many other applications in physics and engineering where a correct quantification of these processes is needed. For instance, in recent years, the need to connect an increasing number of new wind farms is forcing many countries to face the prospect of installing hundreds of kilometres of new cables and hundreds more pylons across the countryside. Therefore the use of underground power cables has now become the only viable alternative. The ageing of underground power cables can be accelerated by overheating and is strongly affected by heat dissipation in soils. Kroener *et al.* (2014*a*) implemented a coupled model for quantification of heat dissipation of underground power cables. This problem is a typical example where heat, liquid water and vapour flow must be coupled for a correct quantification of heat dissipation.

Although the general equations for transport have been described in the previous chapters, we summarize here some of the fundamental transport functions for liquid, heat and vapour flow, to describe which processes are involved in the coupling and the mathematics involved to couple the processes. Moreover, in this chapter, we include a calculations of the soil surface energy balance by including the computation of short-wave radiation, long-wave radiation, sensible heat and latent heat. A description of each term of the energy balance is provided in more detail in Chapter 15.

Soil Physics with Python. First Edition. Marco Bittelli, Gaylon S. Campbell and Fausto Tomei.
© Marco Bittelli, Gaylon S. Campbell and Fausto Tomei 2015. Published in 2015 by Oxford University Press.

12.1   Transport Equations

12.1.1   Liquid Water Transport

Liquid water flow is described by Richards' equation, where the liquid water flux density f_l is determined by the gradient of the water potential (the sum of the matric potential ψ and the gravitational potential $-gz$) and is proportional to the hydraulic conductivity K:

$$f_l = -K\nabla(\psi - gz) \tag{12.1}$$

where K is a nonlinear function of the matric potential ψ.

12.1.2   Heat Transport

Thermal energy flow in soils can be divided into sensible and latent heat. Sensible heat flow is driven by a gradient in temperature T and is proportional to the thermal conductivity λ. Latent heat flow is the thermal energy carried by evaporating or condensing water vapour. Therefore the total heat flow f_h is

$$f_h = -\lambda\nabla T + L_v f_v \tag{12.2}$$

Conservation of mass provides the relationship between flux density of water, both in liquid and vapour form, and change of water content $\rho_l\theta$, where ρ_l is the water density and θ the volumetric water content:

$$\rho_l\frac{\partial\theta}{\partial t} = -\nabla \cdot (f_l + f_v) \tag{12.3}$$

where f_l and f_v are the flux densities of liquid water and water vapour, respectively. At low water potentials, the hydraulic conductivity of liquid water becomes very small and the contribution of vapour flow can be in the same range as liquid water flow or even bigger. The contribution of water vapour to the local water content, however, can be neglected.

The conservation of energy states that the change in specific heat equals the negative divergence of the thermal energy flow. Specific heat is stored in the liquid water, $C_l\rho_l\theta$, the solid part, $C_m\rho_s x_m$ and the air fraction. The changes in the thermal energy stored in the air fraction, however, are so small compared with the changes in thermal energy in the liquid and solid fractions that we can neglect the contribution of the air fraction to the changes in thermal energy:

$$(C_m\rho_s\phi_s + C_l\rho_l\theta)\frac{\partial T}{\partial t} = -\nabla \cdot q_h \tag{12.4}$$

where ρ_s is the density of minerals, x_m is the volumetric fraction of the solid part, C_m is the specific heat of soil minerals and C_l is the specific heat of water.

12.1.3   Vapour Transport

The isothermal, vertical transport of water vapour is given by Fick's law:

$$f_v^i = -D_v \frac{dc_v}{dz} \tag{12.5}$$

where f_v^i is the vapour flux density [kg m^{-2} s^{-1}], D_v is the vapour diffusivity [m^2 s^{-1}], c_v is the vapour concentration [kg m^{-3}] and z is distance [m]. The superscript i indicates isothermal flow. Vapour concentration is obtained as

$$c_v = h_r c_v' \tag{12.6}$$

where h_r is the fractional relative humidity and c_v' is the saturation vapour concentration. If the soil is isothermal, the variation of vapour concentration with depth can be written as

$$\frac{dc_v}{dz} = c_v' \frac{dh_r}{dz} \tag{12.7}$$

The partial pressure of water vapour is related to the soil water potential through the fractional relative humidity, which can be obtained from the water potential by solving eqn (5.44) for the relative humidity:

$$h_r = \exp\left(\frac{M_w \psi}{RT_K}\right) \tag{12.8}$$

where M_w is the molecular weight of water (0.018 kg mol^{-1}), ψ is the soil water potential [J kg^{-1}], R is the gas constant (8.31 J mol^{-1} K^{-1}) and T_K is the temperature [K]. Substituting these expressions into eqn (12.5), the vapour flow can be written as

$$f_v^i = -K_v \frac{d\psi}{dz} \tag{12.9}$$

where K_v is the water vapour conductivity,

$$K_v = \frac{D_v c_v' h_r M_w}{RT_K} \tag{12.10}$$

This description only accounts for matric potential gradients. The vapour concentration gradient is the driving force and can be separated into thermally driven and water-potential-driven flow. A more general analysis of soil evaporation must include the effects of temperature gradients on water transport. Therefore the total water vapour flow f_v^* can be formulated as sum of an isothermal flux component f_v^i and a temperature-driven flux component q_v^T, as described by Bittelli *et al.* (2008a):

$$f_v^* = f_v^i + f_v^T = -D_v c_v' \frac{dh_r}{dz} - D_v h_r s \frac{dT}{dz} \qquad (12.11)$$

where D_v is the vapour diffusivity, c_v' is the saturated vapour concentration, and the relative humidity $h_r = c_v/c_v'$ is the ratio of the vapour concentration c_v to the saturated vapour concentration c_v'. The slope of the saturation vapour concentration function is given by

$$s = \Delta M_w V_m / P \qquad (12.12)$$

where P is the atmospheric pressure, Δ is the slope of the saturation vapour pressure curve and V_m is the molar volume of air. Multiplication by $M_w V_m$ is necessary to convert mol mol^{-1} to kg m^{-3}. The slope of the saturation vapour pressure curve is given by its derivative:

$$\Delta = \frac{4098 e_s}{(T + 237.3)^2} \qquad (12.13)$$

where T is in Celsius and Δ has units of Pa C^{-1}.

12.1.4 Effect of Water Vapour Flow on the Heat Flow Equations

Two corrections can be made to the heat flux equations in Chapter 4 to make them compatible with the vapour flux equations just derived. First, the thermally induced latent heat flux can now be explicitly calculated. Since the latent heat flux is the latent heat of vaporization L_v times the vapour flux, the apparent thermal conductivity becomes the conductivity for sensible heat plus L_v times the vapour conductivity:

$$\lambda_a = \lambda + D_v h_r s L_v \qquad (12.14)$$

The second correction accounts for the latent heat that is transported by water potential gradients. This term is used for correcting the thermal conductivity by changes in the vapor flux. The vapour flux is given by eqn (12.9), so a source term is added to the heat flux equation, which is the latent heat of vaporization multiplied by the divergence in vapour flux.

12.1.5 Comparison of Thermally Induced Liquid and Vapour Flow

It is useful to compare the thermally induced liquid flux and vapour flux in these two phases, to determine when each might be important. The thermal liquid conductivity $Kd\psi/dT$ was computed by Philip and de Vries (1957) assuming that only the surface tension γ of the water changes with temperature:

$$\frac{d\psi_m}{dT} = \frac{d\psi_m}{d\gamma}\frac{d\gamma}{dT} \qquad (12.15)$$

From the capillary equation,

$$\frac{d\psi_m}{d\gamma} = \frac{\psi_m}{\gamma}$$

(12.16)

The temperature dependence of the water potential is therefore predicted to be around $2 \times 10^{-3}\ \psi_m$ at 20 °C. Taylor and Stewart (1960) found $d\psi_m/dT$ to be about 10 times this large for a silt loam soil. If the temperature dependence found by Taylor and Stewart (1960) is assumed, then typical thermal liquid conductivities for a silt loam soil would be 3×10^{-7} kg s^{-1} m^{-1} K^{-1} at $\psi_m = -10$ J kg^{-1}, 5×10^{-8} kg s^{-1} m^{-1} K^{-1} at $\psi_m = -20$ J kg^{-1}, and 6×10^{-10} kg s^{-1} m^{-1} K^{-1} at $\psi_m = -100$ J kg^{-1}. Thermal vapour conductivity is around 4×10^{-8} kg s^{-1} m^{-1} K^{-1}, or about the same as the liquid value at $\psi_m = -20$ J kg^{-1}. Thermally induced liquid flux would therefore be important only in wet soil. At saturation, a temperature gradient of 1 K m^{-1} could produce a flux of 2×10^{-5} kg m^{-2} s^{-1} K, about 0.5 % of the saturated hydraulic conductivity.

12.2   Partial Differential Equations

The transport equations described above for liquid water, heat and water vapour are combined to obtain a coupled system of two partial differential equations, which need to be solved for the two unknowns temperature T and matric potential ψ:

$$\frac{\partial}{\partial t}\theta(\psi) = \nabla \cdot \left[-K(\psi)\nabla(\psi - gz) - D_v h c_v' \frac{M_w}{RT}\nabla\psi - D_v h s \nabla T \right]$$

(12.17)

and the partial differential equation for temperature,

$$[C_m\rho_s x_m + C_w\rho_l\theta(\psi)]\frac{\partial T}{\partial t} = \nabla \cdot \left[-\lambda\nabla T - (TC_w)K(\psi)\nabla(\psi - gz) \right.$$

$$\left. - (L + TC_w)D_v h c_v' \frac{M_w}{RT}\nabla\psi - (L + TC_w)D_v h s \nabla T \right]$$

(12.18)

These coupled differential equations are solved under prescribed boundary conditions.

12.3   Surface Boundary Conditions

Liquid water flow at the upper surface is given by precipitation (input) and evaporation (output). The upper boundary condition for heat transfer is of known temperature (measured air temperature). Vapour flow is given by evaporation, with an evaporation rate E of

$$E = \frac{c_{va} - c_{vs}}{r_v + r_s}$$

(12.19)

where r_v is the aerodynamic resistance for water vapour transfer, r_s is the soil surface resistance for water vapour transfer, c_{va} is the atmosphere vapour concentration at height z_{ref} and c_{vs} is the soil surface vapour concentration, where the surface is at height z_0. Soil surface resistance depends on the water content in the top soil layer (van de Griend and Owe, 1994):

$$r_s = 10e^{0.3563(\theta_{min}-\theta_{top})} \tag{12.20}$$

where $\theta_{min} = 0.15$ is an empirical parameter and θ_{top} is the water content of the top 1 cm layer. The aerodynamic resistance r_v depends on wind speed, level of turbulence, soil surface roughness and thermal stratification of the boundary layer. For the calculation of r_v, the reader should refer to Chapter 15, where atmospheric boundary conditions are described.

12.4   Numerical Implementation

Here we present a series of modules, each containing a part of the program for one-dimensional simulation of coupled flow of liquid water, heat and water vapour. In this program, we also included the computation of the soil energy budget. Each component of the energy balance is computed and saved into an output file, to allow the reader to analyse the orders of magnitude and diurnal oscillations of the different terms.

While in the previous chapters we used simplified upper boundary conditions, in this example we utilize atmospheric boundary conditions. Therefore we provide real weather data for precipitation, global solar radiation, relative humidity, air temperature and wind speed. The data were collected at an experimental station installed in Ozzano Emilia, Bologna, Italy (Bittelli *et al.*, 2012). Wind direction was also collected and it is stored in the data file, but it is not used here.

The weather data used in this example were collected from 23 April until 31 December 2005. The user can select the simulation period in the file `PSP_public.py` as described below. Simulations performed over different periods in the winter and in the summer are useful to analyse the different processes in different seasons and the water balance terms. The project is called `PSP_coupled` and comprises eleven modules and two data files. One data file contains the soil data and the other the weather data.

1. `main.py`
2. `PSP_boundary.py`
3. `PSP_public.py`
4. `PSP_soil.py`
5. `PSP_coupled1D.py`
6. `PSP_readDataFile.py`
7. `PSP_longWaveRadiation.py`
8. `PSP_grid.py`
9. `PSP_plot.py`

10. `PSP_plotEnergy.py`
11. `PSP_ThomasAlgorithm.py`
12. `soil.txt`
13. `weather_ozzano.dat`

In the following, we will not describe the modules `PSP_readDataFile.py` for reading the data, `PSP_grid.py` for building the computational grid and `PSP_ThomasAlgorithm.py` for solving the system of equations, since they have already been described in previous chapters. Moreover, we will not describe the modules `PSP_plot` and `PSP_plotEnergy`, since they are straightforward modules with simple plotting instructions using `matplotlib`. They are written to plot the results of the simulation at runtime and, after the simulation is finished, to plot the energy budget terms as functions of time. When plotting the energy budget by executing the module `PSP_plotEnergy`, it is important to be sure that the generated output file (`energy_balance.dat`) has columns of the same length. If the simulation is interrupted before completion of the last time step, the last row may be incomplete. This anomaly may prevent plotting of the file and the program `PSP_plotEnergy` will give an error. If this happens, it is necessary to manually remove the last row from the file (`energy_balance.dat`) and rerun the program.

12.4.1 Main

The main file `main.py` contains the calls to the following other files:

1. `PSP_public` import
2. `PSP_coupled1D` as coupled
3. `PSP_soil` import readSoil
4. `PSP_readDataFile` import readDataFile
5. `PSP_plot` import *
6. `PSP_longWaveRadiation` import *

The module `PSP_public` is written to define the variables that are available to all modules. The module `PSP_coupled1D` implements the solvers for the numerical solutions. `PSP_soil` import is used here to read the soil data, since the file reader is imported into the module `PSP_soil`, while `PSP_readDataFile` is used to read the data file. The module `PSP_plot` is written to plot the results and imports all the modules contained in this file, by using the asterisk. Finally, the module `PSP_longWaveRadiation` is used to compute the long-wave radiation. A detailed description of each module is presented below.

The module **main** reads the soil data written in the file `soil.txt` and the weather data written in the file `weather.dat`. The weather variables are assigned to arrays and then variables needed for the numerical solution are initialized. The module `coupled.initialize`, which is written in the module `PSP_coupled1D`, is called to initialize the soil properties, the soil water potential and the soil temperature.

The following `while` loop, which runs from `time` to `endTime`, is the main loop. Inside this loop, the boundary conditions are defined, as well as the calls for the numerical solution.

```python
#PSP_coupled
from __future__ import print_function, division
import numpy as np
from PSP_public import *
import PSP_coupled1D as coupled
from PSP_soil import readSoil
from PSP_readDataFile import readDataFile
from PSP_plot import *
from PSP_longWaveRadiation import *

def main():
    isSuccess, mySoil = readSoil("soil.txt")
    if not isSuccess:
        print("warning: wrong soil file.")
        return

    A, isFileOk = readDataFile('weather.dat', 1, '\t', False)
    if not isFileOk:
        print ("Incorrect format in row: ", A)
        return()

    airT = A[:,1]
    prec = A[:,2]
    relativeHumidity = A[:,3]
    windSpeed = A[:,4]
    globalRadiation = A[:,6]
    longWaveRadiation = longWaveRadiationFromWeather(nrDays)

    time = 0
    dt = 300
    sumWaterFlow = 0.
    sumHeatFlow = 0.
    sumEvaporationFlow = 0.
    nrIterations = 0

    coupled.initialize(mySoil, initialPotential, initialTemperature)
    plot_start(endTime)

    while (time < endTime):
        dt = min(dt, endTime - time)
        myBoundary = coupled.boundary.Cboundary()
        i = int(time/3600)
```

```
        myBoundary.time = time
        myBoundary.airTemperature = airT[i]
        myBoundary.precipitation = prec[i]
        myBoundary.relativeHumidity = relativeHumidity[i]
        myBoundary.windSpeed = windSpeed[i]
        myBoundary.globalRadiation = globalRadiation[i]
        myBoundary.longWaveRadiation = longWaveRadiation[i]

        (isBalanceOk, waterFlux, heatFlux, boundaryLayerConductance,
         evaporationFlux, nrIterations,massBalance) = coupled.solver
         (mySoil, myBoundary, isFreeDrainage, dt)

        if isBalanceOk:
            for i in range(coupled.n+1):
                coupled.oldTheta[i] = coupled.theta[i]
                coupled.oldPsi[i] = coupled.psi[i]
                coupled.oldT[i] = coupled.T[i]
                coupled.oldCh[i] = coupled.Ch[i]
            sumWaterFlow += waterFlux * dt
            sumHeatFlow += heatFlux * dt
            sumEvaporationFlow += evaporationFlux * dt
            time += dt
            print("time =", int(time), "\tdt =", dt,
                  "\tnrIterations =", int(nrIterations),
                  "\tsumWaterFlow:", format(sumWaterFlow, '.3f'),
                  "\tsumHeatFlow:", format(sumHeatFlow,'.1f'))
            plot_variables(coupled.z, coupled.theta, coupled.T,
            coupled.psi, time, boundaryLayerConductance,
            sumEvaporationFlow, myBoundary.precipitation,
            myBoundary.airTemperature, myBoundary.relativeHumidity)
            if (float(nrIterations / maxNrIterations) <= 0.1):
                dt = int(min(dt*2, maxTimeStep))

        else:
            print ("time =", int(time), "\tdt =", dt, "\tNo convergence")
            dt = max(int(dt / 2), 1)
            for i in range(coupled.n+1):
                coupled.theta[i] = coupled.oldTheta[i]
                coupled.psi[i] = coupled.oldPsi[i]
                coupled.T[i] = coupled.oldT[i]
    plot_end()
main()
```

12.4.2 Boundary

The module PSP_boundary defines the initial and boundary conditions. The first lines
are written to import the modules needed for this module. The results of the energy
budget computation are then written in the file energy_balance.dat . The class

Cboundary initializes the atmospheric variables by setting them equal to the constant NODATA. The function **boundaryLayerConductance** computes the boundary-layer conductance for transfer of heat and vapour. The theory for the implementation of this function is provided in Chapter 15. The function **soilResistance** is an additional resistance term defined to compute the resistance for vapour flow for bare soils as detailed by Bittelli *et al.* (2008*b*).

The function **evaporationFlux** computes the evaporation flux based on eqn (12.19), while the function **dEvaporationFluxdPsi** computes the changes in evaporation flux as functions of water potential gradients. The function **waterFlux** computes the sum of vapour flux and precipitation flux. Finally, the function **thermalFlux** computes the fluxes of short-wave, long-wave and net radiation, as well as sensible and latent heat.

```
#PSP_boundary
from __future__ import division, print_function
import numpy as np
import PSP_soil as soil
from PSP_public import *

energy_balance = open('energy_balance.dat','w')
energy_balance.write('#Time \t LWUp \t SWDown \t LWDown \t Latent \t
                     Sensible \t G \n')

class Cboundary:
    time = 0
    airTemperature = NODATA
    precipitation = NODATA
    relativeHumidity = NODATA
    windSpeed = NODATA
    globalRadiation = NODATA

def boundaryLayerConductance(windSpeed, Tair, Tsoil):
    z = 2.0
    cp = 29.3
    h = 0.01
    D = 0.77 * h
    zm = 0.13 * h
    zh = 0.2 * zm
    psi_m = 0; psi_h = 0
    vk = 0.4
    TsoilK = Tsoil + zeroKelvin      #[K]
    # molar density of the gas
    ro = 44.6 * (atmPressure / 101.3) * (293.15 / TsoilK)
    # volumetric heat of air (= 1200 J/m^3*K at 20C e sea level)
    Ch = ro * cp
    for i in range(3):
        #friction velocity         #[m/s]
```

```
            ustar = vk * windSpeed / (np.log((z - D + zm) / zm) + psi_m)
            Kh = vk * ustar / (np.log((z - D + zh) / zh) + psi_h)
            Sp = -vk * z * g * Kh * (Tair - Tsoil) / (Ch * TsoilK *
                np.power(ustar,3))
            if (Sp > 0):
                psi_h = 4.7 * Sp
                psi_m = psi_h
            else:
                psi_h = -2 * np.log((1 + np.sqrt(1 - 16 * Sp)) / 2)
                psi_m = 0.6 * psi_h
        return Kh


# soil resistance
def soilResistance(theta):
    return 10. * np.exp(0.3563 * (22. - (theta * 100.)))


# [kg/m2s]
def evaporationFlux(psi, theta, Tsoil, Tair, rhAir,
                    aerodynamicResistance):
    vapourConcAir = soil.vapourConcentration(Tair, rhAir/100.)
    rhSoil = soil.relativeHumidity(psi, Tsoil + zeroKelvin)
    vapourConcSoil = soil.vapourConcentration(Tsoil, rhSoil)
    dVapour = vapourConcSoil - vapourConcAir
    #print(vapourConcAir,rhSoil,vapourConcSoil,
        dVapour,aerodynamicResistance)
    return -(1.0 / (aerodynamicResistance + soilResistance(theta)) *
                    dVapour)


def dEvaporationFluxdPsi(psi, theta, Tsoil, Tair,
                         aerodynamicResistance):
    TKelvinAir = Tair + zeroKelvin
    rhSoil = soil.relativeHumidity(psi, Tsoil + zeroKelvin)
    vapourConcSoil = soil.vapourConcentration(Tsoil, rhSoil)
    return (1.0 / (aerodynamicResistance + soilResistance(theta))
            * Mw/(R*TKelvinAir) * vapourConcSoil)


# [kg/m2s]
def waterFlux(psi, theta, Tsoil, myBoundary, aerodynamicResistance):
    evapFlux = evaporationFlux(psi, theta, Tsoil,
                               myBoundary.airTemperature,
    myBoundary.relativeHumidity, aerodynamicResistance)
    precFlux = myBoundary.precipitation/3600
    return evapFlux + precFlux


def dWaterFluxdPsi(psi, theta, Tsoil, myBoundary,
                   aerodynamicResistance):
    return (dEvaporationFluxdPsi(psi, theta, Tsoil,
    myBoundary.airTemperature, aerodynamicResistance))
```

```
def thermalFlux(psi, theta, Tsoil, myBoundary,
                aerodynamicResistance, isWrite):
    TairK = myBoundary.airTemperature + zeroKelvin
    longWaveSoilTaylorAtmosphericPart =-(((sigma*(TairK)**4 -
    4.0*sigma*(TairK)**3 * myBoundary.airTemperature)))
    sensibleHeatAtmosphericPart = (1200.0 /
    aerodynamicResistance * myBoundary.airTemperature)
    sensibleHeat = (1200.0 /
    aerodynamicResistance * (myBoundary.airTemperature-Tsoil))
    evaporation = evaporationFlux(psi, theta,
    Tsoil, myBoundary.airTemperature,
    myBoundary.relativeHumidity, aerodynamicResistance)
    shortWaveAbsRadiation = (1 - albedo) * myBoundary.globalRadiation
    longWaveAbsRadiation = myBoundary.longWaveRadiation
    netAbsRadiation = shortWaveAbsRadiation + longWaveAbsRadiation
    if isWrite:
        dTemperature = myBoundary.airTemperature - Tsoil
        sensibleHeat = (1200.0 / aerodynamicResistance * dTemperature)
        longWaveSoil = -(sigma*(TairK)**4 -4.0*sigma*(TairK)**3 *
                         dTemperature)
        G=(longWaveAbsRadiation+shortWaveAbsRadiation+
        sensibleHeat+(evaporation*L)+longWaveSoil)
        energy_balance.write(str(myBoundary.time) + '\t'
                        + format(longWaveSoil, ".1f")
                        + '\t' + format(shortWaveAbsRadiation,
                                        ".1f")
                        + '\t' + format(longWaveAbsRadiation,
                                        ".1f")
                        + '\t' + format(evaporation*L, ".1f")
                        + '\t' + format(sensibleHeat, ".1f")
                        + '\t' + format(-G, ".1f") + '\n')
    return (longWaveSoilTaylorAtmosphericPart +
    netAbsRadiation + evaporation*L + sensibleHeatAtmosphericPart)
```

12.4.3   Public

This module contains the variables that are read by all modules. First the information about the study site is defined, such as latitude, longitude, altitude, albedo, atmospheric pressure and clay content. The information included in this example is for the Centonara location, an experimental site of the Department of Agricultural Sciences of the University of Bologna (Bittelli *et al.*, 2012).

   Then information about the initial conditions is included, such as the initial value of the soil matric potential and the initial soil temperature. The selection of the lower boundary condition must also be included here, by setting the variable isFreeDrainage equal to True or False. Clearly, if the variable is set to True, the lower boundary condition is of free drainage; otherwise it is constant potential, equal to the value set as the initial condition. The next block of information regards the size of the area of the interfacial node. Since this example is one-dimensional, this value is not important and can be set equal to 1. However, we have seen its importance in the description of three-dimensional flow. Then the number of simulation days is set, as well as the end time. The user can select the simulation period by modifying the variable nrDays. Since the simulation time step is in seconds, the following variable endTime is obtained by multiplying the number of seconds in an hour by the number of hours in a day and by the number of days. The accepted tolerance for the energy and mass balance errors are then defined both for vapour and for heat. The maximum number of iterations and the maximum time step is defined.

The following block describes some constants. The variables are latent heat of vaporization, gravitational acceleration, density of water, molecular mass of water, gas constant, vapour diffusivity, Stefan–Boltzmann constant, heat capacity of water and the kelvin temperature at zero degrees Celsius.

```
#PSP_public.py
NODATA = -9999.

# Site
latitude = 44.5
longitude = 11.5
altitude = 120
albedo = 0.2
atmPressure = 101.3
clay = 0.2

# Initial Condition
initialPotential = -1000.0
initialTemperature = 10.0
isFreeDrainage = False

# Simulation
area = 1
nrDays = 10
endTime = 3600*24*nrDays
toleranceVapour = 1.e-3
toleranceHeat = 1.e-5
maxNrIterations = 50.
maxTimeStep = 3600.

# Physics
L = 2.45e6
g = 9.8065
waterDensity = 1000.
Mw = 0.018
R = 8.31
```

```
waterVapourDiff0 = 2.12E-5
sigma = 5.67e-8
heatCapacityWater = 4.18e3
zeroKelvin = 273.15
```

12.4.4 Soil

The module `PSP_soil` is written to define the soil properties, as was done in previous chapters. In this version, there are additional functions needed for coupled heat and vapour fluxes. The function **relativeHumidity** implements eqn (12.8), while the function **vapourDiffusivity** implements eqn (3.13). The function **satVapourPressure** implements eqn (5.46). This equation computes the saturation vapour pressure as function of temperature. The saturation vapour concentration is computed by the function **satVapourConcentration**. The actual vapour pressure and concentration are obtained by multiplying the saturation vapour pressure and saturation vapour concentration by the relative humidity as performed in the functions **vapourPressure** and **vapourConcentration**, respectively. Equation (12.13) is implemented in the function **slopeSatVapourConcentration**, which computes the slope of the saturation vapour concentration curve. The conductivity for water vapour described in eqn (12.10) is computed by the function **vapourConductivity**. The function **computeVapour** is used to compute the vapour concentration. The following functions are used to compute parameters needed for heat transfer and were described in Chapter 4. Finally, the class `PropertiesForMassBalance` is written to define properties needed for the mass balance computation. The array `self.K11` is the conductivity term for liquid water plus vapour flow driven by the gradient in water potential. Append is used to append the first node (0) to the other nodes for the upper boundary. The array `self.gravityWaterFlow` is the gravity-driven water flow, while the array `self.K21` is the conductivity term for vapour flow driven by a temperature gradient. The array `self.K12` is the thermal conductivity for a gradient in heat flow driven by a gradient in water potential. Finally, the array `self.K22` contains the thermal conductivity for heat flow determined by temperature gradients. The code is as follows:

```
#PSP_soil
from PSP_readDataFile import readDataFile
from PSP_public import *
import numpy as np

CAMPBELL = 1
RESTRICTED_VG = 2
IPPISCH_VG = 3
VAN_GENUCHTEN = 4

CELL_CENT_FIN_VOL = 1
NEWTON_RAPHSON_MP = 2
NEWTON_RAPHSON_MFP = 3
```

```
LOGARITHMIC = 0
HARMONIC = 1
GEOMETRIC = 2

class Csoil:
    upperDepth = NODATA
    lowerDepth = NODATA
    Campbell_he = NODATA
    Campbell_b = NODATA
    CampbellMFP_he = NODATA
    VG_alpha = NODATA
    VG_n = NODATA
    VG_m = NODATA
    VG_he = NODATA
    VG_alpha_mod = NODATA
    VG_n_mod = NODATA
    VG_m_mod = NODATA
    VG_Sc = NODATA
    VG_thetaR = NODATA
    Mualem_L = NODATA
    thetaS = NODATA
    Ks = NODATA


def readSoil(soilFileName):
    soil = Csoil()
    A, isFileOk = readDataFile(soilFileName, 1, ',', False)
    if ((not isFileOk) or (len(A[0]) < 12)):
        return False, soil

    soil.upperDepth = A[0,0]
    soil.lowerDepth = A[0,1]
    soil.Campbell_he = A[0,2]
    soil.Campbell_b = A[0,3]
    soil.Campbell_n = 2.0 + (3.0 / soil.Campbell_b)
    soil.VG_he = A[0,4]
    soil.VG_alpha = A[0,5]
    soil.VG_n = A[0,6]
    soil.VG_m =  1. - (1. / soil.VG_n)
    soil.VG_alpha_mod = A[0,7]
    soil.VG_n_mod = A[0,8]
    soil.VG_m_mod =  1. - (1. / soil.VG_n_mod)
    soil.VG_Sc = ((1. +
    (soil.VG_alpha_mod *
    abs(soil.VG_he))**soil.VG_n_mod)**(-soil.VG_m_mod))
    soil.VG_thetaR = A[0,9]
    soil.thetaS = A[0,10]
    soil.Ks = A[0,11]
    soil.Mualem_L = 0.5
```

```
        soil.CampbellMFP_he = (soil.Ks *
        soil.Campbell_he / (1.0 - soil.Campbell_n))
        return True, soil

    def airEntryPotential(funcType, soil):
        if (funcType == CAMPBELL):
            return(soil.Campbell_he)
        elif (funcType == IPPISCH_VG):
            return(soil.VG_he)
        elif (funcType == RESTRICTED_VG):
            return(0)
        else:
            return(NODATA)

    def waterPotential(funcType, soil, theta):
        psi = NODATA
        Se = SeFromTheta(funcType, soil, theta)
        if (funcType == RESTRICTED_VG):
            psi = (-(1./soil.VG_alpha)*
            ((1./Se)**(1./soil.VG_m) - 1.)**(1./soil.VG_n))
        elif (funcType == IPPISCH_VG):
            psi = (-((1./soil.VG_alpha_mod)*
            ((1./(Se*soil.VG_Sc))
             **(1./soil.VG_m_mod)-1.)**(1./soil.VG_n_mod)))
        elif (funcType == CAMPBELL):
            psi = soil.Campbell_he * Se**(-soil.Campbell_b)
        return(psi)

    def SeFromTheta(funcType, soil, theta):
        check = np.greater_equal(theta,soil.thetaS)
        if (funcType == CAMPBELL):
            Se = theta / soil.thetaS
        else:
            Se = ((theta - soil.VG_thetaR)
            / (soil.thetaS - soil.VG_thetaR))
        return (1.*check + Se*np.logical_not(check))

    def thetaFromSe(funcType, soil, Se):
        if (funcType == RESTRICTED_VG) or (funcType == IPPISCH_VG):
            theta = (Se * (soil.thetaS
            - soil.VG_thetaR) + soil.VG_thetaR)
        elif (funcType == CAMPBELL):
            return(Se * soil.thetaS)
        return(theta)

    def degreeOfSaturation(funcType, soil, psi):
        if (funcType == IPPISCH_VG):
            Se = ((1./soil.VG_Sc) * pow(1.+pow(soil.VG_alpha_mod
```

```
            * np.maximum(psi,0.), soil.VG_n_mod), -soil.VG_m_mod))
        return Se
    elif (funcType == RESTRICTED_VG):
        Se = (1 / pow(1 +
        pow(soil.VG_alpha * np.maximum(-psi,0.),
        soil.VG_n), soil.VG_m))
        return Se
    elif (funcType == CAMPBELL):
        Se = (pow(np.minimum(psi,soil.Campbell_he)
        / soil.Campbell_he, -1. / soil.Campbell_b))
        return Se

def thetaFromPsi(funcType, soil, psi):
    Se = degreeOfSaturation(funcType, soil, psi)
    theta = thetaFromSe(funcType, soil, Se)
    return(theta)

def dTheta_dPsi(funcType, soil, psi):
    airEntry = airEntryPotential(funcType, soil)
    if (funcType == RESTRICTED_VG):
        check = np.greater(psi,airEntry)
        dSe_dpsi = soil.VG_alpha * soil.VG_n * (soil.VG_m
        * pow(1. + pow(soil.VG_alpha * abs(psi), soil.VG_n),
        -(soil.VG_m + 1.)) * pow(soil.VG_alpha * abs(psi),
                        soil.VG_n - 1.))
        return (0*check + dSe_dpsi * (soil.thetaS
        - soil.VG_thetaR)*np.logical_not(check))
    elif (funcType == IPPISCH_VG):
        check = np.greater(psi,airEntry)
        dSe_dpsi = soil.VG_alpha_mod * soil.VG_n_mod * (soil.VG_m_mod
        * pow(1. + pow(soil.VG_alpha_mod * abs(psi), soil.VG_n_mod),
        -(soil.VG_m_mod + 1.)) *
        pow(soil.VG_alpha_mod * abs(psi), soil.VG_n_mod - 1.))
        dSe_dpsi *= (1. / soil.VG_Sc)
        return (0*check + dSe_dpsi * (soil.thetaS
        - soil.VG_thetaR)*np.logical_not(check))
    elif (funcType == CAMPBELL):
        check = np.greater(psi,airEntry)
        theta = soil.thetaS * degreeOfSaturation(funcType, soil, psi)
        return (0*check + -theta
        / (soil.Campbell_b * psi)*np.logical_not(check))

def hydraulicConductivityFromPsi(funcType, soil, psi):
    if (funcType == RESTRICTED_VG):
        psi = abs(psi)
        num = (1. - pow(soil.VG_alpha * psi, soil.VG_m * soil.VG_n)
        *pow(1. + pow(soil.VG_alpha*psi, soil.VG_n), -soil.VG_m))**2
        denom = (pow(1. + pow(soil.VG_alpha*psi, soil.VG_n),
```

```
            soil.VG_m * soil.Mualem_L))
        k = soil.Ks * (num / denom)
    elif (funcType == IPPISCH_VG):
        k = NODATA
    elif (funcType == CAMPBELL):
        k = soil.Ks * (soil.Campbell_he / psi)**soil.Campbell_n
    return(k)

def relativeHumidity(psi, TKelvin):
    return np.exp(psi * Mw/(R*TKelvin))

def vapourDiffusivity(mySoil, theta, TKelvin, atmPressure):
    binaryDiffCoeff = (waterVapourDiff0 * (101.3 / atmPressure)
    * (TKelvin / 273.15)**1.75)
    gasPorosity = (mySoil.thetaS - theta)
    bg = 0.9
    mg = 2.3
    return (binaryDiffCoeff * bg * gasPorosity**mg)

def satVapourPressure(TCelsius):
    return 0.611 * np.exp(17.502 * TCelsius / (TCelsius + 240.97))

def satVapourConcentration(TCelsius):
    svp = satVapourPressure(TCelsius)
    TKelvin = TCelsius + zeroKelvin
    return svp * 1.e3 * Mw/(R*TKelvin)

def vapourPressure(TCelsius, rh):
    return satVapourPressure(TCelsius) * rh

def vapourConcentration(TCelsius, rh):
    return satVapourConcentration(TCelsius) * rh

def slopeSatVapourConcentration(TCelsius, atmPressure,satVapConc):
    return 4098.0 * satVapConc / (TCelsius+237.3)**2

def vapourConductivity(mySoil, psi, theta, TCelsius, atmPressure):
    TKelvin = TCelsius + zeroKelvin
    rh = relativeHumidity(psi, TKelvin)
    vapDiff = vapourDiffusivity(mySoil, theta, TKelvin, atmPressure)
    satVapConc = satVapourConcentration(TCelsius)
    return vapDiff * satVapConc * rh * Mw/(R*TKelvin)

def computeVapour(mySoil, psi, theta, rh, satVapConc):
    gasPorosity = (mySoil.thetaS - theta)
    return vapourConcentration(rh, satVapConc) * gasPorosity
```

```
def heatCapacity(mySoil, waterContent):
    solidContent = 1.0 - mySoil.thetaS
    return (2.4e6 * solidContent + 4.18e6 * waterContent)

def thermalConductivity(mySoil, waterContent, TCelsius):
    ga = 0.088
    thermalConductivitysolid = 2.5

    q = 7.25 * clay + 2.52
    xwo = 0.33 * clay + 0.078

    porosity = mySoil.thetaS
    solidContent = 1 - porosity
    gasPorosity = np.maximum(porosity - waterContent, 0.0)

    TKelvin = TCelsius + zeroKelvin
    Lv = 45144 - 48 * TCelsius
    svp = satVapourPressure(TCelsius)
    slope = 17.502 * 240.97 * svp / (240.97 + TCelsius)**2.0
    Dv = 0.0000212 * (101.3 / atmPressure) * (TKelvin / 273.16)**1.75
    rhoair = 44.65 * (atmPressure / 101.3) * (273.16 / TKelvin)
    stcor = np.maximum(1 - svp / atmPressure, 0.3)

    thermalConductivityWater = 0.56 + 0.0018 * TCelsius
    #empirical weighting function D[0,1]
    check = np.less(waterContent,0.01*xwo)
    wf = 0*check + (1 / (1 + (waterContent / xwo)**(-q)))*
                    np.logical_not(check)

    thermalConductivityGas = (0.0242 + 0.00007 * TCelsius +
    wf * Lv * rhoair * Dv * slope / (atmPressure * stcor))
    gc = 1 - 2 * ga
    thermalConductivityFluid = (thermalConductivityGas +
    (thermalConductivityWater - thermalConductivityGas)
    * (waterContent / porosity)**2.0)
    ka = (2 / (1 + (thermalConductivityGas / thermalConductivityFluid -
     1) * ga) + 1 / (1 + (thermalConductivityGas /
    thermalConductivityFluid - 1) * gc)) / 3 kw = (2 / (1 +
    (thermalConductivityWater / thermalConductivityFluid - 1) * ga) +
    1 / (1 + (thermalConductivityWater / thermalConductivityFluid -
     1) * gc)) / 3 ks = (2 / (1 + (thermalConductivitysolid /
    thermalConductivityFluid - 1) * ga) + 1 / (1 +
    (thermalConductivitysolid / thermalConductivityFluid - 1) * gc)) / 3

    thermalConductivity = ((kw * thermalConductivityWater *
    waterContent + ka * thermalConductivityGas * gasPorosity + ks *
```

```
        thermalConductivitysolid * solidContent) / (kw * waterContent +
        ka * gasPorosity + ks * solidContent))
        return(thermalConductivity)

class PropertiesForMassBalance:
    def __init__(self,TCelsius,psi,mySoil):
        k = hydraulicConductivityFromPsi(CAMPBELL, mySoil, psi)
        theta = thetaFromPsi(CAMPBELL, mySoil, psi)
        TKelvin = TCelsius + zeroKelvin
        rh = relativeHumidity(psi, TKelvin)
        vapourDiff = vapourDiffusivity(mySoil, theta, TKelvin,
                    atmPressure)
        satVapConc = satVapourConcentration(TCelsius)
        slope = slopeSatVapourConcentration(TCelsius, atmPressure,
            satVapConc)
        vapour = computeVapour(mySoil, psi, theta, rh, satVapConc)
        thermalCond = thermalConductivity(mySoil,theta, TCelsius)

        self.K11 = np.append(0,k + vapourDiff*satVapConc*rh*Mw/
                            (R*TKelvin))
        self.gravityWaterFlow = np.append(0,-k*g)
        self.K21 = np.append(0,vapourDiff*rh*slope)
        self.K12 = np.append(0,L*vapourDiff*satVapConc*rh*Mw/
                            (R*TKelvin))
        self.K22 = np.append(0,thermalCond + L*vapourDiff*rh*slope)
```

12.4.5 Coupling

PSP_coupled1D is the module that implements the solver for the different flux equations. The first eight lines are used to import the modules necessary for the calculation. Then the matrix vectors are defined, as well as the variables for depth, water potential, water content, vapour and heat capacity. The core of the numerical solution is in the function **solver**, where the matrices for vapour, heat and water flow are implemented and solved using the Thomas algorithm. The mathematical formulation of these solutions has been presented in Chapters 3, 4 and 8. Here they are coupled, by including all processes and their non-linear coupling at once by coupling the processes (for instance solving for the thermally driven vapour flow from knowledge of temperature) and solving the matrices within the same time iteration.

```
#PSP_coupled1D.py
from __future__ import print_function, division
import numpy as np
import PSP_soil as soil
from PSP_public import *
import PSP_boundary as boundary
from PSP_ThomasAlgorithm import ThomasBoundaryCondition
import PSP_grid as grid
```

```
n = 10

def initialize(mySoil, psi0, T0):
    global z, vol, dz, psi, oldPsi, dpsi, T, oldT
    global theta, oldTheta, vapour, oldVapour, Ch, oldCh, dHydr2
    z = grid.linear(n, mySoil.lowerDepth)
    dz = np.append(z[1:n+2]-z[0:n+1],z[n+1]-z[n])
    dz[0] = 1.
    vol = (np.append(np.append(0,(z[2:n+1] - z[0:n-1]) / 2.0),
                     (z[n] - z[n-2]) / 2.0))

    theta0 = soil.thetaFromPsi(soil.CAMPBELL, mySoil, psi0)
    Ch0 = soil.heatCapacity(mySoil, theta0)

    psi = np.zeros(n+2,float)+psi0
    oldPsi = np.zeros(n+2,float)+psi0
    dPsi = np.zeros(n+2,float)
    T = np.zeros(n+2,float)+T0
    oldT = np.zeros(n+2,float)+T0
    theta = np.zeros(n+2,float)+theta0
    oldTheta = np.zeros(n+2,float)+theta0
    Ch = np.zeros(n+2,float)+Ch0
    oldCh = np.zeros(n+2,float)+Ch0
    dpsi = np.zeros(n+2,float)

def solver(mySoil, myBoundary, isFreeDrainage, dt):
    global psi, oldPsi, dpsi, T, oldT, theta
    global oldTheta, vapour, oldVapour, Ch, oldCh, dHydr2
    massBalance = 1.
    energyBalance = 1.
    nrIterations = 0
    psi = np.copy(oldPsi)
    theta = np.copy(oldTheta)
    T = np.copy(oldT)
    if (isFreeDrainage):
        psi[n+1] = psi[n]
        theta[n+1] = theta[n]
        T[n+1] = T[n]

    while ((((massBalance > toleranceVapour) or
    (energyBalance > toleranceHeat))
    and (nrIterations < maxNrIterations))):
        Cw = soil.dTheta_dPsi(soil.CAMPBELL, mySoil, psi)
        meanPsi = np.append(0,(psi[2:n+2]+psi[1:n+1])/2.)
        TCelsius = np.append(0,(T[2:n+2]+T[1:n+1])/2.)
        gradPsi = np.append(0,(psi[2:n+2]-psi[1:n+1])/dz[1:n+1])
        gradT = np.append(0,(T[2:n+2]-T[1:n+1])/dz[1:n+1])
        FlowProp = soil.PropertiesForMassBalance(TCelsius[1:n+1],
                   meanPsi[1:n+1],mySoil)
```

```
dHydr = np.append(np.append(0,(theta[1:n+1]-oldTheta[1:n+1]) *
vol[1:n+1]*1.e3 \ + (FlowProp.K11[0:n]*gradPsi[0:n]-
FlowProp.K11[1:n+1]*gradPsi[1:n+1])*dt \ + (FlowProp.K21[0:n]*
gradT[0:n]-FlowProp.K21[1:n+1]*gradT[1:n+1])*dt \ +
(FlowProp.gravityWaterFlow[0:n]-
FlowProp.gravityWaterFlow[1:n+1])* dt),0)
aHydr = -np.append([0,0],FlowProp.K11[1:n+1]/dz[1:n+1]*dt)
bHydr = np.append([0,0],FlowProp.K11[1:n+1]/dz[1:n+1]*dt)\
+ np.append(np.append(0,Cw[1:n+1]*vol[1:n+1]*1e3),0)\
+ np.append(np.append(0,FlowProp.K11[1:n+1]/dz[1:n+1]*dt),0)
cHydr = -np.append(np.append(0,FlowProp.K11[1:n+1]/dz[1:n+1]*
dt),0)
airResistance = (1.0
/ boundary.boundaryLayerConductance(myBoundary.windSpeed,
myBoundary.airTemperature, T[1]))
dHydr[1] -= boundary.waterFlux(psi[1],
theta[1], T[1], myBoundary, airResistance) *dt
bHydr[1] -= (boundary.dWaterFluxdPsi(psi[1],
theta[1], T[1], myBoundary, airResistance) *dt)
massBalance = np.sum(abs(dHydr))
ThomasBoundaryCondition(aHydr, bHydr,cHydr,dHydr, dpsi, 1, n)
psi -= dpsi
theta = soil.thetaFromPsi(soil.CAMPBELL, mySoil, psi)
if (isFreeDrainage):
    psi[n+1] = psi[n]

Ch = soil.heatCapacity(mySoil, theta)
meanPsi = np.append(0,(psi[2:n+2]+psi[1:n+1])/2.)
TCelsius = np.append(0,(T[2:n+2]+T[1:n+1])/2.)
gradPsi = np.append(0,(psi[2:n+2]-psi[1:n+1])/dz[1:n+1])
gradT = np.append(0,(T[2:n+2]-T[1:n+1])/dz[1:n+1])
FlowProp = soil.PropertiesForMassBalance(TCelsius[1:n+1],
            meanPsi[1:n+1],mySoil)
dTherm = np.append(np.append(0,Ch[1:n+1]*oldT[1:n+1]*
vol[1:n+1] \ + (FlowProp.K12[0:n]*gradPsi[0:n]-
FlowProp.K12[1:n+1]*gradPsi[1:n+1])*dt),0)
aTherm = -np.append([0,0],FlowProp.K22[1:n+1]/dz[1:n+1]*dt)
bTherm = np.append([0,0],FlowProp.K22[1:n+1]/dz[1:n+1]*dt) \
+ np.append(np.append(0,Ch[1:n+1]*vol[1:n+1]),0) \
+ np.append(np.append(0,FlowProp.K22[1:n]/dz[1:n]*dt),[0,0])
cTherm = -np.append(np.append(0,FlowProp.K22[1:n]/dz[1:n]*dt),
                              [0,0])
airResistance = (1.0 /
boundary.boundaryLayerConductance(myBoundary.windSpeed,
myBoundary.airTemperature, T[1]))
BoundaryThermalFlux =( boundary.thermalFlux(psi[1],
theta[1], T[1], myBoundary, airResistance, False))
dTherm[1] += BoundaryThermalFlux*dt
```

```
    bTherm[1] += (4.0*sigma*(myBoundary.airTemperature+273.15)**3*dt
    +(1200.0 / airResistance)*dt)
    ThomasBoundaryCondition(aTherm, bTherm, cTherm, dTherm, T, 1, n)

    #Energy balance for the whole profile
    #energyBalance = abs(sum(Ch[1:n+1]*vol[1:n+1]*(T[1:n+1]-
                          oldT[1:n+1]))\
    #- (FlowProp.K12[0]*gradPsi[0]-FlowProp.K12[n]*gradPsi[n])*dt \
    #+ (4.0*sigma*(myBoundary.airTemperature+273.15)**3*dt
    #+(1200.0 / airResistance)*dt) * T[1] \
    #- BoundaryThermalFlux*dt)

    #Energy balance for the surface layer
    energyBalance = (abs(Ch[1]*vol[1]*(T[1]-oldT[1]) +
    FlowProp.K22[1]/dz[1]*dt*(T[1]-T[2])\
    + FlowProp.K12[1]*gradPsi[1]*dt \
    + (4.0*sigma*(myBoundary.airTemperature+273.15)**3*dt
    +(1200.0 / airResistance)*dt) * T[1] \
    - BoundaryThermalFlux*dt))
    print(energyBalance)

    nrIterations += 1
if ((massBalance < toleranceVapour) and (energyBalance <
                                    toleranceHeat)):
    isBalanceOk = True
    aerodynamicResistance = (1.0
    / boundary.boundaryLayerConductance(myBoundary.windSpeed,
    myBoundary.airTemperature, T[1]))
    waterFlux = (boundary.waterFlux(psi[1],
    theta[1], T[1], myBoundary, aerodynamicResistance))
    heatFlux = (boundary.thermalFlux(psi[1],
    theta[1], T[1], myBoundary, aerodynamicResistance, True))
    evaporationFlux = (boundary.evaporationFlux(psi[1],
    theta[1], T[1], myBoundary.airTemperature,
    myBoundary.relativeHumidity, aerodynamicResistance))
    boundaryLayerConductance =
    boundary.boundaryLayerConductance(myBoundary.windSpeed,
    myBoundary.airTemperature, T[1]))
else:
    isBalanceOk = False
    waterFlux = NODATA
    heatFlux = NODATA
    boundaryLayerConductance = NODATA
    evaporationFlux = NODATA
return (isBalanceOk, waterFlux, heatFlux,
boundaryLayerConductance, evaporationFlux, nrIterations,
massBalance)
```

12.4.6   Long-Wave Radiation

The module PSP_longWaveRadiation is written to compute the long-wave radiation component of the radiation balance at the soil surface. A detailed description of the energy budget is provided in Chapter 15. Three functions are written in this module. The function **vaporConcentrationAir** computes the saturated vapour pressure and vapour concentration as function of air temperature and relative humidity. The saturation vapour pressure of the atmosphere $e_s(T)$ is computed by employing eqn (5.46) and is multiplied by the molecular weight of water and divided by the gas constant and temperature to convert the vapour pressure into a vapour concentration. From the ideal gas law, the corresponding concentration [mol m^{-3}] is used to calculate $PV = nRT_K$, where $[P]$ = Pa, $[V]$ = m^3, $[n]$ = mol, $[T]$ = K and $R = 8.314\,\mathrm{J\,mol^{-1}K^{-1}}$. Therefore $n/V = P/RT_K$, with concentration $[n/V]$ = mol m^{-3}. To convert mol m^{-3} into kg m^{-3}, we multiply by the molecular mass of water, $M_w = 0.018\,\mathrm{kg\,mol^{-1}}$. Therefore

$$c_s = \frac{e_s \times 1000 \times M_w}{R(T + 273.15)} \tag{12.21}$$

where T is temperature in °C. Note that the multiplication by one thousand is written to convert from kPa into Pa. Then the actual vapour concentration is computed by multiplying the saturation vapour concentration by the relative humidity. The value of atmospheric vapour concentration is needed to compute the atmospheric emissivity, which is calculated by the function **atmEmissivity**. Its arguments are measured radiation, day, air temperature and relative humidity. Finally, the function **longWaveRadiationFromWeather** computes the long-wave radiation.

```
#PSP_longWaveRadiation
from __future__ import division
import numpy as np

albedo = 0.2
sigma = 5.670e-8   #[W/m2K4]
latitude = 44.5/360.*2.*3.1415
M_w = 0.018
R = 8.31

def vaporConcentrationAir(T,relHum):
    c_vsat = 0.611*1.e3*np.exp(17.27*T/(T+237.3)) * M_w/(R*(T+273.15))
    c_v = relHum*c_vsat
    return c_v

def atmEmissivity(measuredRadiation,day,T,relHum):
  sin_sD = (0.3985 * np.sin(4.869+0.0172*day
  +0.03345*np.sin(6.224+0.0172*day)))
  cos_sD = np.sqrt(1-sin_sD*sin_sD)
  h_s = np.arccos(-np.tan(latitude)*(sin_sD/cos_sD))
```

```
potentialRadiation = (117.5e6*(h_s*np.sin(latitude)*sin_sD
+np.cos(latitude)*cos_sD*np.sin(h_s))/3.1415)
T_t = measuredRadiation/potentialRadiation
c_1 = 2.33-3.33*T_t
if(c_1 < 0):
   c_1 = 0.
elif(c_1 > 1):
   c_1 = 1.
c_va = vaporConcentrationAir(T,relHum)*1.e3
epsilon_a = 0.58*np.power(c_va,1./7.)
emissivity = (1.-0.84*c_1)*epsilon_a + 0.84*c_1
return emissivity

def longWaveRadiationFromWeather(nrDays):
  longWaveRadiation = np.zeros(nrDays*24)
  hour,temp,prec,hum,wvl,rad = (np.loadtxt('weather.dat',
  usecols=(0,1,2,3,4,6),unpack=True))
  for k in range(nrDays):
   dailyRad = sum(rad[k*24:(k+1)*24])*3600.
   for j in range(24):
     i = k*24+j
     longWaveRadiation[i] = (atmEmissivity(dailyRad,k,
     temp[i],hum[i]/100.)
     *sigma*np.power(temp[i]+273.15,4.))
  return longWaveRadiation
```

Figure 12.1 shows the output of the simulation. In the left-hand column of plots, experimental weather variables are shown with precipitation, air temperature and relative humidity (from top to bottom). The middle column depicts soil water content, soil water potential and soil temperature as functions of depth (from top to bottom). Finally, the right-hand column shows the boundary-layer conductance, the cumulative evaporation, and the air (dots) and soil surface (triangles) temperatures. This last column is shown to observe the heat and vapour transport dynamics at the soil–atmosphere interface. Evaporation is depicted with a negative sign, since it is a loss in the water budget computation.

Note that heat and water are now coupled so the variation of soil water content is affected not only by liquid water transport, but also by transport of water vapour and condensation. Temperature is affected not only by sensible heat transport, but also by latent heat transport associated with vapour transport.

The simulation is performed for a period of 10 days, using weather data from 23 April until 2 May, 2006. Note that the weather data contained in the file weather.dat starts on 23 April 2006. As depicted in Fig. 12.1, during this period, there were some days where precipitation occurred (note the graph on the upper left side). Moreover, after about 90 hours, the relative humidity reached values of 100 % and the soil surface temperature decreased (lower right graph). Since the relative humidity was at or close to 100 %, the vapour pressure deficit was very small and evaporation was negligible. The middle graph in the right-hand column depicted no cumulative evaporation from approximately hour 90 to hour 130.

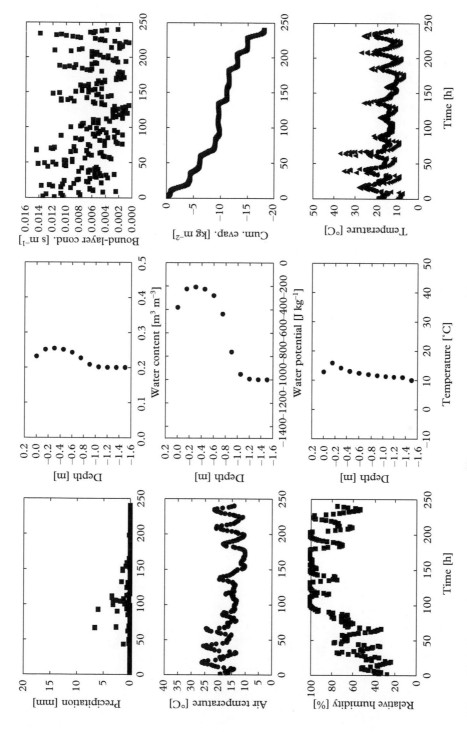

Figure 12.1 *Output of the coupled model.*

The model generates an output file called `energyBalance.dat`, where the energy budget terms are saved at each time step. Specifically, the variables are called `Time`, `RadEmitFromSoil`, `NetRadOnSoil` `latentHeat` and `sensibleHeat` and `G`, with units of W m^{-2}. Specifics about the energy budget computation are provided in Chapter 15. The following is an example of output for a few time steps of simulation:

```
Time      RadEmitFromSoil     NetRadOnSoil    latentHeat      sensibleHeat
0 -340.9 285.7 42.3 59.1
600 -342.7 285.7 37.9 54.3
1800 -343.7 285.7 35.2 51.5
3000 -344.3 285.7 33.5 49.8
4200 -344.7 285.2 31.1 50.7
5400 -345.1 285.2 30.3 49.8
```

Note that the time step is in seconds. Typical values of the surface energy budget terms are shown in Fig. 15.1 in Chapter 15. We implemented a module called `PSP_plotEnergy` that is used to plot the results of the energy budget computation. Figure 12.2 shows the energy budget terms as functions of time for this simulation. As expected, the long-wave upward radiation is always negative, since it is a loss of energy, during both daytime and nighttime. The long-wave downward radiation is always positive. The short-wave (solar radiation) is positive during the day and zero at night. The other fluxes of energy may change depending on the soil and atmospheric boundary-layer conditions. In these weather conditions, the latent heat is negative during the day, since evaporation is predominant during the day; however, there could be latent heat flux during the night as well.

The sensible heat is usually predominant during the daytime, since the soil temperature is larger than air temperature; therefore, heat left the soil surface towards the

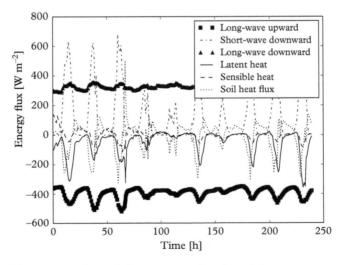

Figure 12.2 *Energy balance terms as functions of time.*

atmosphere and the flux is negative. When meteorological or micro-meteorological experiment are conducted the signs are effectively inverted since the reference framework is now the atmospheric surface layer rather than the subsoil ground medium in regards to turbulent fluxes. Under particular weather conditions, the soil temperature could be lower than the air temperature and therefore sensible heat flux can be positive. Note how, during the period between hours 90 and 120, relative humidity was at or close to 100 % and the latent heat close to zero.

These results are in agreement with the discussion above, where cumulative evaporation was close to zero. Clearly, since evaporation is negligible, latent heat transport is also negligible. During this period, there is also a smaller value of short-wave radiation, indicating overcast meteorological conditions. Overall, the coupled simulations of heat, liquid water and vapour, and the energy budget computation allow for a detailed description of the water and heat transport in the soil–atmosphere system.

12.5 EXERCISES

12.1. Using the program PSP_coupled, compute the evaporation rate for a period of 5 days.
12.2. Select a different period, by changing the start and end times of the simulation (for example by choosing a period during the summer), and discuss the results. Analyse the soil water and soil temperature dynamics, as well as the energy budget terms.

13

Solute Transport in Soils

The ability to predict the rate of movement of solutes in soils and the effect of the soil on this rate is useful in many ways. Knowing the rates of loss of various solutes carried in water moving out of the soil can assist in forming more accurate nutrient budgets and increasing understanding of nutrient cycling. Also, there is interest in knowing the amount and concentration of fertilizer nutrients below the root zone to design management schemes that minimize fertilizer losses and keep concentrations of certain solutes, such as nitrate, within acceptable water quality limits. In addition, understanding the movement of pesticides and other toxic compounds, heavy metals, viruses, and radioactive materials, once they have been applied or spilled onto soils, is essential for predicting their impact on the environment. Knowledge of transport in soils is also useful to reclaim salt-affected soils and to devise management practices for irrigated cropland to prevent excessive salt accumulation in soil. Finally, an understanding of solute transport to roots allows better prediction of the effect of soil on plant nutrition.

Equations describing the movement of a solute through a porous medium such as soil have been derived and investigated (Jury and Roth, 1990). These equations use both analytical and numerical techniques to describe non-interacting as well as interacting solute transport. A particularly good summary of this work is given by Bresler *et al.* (1982). Software with numerical solutions for simulation of advective–dispersive transport for multiple solutes have been developed (Simunek and van Genuchten, 1994; Simunek *et al.*, 2005) and extensive research has been performed in this field, including the application of stochastic methods (Dagan and Neuman, 2005). Moreover, solute transport has been simulated by random-walk particle tracking as an alternative to grid-based Eulerian approaches (Bechtold *et al.*, 2011). Overall, a large body of research and numerical solutions have been performed and developed. It is beyond the scope of this book to provide an exhaustive discussion of the many problems and possible solutions arising when dealing with solute transport, but this chapter aims at providing the fundamental concepts and equations for simple cases of one-dimensional transport. Following this, some consideration will be given to situations where solutes interact with the soil.

Soil Physics with Python. First Edition. Marco Bittelli, Gaylon S. Campbell and Fausto Tomei.
© Marco Bittelli, Gaylon S. Campbell and Fausto Tomei 2015. Published in 2015 by Oxford University Press.

13.1 Mass Flow

Chemical species dissolved in water are transported with the water. This type of transport is termed the *mass flow* of solutes. It is also called *advective transport,* which is transfer of energy or matter by the movement of fluids. The change in concentration with time, at a given point in the soil, of a solute moving by steady mass flow can be described by the equation

$$\rho_b \frac{\partial C}{\partial t} = -f_w \frac{\partial c}{\partial z} \tag{13.1}$$

where c is the solute concentration in the soil solution [kg kg^{-1}], C is the solute present per unit mass of soil [kg kg^{-1}], f_w is the water flux density [kg m^{-2} s^{-1}], z is the soil depth [m], ρ_b is the bulk density [kg m^{-3}] and t is time [s].

A very simple model of solute transport is obtained by assuming that flow through all the soil pores is of uniform velocity (steady state). In this case, the flux of a solute into and out of a soil can be described by eqn (13.1) if the initial solute concentration and water flux density in the soil are known. A typical method used to compare calculated solute flow with actual solute flow is to establish a saturated column of soil in the laboratory and initiate water flow in the column at a rate f_w. A slug of solute is then introduced at the top of the column and the amount of water moving through the column as well as the solute concentration in the water at the bottom of the column are measured. If the solute moved only through mass flow and the flow of all the water in all pores were uniform, then the breakthrough curve of the solute would appear as a piston flow response (Fig. 13.1). The amount of time required for the solute to appear in the outflow is equal to the amount of time taken to replace all the water in the soil (one pore volume) with the introduced water. This depends on f_w, the water flux density in the soil column. Increasing or decreasing the hydraulic conductivity would increase or decrease the time taken for the solute to appear, but would not change the shape of the breakthrough curve.

It is, of course, not realistic to assume that all the water in all the pores will move at the same speed. If we accept that Poiseuille's equation, developed to describe the flux of water in cylindrical tubes, can at least qualitatively describe the influence of viscosity on water flow in non-uniform soil channels, then we can assume that water flux density will be directly proportional to the square of the pore radius as we discussed in Chapter 6 and 8. In other words, water, along with the solutes dissolved in it, will move fastest through the largest pores. Solutes flowing in channels of average pore size would break through after one pore volume. Solute in larger channels would break through sooner than the average and water in small pores would come through well behind the average. The result would be a breakthrough curve like the dispersion curve in Fig. 13.1.

Furthermore, these curves cannot be expected to be linearly correlated with the pore size; the variations around the average hydraulic conductivity will not have a Gaussian distribution (van de Pol *et al.*, 1977). Solute transport due to mass flow, therefore, might be more precisely described by the equation

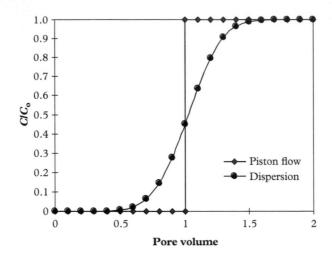

Figure 13.1 *Advection with and without dispersion. The initial concentration of the solute is low. At time 0, solute with concentration c_0 is introduced at the top of the column. c/c_0 is the ratio of outflow to inflow concentration.*

$$\rho_b \frac{\partial C}{\partial t} = -\frac{\partial c}{\partial z} \sum f_{wi} \tag{13.2}$$

where

$$f_{wi} = -K_i \frac{d\psi}{dt} \tag{13.3}$$

is a function of both pore size and the number of pores in a pore size class, and f_{wi} is the water flux density per pore.

Laminar flow through a porous medium further complicates the mass flow of solutes because flow within a single pore is not uniform (i.e. there is a gradient in velocity within the pore). According to Newton's law of viscosity, this gradient in velocity is proportional to the shear force in the flowing liquid, divided by the dynamic viscosity. The water flux density at any one point in the pore is therefore a function of distance from the centre of the pore:

$$f_w = \rho_l \frac{a^2 - r^2}{4\eta} \frac{\Delta P}{\Delta x} \tag{13.4}$$

where a is the radius of the pore, r is the radial distance from the centre of the pore, η is the dynamic viscosity and P is the pressure. In other words, solutes dissolved in water near the side of the pore will flow more slowly than those in the middle of the pore. Since the water containing the solutes is not moving at the same velocity within the pore, mixing occurs along the flow path.

13.2   Diffusion

Solutes move in response to concentration gradients. This transport mechanism is called *diffusion*. Chemical diffusion in free water is described by Fick's law:

$$f_i = -D_i \frac{dC_i}{dx} \qquad (13.5)$$

where f_i is the diffusive flux of the solute i in response to a concentration gradient and D_i is the diffusion coefficient in free water.

Diffusion coefficients for various solutes in free water are available in handbooks. A discussion of solute diffusion, diffusion coefficients and experimental methods has been presented by Flury and Gimmi (2002). To predict the diffusion of an ion in soil, we must use a diffusion coefficient that accounts for both the tortuosity in a porous medium such as soil and the cross-sectional area of water available for diffusion (which is a function of the water content of the soil). Several equations have been proposed for calculating the diffusion of an ion in soil. One commonly used equation was proposed by Bresler (1973):

$$D_m = D_o a e^{b\theta} \qquad (13.6)$$

where D_o is the diffusion coefficient of the ion in water [m^2 s^{-1}] and a and b are soil-dependent parameters. Papendick and Campbell (1980) later proposed the equation

$$D_m = D_o a \theta^3 \qquad (13.7)$$

where a is a parameter to account for tortuosity and θ is the volumetric water content. The equations proposed by Bresler (1973) and Papendick and Campbell (1980) give similar results except in dry soil, where the equation proposed by Papendick and Campbell appears to give more realistic values. Diffusivity is predicted to decrease rapidly with decreasing water content by both equations, because the tortuosity of the flow path increases.

13.3   Hydrodynamic Dispersion

As mentioned earlier when describing the process of *mass flow*, the fact that the convective transport of a solute cannot be accurately described by an average water flux density has led to the inclusion in the diffusion coefficient of a term for hydrodynamic dispersion. The reasons for hydrodynamic dispersion are (a) the different pore sizes, (b) the different distribution of velocities, which changes depending upon the distance from the pore walls, and (c) the different path lengths of the streamlines due to the tortuosity of the porous media.

Bresler (1973) suggests that the dispersion coefficient can be calculated from

$$D = D_m + \frac{\vartheta f_w}{\rho_l \theta} \tag{13.8}$$

where ϑ [m] is a scale-dependent constant, called dispersivity. For laboratory columns, ϑ has values that depend on the porous medium. The dispersion effect (the second term in eqn (13.8)) is a function of f_w and at most flow rates is more important than molecular diffusion. Now using this definition where the parameter D includes both the diffusion and the hydrodynamic dispersion effects, we can write

$$f_s = f_w c - D \frac{\partial c}{\partial z} \tag{13.9}$$

where f_s is the solute flux. The first term on the right-hand side is the mass flow and the second term is the diffusion term (including hydrodynamic dispersion).

13.4   Advection–Dispersion Equation

The equations for mass flow, diffusion and hydrodynamic dispersion can be combined into a general mass conservation equation:

$$\rho_b \frac{\partial C}{\partial t} = \frac{\partial}{\partial z} \left(\rho_l D \frac{\partial c}{\partial z} \right) - \frac{\partial (c f_w)}{\partial z} \tag{13.10}$$

This equation is called the *advection–dispersion equation* and is a conservation equation applied to solutes. The first term on the right-hand side accounts for diffusion and hydrodynamic dispersion, while the second accounts for mass flow, as described above. In some cases, however, the convective component can dominate. It is often convenient to analyse the contribution of mechanical dispersion and diffusion to the solute transport problem. The *Péclet number* is a dimensionless number defined as the ratio of the rate of advection to the rate of diffusion, of the same quantity driven by a gradient. It is defined as

$$Pe = \frac{v_z L}{D} \tag{13.11}$$

where v_z is the advective velocity (in the z vertical direction), L is a characteristic flow length and D is the coefficient of molecular diffusion. As shown in eqn (13.11), the Péclet number increases when the advective component of the flux increases, while it decreases when the diffusive part increases. Overall, the transport of a solute is determined by advection and dispersion–diffusion. The relative importance of these mechanisms depends on the magnitude of the advective component. On the other hand, diffusion becomes dominant at low flow rates.

13.5   Solute–Soil Interaction

In eqns (13.1) and (13.9), the flux of solute is determined by the concentration in solution, c, while the storage at any point in the soil is shown by changes in the variable C. Before the transport equation can be solved, the relationship between c and C must be specified. This relationship may be extremely complex. The presence of each of these can depend on the concentrations of other solutes as well as the exchange characteristics of the soil. Equilibrium and non-equilibrium chemistry models are available for determining the concentrations of a suite of ions and complexes in soil, and such models have been used in conjunction with water flow models to predict reactive transport of a large number of inorganic and organic molecules (Seaman *et al.*, 2012).

Often the absorbed and solution-phase solute concentrations can be related by Langmuir's equation:

$$N = \frac{k_l Q c}{1 + k_l c} \tag{13.12}$$

where N is the coverage of absorption and k_l and Q are constants for a given soil and solute. Langmuir's equation is an equilibrium model of adsorption that describes the adsorption of a monolayer of adsorbate over an homogeneous surface. These constants may depend on the cation exchange capacity (CEC) of the soil and concentrations of other ions in the soil solution. However, Langmuir's equation can also be used for solutes that do not sorb by cation exchange. Values for k_l and Q are normally found from experiments or the more complete equilibrium chemistry models mentioned previously. Values from Smith (1991) for ammonium, phosphorus and potassium are given in Table 13.1

The total amount of solute in the soil is the sum of the adsorbed phase and the solution phase and is given by

$$C = N + wc \tag{13.13}$$

where w is the mass-basis water content [kg kg^{-1}]. Solute concentrations in soil are often sufficiently low that the denominator of eqn (13.12) is near unity. For these conditions, eqn (13.13) becomes

$$C = (k_l Q + w)c \tag{13.14}$$

Table 13.1 *Values of the coefficients for eqn (13.12) (from Smith (1991)*

Solute	k_l [kg water kg^{-1} salt]	Q [kg salt kg^{-1} soil]	$k_l Q$ [kg water kg^{-1} soil]
Phosphorus	3×10^5	2×10^{-4}	60
Potassium	8×10^2	8×10^{-3}	6.4
Ammonium	2×10^3	3×10^{-3}	5.6

This can be written in another form that is often used in solute transport studies. Multiplying both sides by the ratio of bulk density to water density gives

$$\frac{\rho_b C}{\rho_l} = \frac{\rho_b k_l Q}{\rho_l \theta} \theta C = R_f \theta c \qquad (13.15)$$

Here R_f is the retardation factor, which describes the slowing of the solute movement due to adsorption. The term kQ/ρ_l is sometimes called the linear coefficient k_d [m^3 kg^{-1}] for the solute.

Equation (13.15) allows comparison of quantities of adsorbed and solution-phase solutes. For the soil represented by the values in Table 13.1, at a water content of 0.2, the adsorbed phase contains 300 times as much phosphorus, 32 times as much potassium and 28 times as much ammonium as the solution phase. Changes in soil pH and CEC will influence these values, but these numbers give some idea of the amounts of these nutrients in the soil solution compared with total soil amounts. For non-interacting solutes such as Cl$^-$, NO$_3^-$ or SO$_4^{2-}$, eqn (13.14) can still be used, but k_l is set to zero.

13.6   Sources and Sinks of Solutes

In addition to dissolution–precipitation reactions, which are treated by equilibrium chemistry models, biological sources and sinks of solutes must be considered. Nitrification consumes ammonium and produces nitrite and nitrate. Ammonium is a product of decomposition. Microbes and roots take up nutrients from the soil solution.

13.7   Analytical Solutions

For simple conditions, such as steady-state water flow and uniform soil conditions, the advection–dispersion equation can be solved analytically. A detailed description of different analytical solutions was presented by Jury and Roth (1990). Here we describe an analytical solution of the advection–dispersion equation to calculate concentration for a pulse of solute at the inlet. The solute having concentration c_0 is applied at time zero t_0. This condition, of a narrow input pulse, is represented mathematically by the Dirac delta function. The derivation of this function begins with the definition of the Gauss normal distribution, which we have already discussed in previous chapters (eqn (2.26)):

$$f(x) = \frac{1}{\sigma (2\pi)^{1/2}} \exp\left(-\frac{t^2}{2\sigma^2}\right) \qquad (13.16)$$

where σ^2 is the variance and σ is the standard deviation. The distribution is centred at $t = 0$. The area under the curve (integral) is always equal to 1. The *Dirac delta function* is the limit of the distribution as the standard deviation approaches zero, $\sigma \to 0$:

$$\delta(t) = \lim_{\sigma \to 0} f(t, \sigma) \qquad (13.17)$$

This equation represents an infinite spike at $t = 0$. This formulation is used to represent sudden events such as a voltage peak. It is also used to represent the spike of concentration of a solute applied to a soil profile at the upper boundary condition. This condition of a short spike can be represented by an upper boundary condition of the form

$$c_0(t) = \frac{m}{f_w}\delta(t) \tag{13.18}$$

where f_w is the water flux density and m is the solute mass applied per unit area of soil [kg m^{-2}]. The analytical solution for a Dirac delta function for this boundary condition is (Jury and Roth, 1990)

$$c(z, t) = \frac{m}{f_w}\left\{ \frac{v}{\sqrt{\pi Dt}}\exp\left[-\frac{(z - vt)^2}{4Dt}\right] - \frac{v^2}{2D}\exp\left(\frac{vz}{D}\right)\text{erfc}\left(\frac{z + vt}{\sqrt{4Dt}}\right)\right\} \tag{13.19}$$

where f_w is the water flux density, m is the mass of the solute, v is the velocity, D is the dispersion coefficient, z is depth, t is time and erfc is the complementary error function.

13.7.1   Flux and Resident Concentrations

When measuring solute transport and concentration, it is possible to distinguish between *resident* and *flux* concentration. If a soil sample is collected and the concentration of given solute is measured by extracting the soil solution, the measurement is a resident concentration. The resident concentration can also be divided into components depending on the contribution of the pore space to solute transport. Therefore, in some cases, the resident concentration can be divided into mobile and immobile liquid phase. For solute transport studies, often the concentration is measured during a flux experiment, for instance by a fraction collector connected to the outflow of a soil column. In this case, it is called *flux* concentration. These two concentrations may not be the same, since the latter represents solutes moving within macropores and therefore may not represent the solute concentration in micropores. The difference between *resident* and *flux* concentrations is usually larger for high water flux rates, since the contribution of macropore flow is larger.

13.7.2   Implementation

The program `PSP_soluteTransportAnalytical` implements the solution presented in eqn (13.19). The solution was obtained for a water flux density $f_w = 2.5\,\text{g m}^{-2}\,\text{h}^{-1}$, velocity $v = 5\,\text{cm h}^{-1}$, dispersion coefficient $D = 2.5\,\text{cm}^2\,\text{h}^{-1}$ and solute mass $m = 3\,\text{g}$.

```
#PSP_soluteTransportAnalytical
import numpy as np
from scipy import special as sp
import matplotlib.pyplot as plt
```

```
def ade_pulse():
    tmax = 11
    nrHours = 10
    n = 100
    D = 1.
    fw = 2.5
    v = 5.0
    mass = 3.0
    z = np.zeros(n)
    conc = np.zeros(n)
    sqrtPi =    np.sqrt(np.pi)
    for i in range(n):
        z[i] = i

    plt.ion()
    conc[0] = 0.0

    for t in range(1, nrHours):
        for i in range(1, n):
            a = v / (sqrtPi*np.sqrt(D*t)) * np.exp(-((z[i] - v*t)**2)
                / (4.0*D*t))
            b = (v**2)/(2*D) * np.exp((v*z[i])/D) * sp.erfc((z[i]+v*t)
                / np.sqrt(4.0*D*t))
            conc[i]= mass/fw * (a - b)

        plt.clf()
        plt.xlabel('Concentration [g cm$^{-2}$]',fontsize=20,
                labelpad=8)
        plt.ylabel('Depth [cm]',fontsize=20,labelpad=8)
        plt.xlim(0, 2)
        plt.plot(conc, -z, 'k-')
        plt.draw()

    plt.ioff()
    plt.show
    a = input()
    #plt.savefig("analytical.eps", transparent = True)

def main():
    ade_pulse()
main()
```

Figure 13.2 shows the results of the program, with solute concentration plotted as a function of time after 10 h. As the peak moves down, it becomes more dispersed, depending on the value of the dispersion coefficient. The reader can analyse the flux for these boundary conditions, as a function of time and depth, by modifying the parameters of the solution.

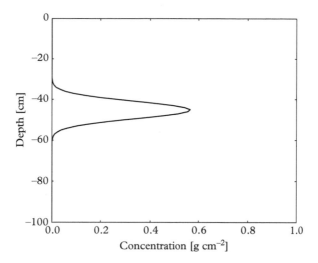

Figure 13.2 *Analytical solution for the solute transport problem with Dirac delta upper boundary condition.*

13.8 Numerical Solution

Solving the solute transport equations by numerical methods requires an approach similar to those used previously for other transport problems. As a first step, eqn (13.9) is formulated as a mass balance for solutes at node i:

$$\frac{\rho_b\left(C_i^{j+1} - C_i^j\right)}{\Delta t} V_i = A_{i-1} f_{i-1}[(1 - \epsilon_{i-1})\bar{c}_{i-1} + \epsilon_{i-1}\bar{c}_i] - A_i f_i[(1 - \epsilon_i)\bar{c}_i + \epsilon_i\bar{c}_{i+1}]$$

$$-\frac{\rho_l A_{i-1} D_{i-1} (\bar{c}_i - \bar{c}_{i-1})}{z_i - z_{i-1}} + \frac{\rho_l A_i D_i (\bar{c}_{i+1} - \bar{c}_i)}{z_{i+1} - z_i} \tag{13.20}$$

where V_i is the volume of cell i, A_i is the area of the interface of cell i and cell $i + 1$, f_i is the water flow from cell i into cell $i + 1$, C is solute present per unit mass of soil and c is the solute concentration in the solution. The left-hand side of the equation is the storage term, the total mass of solute per unit volume. The first term on the right-hand side is the convective solute flux transported by the solution across the interface of cells $i - 1$ and i and the interface of cells i and $i + 1$. The second term on the right-hand side is the diffusive solute flux driven by a gradient in solute concentration in the solution.

Equation (13.14) provides a relation between solute per unit mass of soil C (Conc) and solute concentration in water c (concWat). The dispersion coefficient is calculated using eqn (13.8).

The concentrations are mean values over the time step, calculated from

$$\bar{c}_i = \eta c_i^{j+1} + (1 - \eta)c_i^j \tag{13.21}$$

where η is a weighting factor as described in Chapter 9.

The convective solute transport across two cells is determined by the amount of water flowing across the interface as well as by the solute concentration in the water crossing the interface. If water flows from cell i into cell $i + 1$, then the solute concentration in the water that is crossing the interface is mainly determined by the concentration in cell i. Instead, if water flows from cell $i + 1$ into cell i, then the solute concentration at the interface is mainly determined by the concentration in cell $i + 1$. In the numerical program, the parameter ϵ is set to 0 if water is flowing downwards and to 1 if water is flowing upwards:

$$\begin{aligned} \epsilon = 0 \quad &\text{for} \quad f \geq 0 \\ \epsilon = 1 \quad &\text{for} \quad f < 0 \end{aligned}$$

(13.22)

According to eqn. (13.20), ϵ then determines if the convective solute transport across the interface of cells i and $i + 1$ is calculated based on the concentration in cell i or in cell $i + 1$. This numerical discretization method is called an upwind scheme (Patankar, 1992).

For a simulation of water and solute movement, the equations from this chapter are combined with the water flow programs of Chapter 8. Here we are describing a program where the infiltration and redistribution program presented in Chapter 8 is used to compute the water flow velocity and the solute transport solution is added.

A useful dimensionless number used in numerical solutions is the Courant number C_r, which is used to assess the stability of the solution and was described in Chapter 10:

$$C = \frac{v_x \Delta t}{\Delta x}$$

(13.23)

Specifically the Courant number provides information about the value of the time discretization with respect to the space discretization. For a given spatial discretization, the time step must be selected such that Courant number is always ≤ 1. The number is used to ensure that for a given flow velocity, the time step is not larger than the time necessary for the travelling mass to cover the distance between two nodes. For instance, if the flow is $10 \, \text{ms}^{-1}$ and the spatial discretization is $10 \, \text{m}$, the time step must be ≤ 1 s; otherwise the solute will have travelled a distance larger than the cell size before the end of the time step.

13.9 Numerical Implementation

The project `PSP_soluteTransportNumerical` comprises eight files:

1. `main.py`
2. `PSP_grid.py`
3. `PSP_readDataFile.py`

4. `PSP_soil.py`
5. `PSP_Infiltration1D.py`
6. `PSP_Diffusion1D.py`
7. `PSP_ThomasAlgorithm.py`
8. `soil.txt`

The first four files are the same as in the project `PSP_Infiltration1D` and therefore are not described here. The file `PSP_Infiltration1D.py` was slightly modified by choosing a linear grid and by selecting the matric flux potential solution (MFP), since this provides faster convergence. Moreover, a code line has been added at the end of the code to calculate the Courant number and to check if it is smaller than the provided limit. The file `main.py` was also modified by selecting only the Campbell parameterization for the soil hydraulic properties. To focus on the diffusion process only, we simplified the infiltration process by choosing a constant water potential ψ ($\theta = 0.2$) as the initial potential and also as the upper boundary condition. If free drainage is chosen as the lower boundary condition, the water content in the domain remains constant with a constant gravity flow field. A flow boundary condition corresponding to the precipitation or evaporation rate at the upper boundary is more realistic; however, the simplified conditions allow for visualization of the diffusion process, and the program can be easily modified by using the atmospheric boundary conditions presented in Chapter 12.

The file `PSP_diffusion1d.py` is the only new file in this project. It has a similar structure to the file `PSP_heat.py`. At the beginning, memory is allocated for the arrays. The variable `Conc` is the solute concentration per unit mass of soil at the end of the time step, `oldConc` is the solute concentration per unit mass of soil at the beginning of the time step, `D` is the diffusion coefficient, `fd` is the diffusive part of solute flux, while `fc` is the convective part of solute flux and `epsilon` is used to check the direction of the flow, as described in eqn (13.22). Afterwards, a function initializes the solute concentration array with a solute pulse located at the surface. Finally, a cell-centred finite volume scheme is implemented to set up the matrix that is solved using the Thomas algorithm. The variable `factor` allows for choosing between explicit (factor = 0) and implicit (factor = 1).

In each time step, water content and flux density are passed from `PSP_infiltration1D.py` as arguments to the file `PSP_diffusion1d.py`. The module is as follows:

```
#PSP_diffusion1D
from __future__ import print_function, division
import numpy as np
from PSP_ThomasAlgorithm import ThomasBoundaryCondition

waterDensity = 1000.
bulkSoilDensity = 2650.
```

```
area = 1
Theta = 0.01
D_0   = 1e-9
a_const = 2.8
klQ = 0.

n = 100
z = np.zeros(n+2, float)
vol = np.zeros(n+2, float)
a = np.zeros(n+2, float)
b = np.zeros(n+2, float)
c = np.zeros(n+2, float)
d = np.zeros(n+2, float)

dz = np.zeros(n+2, float)
Conc = np.zeros(n+2, float)
oldConc = np.zeros(n+2, float)
concWat = np.zeros(n+2, float)
D = np.zeros(n+2, float)

fd = np.zeros(n+2, float)
fc = np.zeros(n+2, float)
epsilon = np.zeros(n+2, float)

def initializeDiffusion(n_, z_,dz_,vol_, theta):
    # vector depth [m]
    for i in range(n+2):
        z[i] = z_[i]
        dz[i] = dz_[i]
        vol[i] = vol_[i]
    for i in range(1, n+2):
        if(z[i]>0.05 and z[i]<0.2):
            concWat[i] = 1e-3
        else:
            concWat[i] = 0
        oldConc[i] = concWat[i] * (klQ + (waterDensity*theta[i]) /
                                        bulkSoilDensity)
        Conc[i]    = oldConc[i]

def cellCentFiniteVolWater(dt, waterFluxDensity, theta, meanType,
                           factor):
    g = 1.0 - factor
    fd[0] = 0
    fc[0] = 0
    for i in range(1, n+1):
        theta_mean = 1/2*(theta[i]+theta[i+1])
```

```
      D[i]    = D_0 * a_const * np.power(theta_mean,3.) \
               + (Theta * waterFluxDensity[i]) /
                 (waterDensity*theta_mean)
      fd[i] = area * D[i] / dz[i]
      fc[i] = area * waterFluxDensity[i]
      if(waterFluxDensity[i] < 0):
           epsilon[i] = 1
  for i in range(1, n+1):
      a[i] = -fd[i-1]*factor - fc[i-1]*(1-epsilon[i-1])*factor
      if (i == 1):
          b[i] =((klQ + (waterDensity*theta[i])/bulkSoilDensity) *
                  vol[i]/dt + fd[i]*factor + fc[i]*(1-epsilon[i])*
                  factor - fc[i-1]*epsilon[i-1]*factor)
          c[i] = -fd[i]*factor + fc[i]*epsilon[i]*factor
          d[i] = (oldConc[i] * vol[i]/dt + g * ((fd[i]+fc[i]*
                  (1-epsilon[i])+ fc[i-1]*epsilon[i-1])*oldConc[i] -
                  fc[i]* epsilon[i]*oldConc[i+1]))
      elif (i < n):
          b[i] =(((klQ + (waterDensity*theta[i])/bulkSoilDensity) *
                 vol[i]/dt + fd[i-1]*factor + fd[i]*factor + fc[i]*
                 (1-epsilon[i])*factor) - fc[i-1]*epsilon[i-1]*factor)
          c[i] = -fd[i]*factor + fc[i]*epsilon[i]*factor
          d[i] = (oldConc[i] * vol[i]/dt + g * ((fd[i-1]+fc[i-1]*
                  (1-epsilon[i-1]))* oldConc[i-1] -
                  (fd[i-1]+fd[i]+fc[i]*(1-epsilon[i])-
                  fc[i-1]*epsilon[i-1])* oldConc[i] -
                  fc[i]*epsilon[i]*oldConc[i+1]))

      elif (i == n):
          b[i] = ((klQ + (waterDensity*theta[i])/bulkSoilDensity) *
                 vol[i]/dt + fd[n-1]*factor + fc[i]*(1-epsilon[i])*
                 factor - fc[i-1]*epsilon[i-1]*factor)
          c[i] = 0
          d[i] = (oldConc[i] * vol[i]/dt  + g * ((fd[i-1]+fc[i-1]*
                  (1-epsilon[i-1]))* oldConc[i-1] -
                  (fd[i]+fc[i]*(1-epsilon[i])+
                  fc[i-1]*epsilon[i-1])*oldConc[i]))

  ThomasBoundaryCondition(a, b, c, d, concWat, 1, n)
  return True
```

Figure 13.3 shows the results of the simulation. The water content changes over time are shown in Fig. 13.3(a) the infiltration flux at the upper boundary in (b) and the concentration of solute in (c). As described above, we set a gravity flow boundary condition so that the water content as a function of depth is constant across the soil profile.

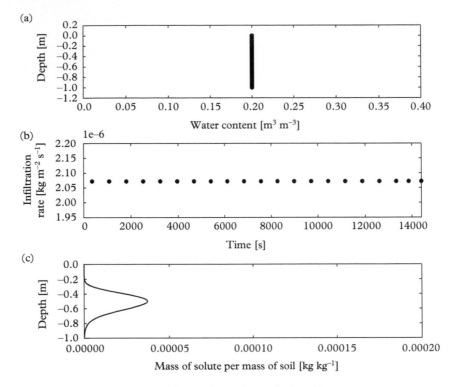

Figure 13.3 *Advection and diffusion for a solute pulse in soil.*

· ·

13.10 EXERCISES

13.1. Change the value of the dispersivity coefficient in the program `PSP_soluteTransportAnalytical`. Discuss the effect of dispersivity on the shape of the curve.

13.2. Modify the upper boundary condition to set a prescribed flux given by precipitation and observe the changes in solute concentration in the soil profile as a function of time. Discuss the difference between the condition of gravity flow and the results of the simulation.

13.3. Select different values for the Courant number and note how the shape of the pulse depends on it.

13.4. Modify the program to impose atmospheric boundary conditions and discuss the results.

14

Transpiration and Plant–Water Relations

Now that models have been developed for heat, water and solute transport in soil, consideration can be given to the interaction between the physical environment described by such models and biological systems that are influenced by that environment. A very important concern that fits into this category is that of water transport and water potentials in the soil–plant–atmosphere–continuum (SPAC). Figure 14.1 shows a scheme of the SPAC. Water flows from soil to roots, through xylem, through mesophyll cells and cell walls, and finally evaporates into sub-stomatal cavities. It then diffuses out of the stomates, through the leaf and canopy layers, and is mixed with the bulk atmosphere. Water potential is highest in the soil, and decreases along the transpiration path. This potential gradient provides the driving force for water transport from the soil to the atmosphere. Water potentials, water fluxes and resistances to flow in the SPAC will be considered in this chapter.

14.1 Soil Water Content and Soil Water Potential under a Vegetated Surface

Plant transpiration and plant water uptake decrease the water content in the soil profile. For many natural and cultivated plants, the transpiration component of the water balance is very important, often accounting for 50–80 % of the yearly precipitation (Pieri *et al.*, 2007), indicating the importance of adequate transpiration quantification in models.

The behaviour described above can be detected in field-measured values of soil water content and soil matric potential as functions of time and soil depth. Figure 14.2 shows soil volumetric water content (dots) and matric potential (triangles) measured over a two-year period (data from Bittelli *et al.*, 2012). Variables were measured at −0.2 and −0.4 m depth. Figure 14.2(a) shows daily precipitation. The area where the data were collected is covered by dense natural vegetation (various species of shrubs and herbaceous plants), resulting in high plant water uptake, with roots that can reach down to 1.5 m depth. Therefore the decrease in soil water content from the beginning of May to the end of summer is due mostly to plant transpiration. The soil profile is recharged by

Soil Physics with Python. First Edition. Marco Bittelli, Gaylon S. Campbell and Fausto Tomei.
© Marco Bittelli, Gaylon S. Campbell and Fausto Tomei 2015. Published in 2015 by Oxford University Press.

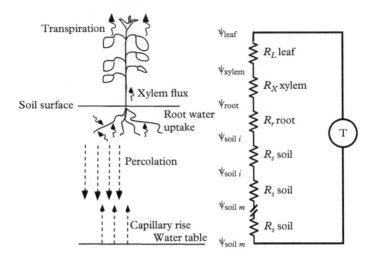

Figure 14.1 *Scheme of the SPAC.*

precipitation in the autumn, with soil water content displaying small variations during the autumn and winter.

The soil water volumetric content and matric potential data exhibit the typical inverse relationship described in Chapter 5, with soil water content decreasing when the matric potential increases (in absolute value), representing a dynamic field-measured soil water retention curve. Precipitation increased soil water content, which is particularly evident at the −0.2 m depth profile. Overall, the sharp decrease in soil water content is due in large part to plant water uptake, which stresses the importance of using correct plant models.

14.2   General Features of Water Flow in the SPAC

Figure 14.3 shows an electrical analogue of the SPAC and indicates where the most important resistances and potentials are located. Evaporation from the leaf is shown as a current source, since the plant water potential has no direct effect on the flux (relative humidity in sub-stomatal cavities is always near 1.0). The circuit analogue equation is written as

$$E = \frac{\psi_{xL} - \psi_L}{R_L} = \frac{\psi_{xr} - \psi_{xL}}{R_x} = \frac{\psi_r - \psi_{xr}}{R_r} = \frac{\psi_s - \psi_r}{R_s} \tag{14.1}$$

where E is transpiration rate [kg m^{-2} s^{-1}], R is resistance [m^4 kg^{-1} s^{-1}] and ψ is water potential [J kg^{-1}]. The variables R_L, R_x, R_r and R_s are the resistances in the leaf, xylem, root and soil, respectively. The gradient is determined by the water potential values at the sub-stomatal cavities (ψ_L), the mesophyll cell and cell walls in the leaf (ψ_{xL}), the xylem (ψ_{xr}), the root (ψ_r), and the soil (ψ_s).

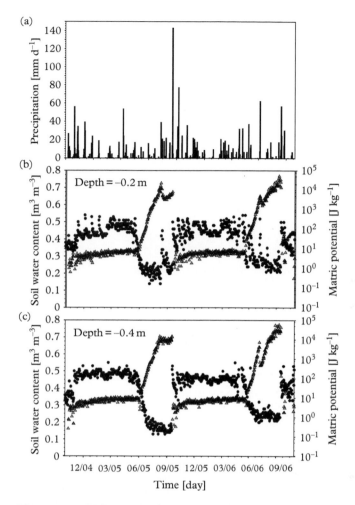

Figure 14.2 *Field-measured soil volumetric water content (dots) and matric potential (triangles) for a two-year period.*

The equation in this form is similar to the steady-state oxygen diffusion problem considered in Chapter 3. All of the potentials can be predicted if the flux and the soil water potential are known. Before actually doing that, however, some simple predictions from eqn (14.1) will be considered.

If an overall resistance is defined as the series combination of all resistances in Fig. 14.3, then

$$E = \frac{\psi_s - \psi_L}{R} \tag{14.2}$$

This equation can be rearranged to predict the leaf water potential:

$$\psi_L = \psi_s - ER \tag{14.3}$$

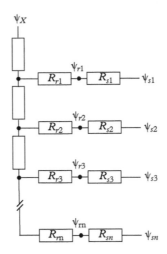

Figure 14.3 *Electrical analogue of the soil–root system, showing water potential with depth in the soil and soil and root resistances to water uptake.*

Equation (14.3) predicts that leaf water potential will be below soil water potential by an amount determined by transpiration rate and resistance to transport. When soil moisture is not limiting, E is independent of ψ_L, and R is about constant, so ψ_L just varies with E. Figure 14.4 shows a simulation of leaf water potential as it responds to fluctuations in E. Until soil water becomes limiting, this response is apparently passive and has little, if any, direct effect on photosynthesis of leaves.

The turgor potential pressure in leaves is $\psi_p = \psi_L - \psi_o$, where ψ_p is the turgor potential and ψ_o is the osmotic potential of the leaf. This value fluctuates with E. Decreased turgor does slow cell expansion, and therefore may indirectly affect plant photosynthesis over time by reducing the photosynthetic area. Reduced turgor apparently does not affect photosynthesis of leaves that are already in place until stomatal closure begins.

It seems reasonable to assume that water stress responses in plants depend on some integrated value of the turgor pressure, since the turgor pressure combines factors of soil water supply, atmospheric demand and plant osmotic potential. This has not yet been proven, however. It also seems apparent that a plant's response to water stress is such that some turgor is maintained. The response to short-term stress is to close stomates, reducing E so that ψ_L is increased. For many species, the response to prolonged stress is to reduce leaf area index so that the leaves that remain are operating at maximum efficiency.

With these ideas in mind, each of the resistances in the transpiration stream will be considered and the relationship between transpiration rate and leaf water potential will be determined.

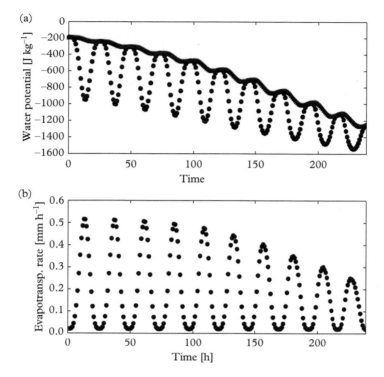

Figure 14.4 *Simulated diurnal fluctuations in soil and leaf water potentials (a) and plant evapotranspiration rate (b) over a 10-day period. In (a) the dots represent leaf water potential and the triangles soil water potential.*

14.3 Resistances to Water Flow within the Plant

From Chapter 6, we know that soil conductance is high when soil is wet. Measurements of water potentials and fluxes for transpiring plants growing in moist soil can therefore be used to find plant resistances. Measurements typically show that xylem resistance is negligible compared with other resistances, although notable examples to the contrary are found in diseased plants and spring-sown cereals (Passioura, 1972). The major resistances within the plant are at the endodermis, where water enters the root steele, and in the leaf, perhaps at the bundle sheath. Water may flow in cell walls and xylem vessels through most of the transpiration stream, but must cross membranes at the endodermis and perhaps at the bundle sheath. For typical plants growing in moist soil, the potential drop across the endodermis is 60–70 % of the total. Between 30 and 40 % of the potential drop is at the leaf. This may be demonstrated by stopping transpiration on a leaf by covering it with aluminium foil and a polyethylene bag and measuring its water potential after allowing a few hours for equilibration. The water potential of the covered leaf is equal to the water potential of the xylem. If the soil resistance and water potential are negligible, then the ratio of xylem water potential to water potential of a freely transpiring leaf is equal to the ratio of root resistance to total resistance, eqn (14.1).

Rough estimates of plant resistance can be made by considering typical potentials and transpiration rates. In a potato crop, for example, after canopy closure, typical leaf water potentials are around -1200 J kg^{-1} when the transpiration rate is 2.4×10^{-4} kg m^{-2} s^{-1}, so a typical total resistance would be around 5×10^{6} m^{4} s^{-1} kg^{-1} and the leaf resistance would be 2×10^{6} m^{4} s^{-1} kg^{-1}. The root resistance is the difference, or 3×10^{6} m^{4} s^{-1} kg^{-1}. The xylem resistance is perhaps a factor of 10 smaller than these in a healthy plant.

For modelling purposes, it is often useful to have an estimate of the resistance per unit length of root. This can be used with estimates of root density to find the resistance to water uptake for any soil layer. Data that provide rooting density estimates, transpiration measurements and water potentials are difficult to find, but some values are given by Bristow *et al.* (1984) for sunflower. They found values of resistance per unit length of root of 2.5×10^{10} m^{3} s^{-1} kg^{-1} To compare this against the calculation just made for potato, a rooting depth of 0.3 m and a root density of 3×10^{4} m^{-2} (metres of root per cubic metre of soil) might be assumed. The resistance per metre of root divided by the length of root per square metre of ground area gives the root resistance. Using the depth and density just assumed, Bristow *et al.* (1984) give $R_r = 2.8 \times 10^{6}$ m^{4} s^{-1} kg^{-1}, which is close to the estimate given above for potato.

14.4   Effect of Environment on Plant Resistance

It is difficult to find quantitative information on the relationship between resistances to water flow and environmental variables for plants growing in field conditions. Several qualitative observations are possible, however. Since water crosses living cells at the endodermis, anything that affects cell metabolism affects resistance. Decreases in temperature, decreases in oxygen supply and decreases in root water potential all decrease root permeability and therefore increase resistance to flow across the endodermis.

14.5   Detailed Consideration of Soil and Root Resistances

The resistances encountered by water as it flows from the bulk soil to the root xylem are highly variable in both space and time. These resistances are the ones most susceptible to alteration by management practices, so a knowledge of their response to environment is important if the soil is managed for maximum crop production.

In the electrical analogue of the root–soil system shown in Fig. 14.3, at each level in the soil, a soil and a root resistance are in series. Axial resistances within the root are also shown, though for computations these are assumed to be negligible compared with soil and root resistances. An interfacial resistance for coarse-textured soil (Bristow *et al.*, 1984) can be added to the soil resistance to simulate water uptake in sandy soils.

The resistance to water uptake in any soil layer, i, is inversely proportional to the length of root in that layer. Therefore, for the root resistance in layer i,

$$R_{r_i} = \frac{\rho_r}{L_i \Delta z} \tag{14.4}$$

where ρ_r is the resistance per unit length of root, L_i is the density of the root system in layer i and Δz is the depth of the layer. If we assume root density to decrease linearly with depth, then

$$L_i = L_0 \left(1 - \frac{z_i}{z_{rd}} \right) \tag{14.5}$$

where L_i is the root density [m^{-2}] (length of root per unit volume of soil) at depth i, L_0 is the rooting density at the surface and z_{rd} is the rooting depth.

The resistance of the soil for water flow to roots is determined by soil hydraulic conductivity, root density and water uptake rate (Cowan, 1965; Gardner, 1960). An equation for this resistance is obtained by considering uptake of water from soil by a cylindrical root. If the rate of water uptake by a root is q [kg s^{-1}], then

$$\frac{q}{A} = -K \frac{d\psi_m}{dr} \tag{14.6}$$

where K is unsaturated hydraulic conductivity and r is the radial distance from the centre of the root. If we assume steady flow, then q is constant with r. The area is $A = 2\pi r l$, where l is the length of root. Substituting eqn (6.35) into (14.6) and integrating from the root surface r_1 to a distance r_2 that represents the mean distance between roots gives

$$\frac{q}{2\pi l} \ln \left(\frac{r_2}{r_1} \right) = \frac{K_2 \psi_2 - K_1 \psi_1}{1 - n} \tag{14.7}$$

This is obtained by using the Campbell formulation of the hydraulic conductivity. The water uptake per unit length of root can be related to root density, soil depth and water extraction rate by

$$\frac{q}{l} = \frac{E}{L_i \Delta z_i} \tag{14.8}$$

where E_i is the uptake rate from soil layer i, L_i is the rooting density and z_i is the thickness of the soil layer. The mean distance between roots is

$$r_2 = \frac{1}{\sqrt{\pi L_i}} \tag{14.9}$$

Making these substitutions and solving for E_i,

$$E_i = \frac{K_{r_i} \psi_{r_i} - K_{s_i} \psi_{s_i}}{B_i} \tag{14.10}$$

where

$$B_i = \frac{(1-n)\ln(r_1^2 L_i)}{4\pi L_i \Delta z_i} \tag{14.11}$$

It is interesting to note that eqn (14.10) predicts a maximum rate at which water can be taken up by the plant. If ψ_{r_i} were $-\infty$, K_{r_i} would approach zero and E_i would equal $-K_{s_i}\psi_{s_i}/B_i$, which is constant for a given soil water potential. The soil resistance to water uptake is

$$R_{s_i} = \frac{\psi_{s_i} - \psi_{r_i}}{E_i} \tag{14.12}$$

so, using eqn (14.10),

$$R_{s_i} = B_i \frac{\psi_{s_i} - \psi_{r_i}}{K_{r_i}\psi_{r_i} - K_{s_i}\psi_{s_i}} \tag{14.13}$$

Usually root resistance is much larger than soil resistance and therefore eqn (14.13) can be simplified to

$$R_{s_i} = \frac{B_i}{K_i} \tag{14.14}$$

Using these equations for soil and root resistance, and returning to Fig. 14.3, the analysis of water uptake can be continued. The transpiration rate must be the sum of the rates of water extraction for each of the soil layers, $E = \sum E_i$, so

$$E = \sum \frac{\psi_{si} - \psi_{xr_i}}{R_{s_i} - R_{r_i}} \tag{14.15}$$

If it can be assumed that the axial resistances are small compared with other resistances, then eqn (14.15) can be solved for ψ_{xr}:

$$\psi_{xr} = \frac{-E + \sum \dfrac{\psi_{s_i}}{R_{s_i} - R_{r_i}}}{\sum \dfrac{1}{R_{s_i} + R_{r_i}}} \tag{14.16}$$

The leaf water potential is given by (assuming negligible xylem resistance)

$$\psi_L = \psi_{xr} - E R_L \tag{14.17}$$

and so, using eqn (14.16),

$$\psi_L = \frac{\sum \dfrac{\psi_{s_i}}{R_{s_i} + R_{r_i}}}{\sum \dfrac{1}{R_{s_i} + R_{r_i}}} - \frac{E}{\sum \dfrac{1}{R_{s_i} + R_{r_i}}} - ER_L \tag{14.18}$$

The first term on the right-hand side in eqn (14.18) can be considered a weighted mean soil water potential, $\overline{\psi_s}$. This can be seen by noting that when $E = 0$, ψ_L equals $\overline{\psi_s}$. The last term is the transpiration rate multiplied by the sum of two resistances: R_L and a weighted mean root–soil resistance $\overline{R_{sr}}$. Equation (14.18) must be solved simultaneously with another that expresses the effect of ψ on E (through stomatal conductance). That relationship can be written as

$$E = \frac{E_p}{1 + \left(\dfrac{\psi_a}{\psi_L}\right)^a} \tag{14.19}$$

where E_p is the potential transpiration rate, ψ_c is the leaf water potential for stomatal closure and a is a species-dependent constant with a typical value of around 10. When $\psi_L = \psi_c$, $E = E_p/a$.

14.6   Numerical Implementation

The equations that we have developed can now be combined in a computer program that will simulate water uptake and loss by plants. The project `PSP_transpiration` comprises eight files:

1. `main.py`
2. `PSP_grid.py`
3. `PSP_readDataFile.py`
4. `PSP_soil.py`
5. `PSP_vapor1D.py`
6. `PSP_plant.py`
7. `PSP_ThomasAlgorithm.py`
8. `soil.txt`

The modules `PSP_grid.py`, `PSP_readDataFile.py`, `PSP_soil.py` have been described in the previous chapters. The module `PSP_ThomasAlgorithm.py` is also described in detail in *Appendix B*.

The **main** function imports the file `PSP_vapor1D.py`, which is very similar to the program presented in Chapter 11. This file is included here to compute infiltration, redistribution and vapour flow, then in this code the plant transpiration is added.

The **main** function begins with the import of the necessary modules as shown in the previous programs. Then it prints on screen the selection of different numerical solutions that the user can choose for liquid water transport, as well as the selection of lower boundary conditions. In this example, the simulation was performed for a period of 10 days (240 hours), with the simulation setup selected at the prompt:

```
1   Cell-Centered Finite Volume
2   Matric Potential with Newton-Raphson
Select solver: 1

]0, 0.46] initial water content (m^3 m^-3):0.2

1: Free drainage
2: Constant water potential
Select lower boundary condition:1

Nr of simulation hours:240
```

The following lines of code are written to set up the simulation length and the time step. The initial condition are then set by calling the functions **initializeWater** and **InitTransp** as **vap.initializeWater** and **vap.plant.InitTransp**. Then the loop over time begins with the instruction while (time < endTime). The following function computes potential evapotranspiration and partitions it between potential evaporation and potential transpiration by simply assigning 10 % to evaporation and the rest to transpiration. More sophisticated subroutines for this will be given in Chapter 15. After the computation of potential evapotranspiration, the solvers are called for solution of liquid water and vapour flow.

```
#PSP_transpiration
from __future__ import print_function, division
from PSP_public import *
import matplotlib.pyplot as plt
import PSP_vapor1D as vap

def main():
    isSuccess, soil = vap.readSoil("soil.txt")
    if not isSuccess:
        print("warning: wrong soil file.")
        return

    funcType = vap.CAMPBELL

    print (vap.CELL_CENT_FIN_VOL,' Cell-Centered Finite Volume')
    print (vap.NEWTON_RAPHSON_MP,' Matric Potential with
           Newton-Raphson')
    solver = int(input("Select solver: "))
```

```
myStr = "]0, " + format(soil.thetaS, '.2f')
myStr += "] initial water content (m^3 m^-3):"
thetaIni = vap.NODATA
print()
while ((thetaIni <= soil.VG_thetaR) or (thetaIni > soil.thetaS)):
    thetaIni = float(input(myStr))


vap.initializeWater(funcType, soil, thetaIni)
vap.plant.InitTransp(soil, vap.z)


print()
print ("1: Free drainage")
print ("2: Constant water potential")
boundary = int(input("Select lower boundary condition:"))
if (boundary == 1):
    isFreeDrainage = True
else:
    isFreeDrainage = False


simulationLenght = int(input("\nNr of simulation hours:"))
endTime = simulationLenght * 3600
dt = maxTimeStep / 10
time = 0
sumETr = 0
totalIterationNr = 0


plt.ion()
f, myPlot = plt.subplots(4, figsize=(8, 9), dpi=80)
f.subplots_adjust(hspace=.35)
myPlot[0].tick_params(axis='both', which='major',
                      labelsize=14,pad=4)
myPlot[1].tick_params(axis='both', which='major',
                      labelsize=14,pad=4)
myPlot[2].tick_params(axis='both', which='major',
                      labelsize=14,pad=4)
myPlot[3].tick_params(axis='both', which='major',
                      labelsize=14,pad=4)


myPlot[1].set_xlim(0, simulationLenght)
myPlot[1].set_ylim(0, 0.8)
myPlot[1].set_ylabel("Evapotransp. Rate [mm h$^{-1}$]")
myPlot[1].set_xlabel("Time [h]")


while (time < endTime):

    dt = min(dt, endTime - time)
    TimeOfDay = float(time%(3600*24))/3600.
    tp = vap.fi * vap.ETp * 2.3 * vap.np.power(0.05
                + vap.np.sin(0.0175 * 7.5 * TimeOfDay),4.)
```

```
vap.plant.PlantWaterUptake(tp, funcType, soil,
                           vap.theta, vap.psi, vap.z)
#print("leaf pot",leafPot)

if (solver == vap.CELL_CENT_FIN_VOL):
    success, nrIterations, flux = (
    vap.cellCentFiniteVolWater(funcType, soil, dt,
                               isFreeDrainage))
elif (solver == vap.NEWTON_RAPHSON_MP):
    success, nrIterations, flux = (
    vap.NewtonRapsonMP(funcType, soil, dt, isFreeDrainage))
totalIterationNr += nrIterations

transp, leafPot, soilPot=vap.plant.PlantWaterUptake(tp,
                funcType, soil, vap.theta, vap.psi, vap.z)
print ("leafPot",leafPot)

if (success):
    for i in range(vap.n+1):
        vap.oldTheta[i] = vap.theta[i]
        vap.oldvapor[i] = vap.vapor[i]
        vap.oldpsi[i] = vap.psi[i]
    sumETr += flux * dt
    time += dt

    print("time =", int(time), "\tdt =", int(dt),
          "\tIter. =", int(nrIterations),
          "\tsum ETr:", format(sumETr, '.3f'))

    myPlot[0].clear()
    myPlot[0].set_xlabel("Water content [m$^3$ m$^{-3}$]")
    myPlot[0].set_ylabel("Depth [m]")
    myPlot[0].set_xlim(-0, soil.thetaS)
    myPlot[0].plot(vap.theta[1:vap.n+1], -vap.z[1:vap.n+1], 'ko')

    myPlot[1].plot(time/3600., flux*3600., 'ko')
    myPlot[3].plot(time/3600., leafPot, 'ko')
    myPlot[3].plot(time/3600., soilPot, 'k^')

    myPlot[2].clear()
    myPlot[2].set_xlabel("Root Water Uptake [-]")
    myPlot[2].set_ylabel("Depth [m]")
    myPlot[2].set_xlim(-0.5e-5,1.5e-5)
    myPlot[2].plot(vap.plant.t[1:vap.n+1], -vap.z[1:vap.n+1], 'ko')

    myPlot[3].set_xlabel("Time [-]")
    myPlot[3].set_ylabel("Water potential [J kg$^{-1}$]")
    myPlot[3].set_xlim(0, simulationLenght)
```

```
            plt.draw()

            if (float(nrIterations/vap.maxNrIterations) < 0.25):
                    dt = min(dt*2, maxTimeStep)
        else:
            dt = max(dt / 2, 1)
            for i in range(vap.n+1):
                vap.theta[i] = vap.oldTheta[i]
                vap.vapor[i] = vap.oldvapor[i]
                vap.psi[i] = vap.waterPotential(funcType, soil,
                                            vap.theta[i])
            print ("dt =", dt, "No convergence")
            print(time)
            break

    plt.ioff()
    plt.show()
main()
```

Similar to other programs, the file PSP_public is written to define constants, simulation properties and variables. The program is as follows:

```
#PSP_public.py
NODATA = -9999.

rootDepth = 0.6
rootMin = 0.02
TKelvin = 293
svp = 0.017
ETp = 0.2/3600.
rhAir = 0.5
fi = 0.9

n = 50
area = 1
maxNrIterations = 100
tolerance = 1e-3
maxTimeStep = 3600

waterDensity = 1000.
L = 2.45e6
g = 9.8065
Mw = 0.018
R = 8.31
waterVapourDiff0 = 2.12E-5
zeroKelvin = 273.15
```

The file `PSP_vapor1D` computes liquid water and vapour flow by implementing the solvers for the numerical solution. A description of this code was provided in Chapters 3 and 8. Here water uptake for each layer becomes a sink–source term for the mass balance equation as detailed here.

```
d[n] = C[n] * oldpsi[n] + f[n]*psi[n+1] -
area * (f[i]*dz[i]-f[i-1]*dz[i-1]) * g - plant.e[i]
```

The code with the numerical solution is as follows:

```
#PSP_vapor1D
from __future__ import print_function, division
from PSP_public import *
import PSP_grid as grid
import PSP_plant as plant
from PSP_ThomasAlgorithm import ThomasBoundaryCondition
from PSP_soil import *

z = np.zeros(n+2, float)
vol = np.zeros(n+2, float)
a = np.zeros(n+2, float)
b = np.zeros(n+2, float)
c = np.zeros(n+2, float)
d = np.zeros(n+2, float)

dz = np.zeros(n+2, float)
psi = np.zeros(n+2, float)
oldpsi = np.zeros(n+2, float)
dpsi = np.zeros(n+2, float)
theta = np.zeros(n+2, float)
oldTheta = np.zeros(n+2, float)
vapor = np.zeros(n+2, float)
oldvapor = np.zeros(n+2, float)
C = np.zeros(n+2, float)
k = np.zeros(n+2, float)
k_mean = np.zeros(n+2, float)

u = np.zeros(n+2, float)
du = np.zeros(n+2, float)
f = np.zeros(n+2, float)
H = np.zeros(n+2, float)
H0 = np.zeros(n+2, float)

def initializeWater(funcType, soil, theta_0):
    global z
```

```
    # vector depth [m]
    z = grid.linear(n, soil.lowerDepth)
    vol[0] = 0
    for i in range(n+1):
        dz[i] = z[i+1]-z[i]
        if (i > 0): vol[i] = area * (z[i+1] - z[i-1]) / 2.0

    #initial conditions
    psi_0 = waterPotential(funcType, soil, theta_0)
    k_0 = (hydraulicConductivityFromTheta(funcType, soil, theta_0)
            + vaporConductivityFromPsiTheta(soil, psi_0, theta_0))

    psi[0] = 0
    for i in range(1, n+2):
        oldTheta[i] = theta_0
        theta[i] = theta_0
        oldvapor[i] = vaporFromPsi(funcType, soil, psi[i], theta[i])
        vapor[i] = vaporFromPsi(funcType, soil, psi[i], theta[i])
        oldpsi[i] = psi_0
        psi[i] = psi_0
        H[i] = psi[i] - z[i]*g
        k[i] = k_0
    return

def NewtonRapsonMP(funcType, soil, dt, isFreeDrainage):
    if (isFreeDrainage):
        psi[n+1] = psi[n]
        theta[n+1] = theta[n]
        k[n+1] = k[n]
    nrIterations = 0
    massBalance = 1.
    while ((massBalance > tolerance) and (nrIterations
                                    < maxNrIterations)):
        massBalance = 0
        for i in range(1, n+1):
            k[i] = (hydraulicConductivityFromTheta(funcType, soil,
                    theta[i]) + vaporConductivityFromPsiTheta
                    (soil, psi[i], theta[i]))
            k_mean[i] = kMean(LOGARITHMIC, k[i], k[i+1])
            u[i] = g * hydraulicConductivityFromTheta(funcType, soil,
                                                    theta[i])
            du[i] = -u[i] * soil.Campbell_n / psi[i]
            capacity = dTheta_dPsi(funcType, soil, psi[i])
            capacity_vapor = dvapor_dPsi(funcType, soil, psi[i],
                                    theta[i])
            C[i] = (vol[i] * (waterDensity*capacity +
                            capacity_vapor)) / dt
```

```
        for i in range (1, n+1):
            f[i] = (k_mean[i]*(psi[i+1]-psi[i])/dz[i]) - u[i]
            if (i == 1):
                a[i] = 0.
                c[i] = -k_mean[i] / dz[i]
                b[i] =  k_mean[i] / dz[i] + C[i] + du[i]
                d[i] = ((evaporation_flux(psi[i]) - f[i] +
                vol[i]*(waterDensity * (theta[i] - oldTheta[i])
                + (vapor[i]-oldvapor[i])) /dt + plant.t[i]))
            else:
                a[i] = -k_mean[i-1] / dz[i-1] - du[i-1]
                c[i] = -k_mean[i] / dz[i]
                b[i] = k_mean[i-1] / dz[i-1] + k_mean[i] / dz[i] +
                        C[i] + du[i]
                d[i] = ((f[i-1] - f[i] +
                        vol[i]*(waterDensity * (theta[i] - oldTheta[i])
                            + (vapor[i]-oldvapor[i])) /dt + plant.t[i]))
                massBalance += abs(d[i])
        ThomasBoundaryCondition(a, b, c, d, dpsi, 1, n)
        for i in range(1, n+1):
            psi[i] -= dpsi[i]
            theta[i] = thetaFromPsi(funcType, soil, psi[i])
            vapor[i] = vaporFromPsi(funcType, soil, psi[i], theta[i])
        nrIterations += 1
        if (isFreeDrainage):
            psi[n+1] = psi[n]
            theta[n+1] = theta[n]
            k[n+1] = k[n]
    if (massBalance < tolerance):
        flux = evaporation_flux(psi[1]) + sum(plant.t)
        return True, nrIterations, flux
    else:
        return False, nrIterations, 0

def cellCentFiniteVolWater(funcType, soil, dt, isFreeDrainage):
    if (isFreeDrainage):
        psi[n+1] = psi[n]
        theta[n+1] = theta[n]
        k[n+1] = k[n]
    for i in range(1, n+1):
        theta[i] = oldTheta[i]
    massBalance = 1.
    nrIterations = 0
    while ((massBalance > tolerance) and (nrIterations <
            maxNrIterations)):
        for i in range(1, n+1):
            k[i] = (hydraulicConductivityFromTheta(funcType, soil,
                    theta[i]) + vaporConductivityFromPsiTheta(soil,
```

```
                psi[i], theta[i]))
    capacity = dTheta_dPsi(funcType, soil, psi[i])
    capacity_vapor = dvapor_dPsi(funcType, soil, psi[i],
    theta[i])
    C[i] = (vol[i] * (waterDensity*capacity +
    capacity_vapor)) / dt
f[0] = 0
for i in range(1, n+1):
    f[i] = area * kMean(LOGARITHMIC, k[i], k[i+1]) / dz[i]
for i in range(1, n+1):
    a[i] = -f[i-1]
    if (i == 1):
        b[i] = C[i] + f[i]
        c[i] = -f[i]
        d[i] = (-evaporation_flux(psi[1]) +
                C[i] * oldpsi[i] - area * f[i]*dz[i] *
                g - plant.t[i])
    elif (i < n):
        b[i] = C[i] + f[i-1] + f[i]
        c[i] = -f[i]
        d[i] = (C[i] * oldpsi[i]
                - area * (f[i]*dz[i]-f[i-1]*dz[i-1]) *
                g - plant.t[i])
    elif (i == n):
        c[n] = 0
        if (isFreeDrainage):
            b[n] = C[n] + f[n-1]
            d[n] = (C[n] * oldpsi[n] -
            area * (k[n] - f[i-1]*dz[i-1]) * g - plant.t[n])
        else:
            b[n] = C[n] + f[n-1] + f[n]
            d[n] = (C[n] * oldpsi[n] + f[n]*psi[n+1]
                    - area * (f[i]*dz[i]-f[i-1]*dz[i-1]) *
                    g - plant.t[i])
ThomasBoundaryCondition(a, b, c, d, psi, 1, n)
for i in range(1, n+1):
    theta[i] = thetaFromPsi(funcType, soil, psi[i])
    vapor[i] = vaporFromPsi(funcType, soil, psi[i], theta[i])
if (isFreeDrainage):
    psi[n+1] = psi[n]
    theta[n+1] = theta[n]
    k[n+1] = k[n]
newSum = 0
for i in range(1, n+1):
    newSum += C[i]*(psi[i]-oldpsi[i])
if (isFreeDrainage):
    massBalance = abs(newSum + evaporation_flux(psi[1]) +
    area*k[n]*g)
```

```
    else:
        massBalance = (abs(newSum + evaporation_flux(psi[1])
                      + f[n]*(psi[n]-psi[n+1]) + area*f[i]*dz[i]*g))
        nrIterations += 1
  if (massBalance < tolerance):
      flux = evaporation_flux(psi[1])+ sum(plant.t)
      return True, nrIterations, flux
  else:
      return False, nrIterations, 0
```

The file `PSP_plant` is the module that computes the initial transpiration by defining the function **InitTransp** and plant water uptake with the function **PlantWaterUptake**.

The soil resistance is a nonlinear function of the water potential at the root surface, which in turn depends in a complex way on the rate of uptake of water from each soil layer and on the rate of water loss from the plant. An iterative method that finds the correct soil resistances, water uptake rates and water potentials is possible, but difficult and slow to run on a computer. Since soil resistance is seldom large enough to contribute significantly to the overall resistance, an approximate soil resistance is used. This is calculated for each layer by assuming that the conductivity is constant in the rhizosphere. Several works have shown that conductivity and hydraulic properties change in the rhizosphere (Carminati *et al.*, 2010, 2011) and a model of water transport in the rhizosphere has been presented (Carminati, 2012); however, these processes are not included here. Moreover, Kroener *et al.* (2014*b*) have shown that the hydraulic properties of the rhizosphere are altered by mucilage, a polymeric gel exuded by the roots, and that the difference in drying and wetting rates of mucilage results in non-equilibrium relations between water content and water potential in the rhizosphere. Therefore, additional resistances may be necessary to model the contribution of the rhizosphere.

The instructions following the `if (pl > pb):` statement are written to calculate the actual transpiration rate from the potential transpiration rate and the water uptake rate. The equations are based on the assumption that transpiration rate is inversely related to stomatal resistance and that stomatal resistance varies with leaf water potential. An improved equation will be derived in Chapter 15 using the Penman–Monteith calculation for potential evapotranspiration. This section of code searches for a stomatal resistance and a leaf water potential that balance supply and demand. Since stomatal resistance is a nonlinear function of leaf water potential, iteration must be used. In this case, a Newton–Raphson procedure is used. Finally, the `for` loop at the end of the function calculates the water uptake at each node. Time steps are short enough that changes in soil water potential during the time step are assumed not to influence uptake during a time step. The variable `rw` is the resistance per unit length of root [$m^3 kg^{-1} s^{-1}$], `rl` is the root radius [m]. The variable `L` is defined in eqn (14.5), with $L_0 = 40\,000$ m^{-2}. The `if` statement is used to check if there are roots. Where no roots are present, the resistance `Rr` is set to a very large number to avoid overflow problems in later computations. The function **PlantWaterUptake** is the function that computes the plant water uptake. The variable `pc` is the water potential for stomatal closure, `RL` is the resistance for liquid water flow to the leaf, sp is the stomatal conductance power, pb is a weighted mean soil water potential and rb is the soil–root resistance. Then a for loop is run over the nodes. The first computation is the one described in eqn (14.14), after which the weighted mean soil water potential pb and the soil–root resistance rb are computed. The soil resistance is a function of the water potential at the root surface, eqns (14.13) and (14.13), which in turn depends on the water uptake rate from the layer and therefore on the total loss from the plant.

The while loop is for computation of mass balance. The variable xp is an argument of the stomatal function, while sl is the derivative of the stomatal function. The term *f* is the transpiration mass balance, while *pl* is a Newton–Raphson estimate of new leaf potential. When the computation is finished, the transpiration rate is computed by the function tr. The final for loop over the nodes computes the water uptake t from each soil layer. The function returns the transpiration rate.

```
#PSP_plant.py
from __future__ import division
from PSP_public import *
from PSP_soil import *

t = np.zeros(n+2)
Rr = np.zeros(n+2)
rs = np.zeros(n+2)
bz = np.zeros(n+2)
leafPot=np.zeros(n+2)
#pb=np.zeros(n+2)

def InitTransp(soil, z):
    rw = 25000000000
    rl = 0.001
    for i in range(1, n+1):
        if (z[i] > rootMin and z[i] < rootDepth):
            L = 40000 * (rootDepth - z[i]) / rootDepth
            Rr[i] = 2 * rw / (L * (z[i+1] - z[i-1]))
            bz[i] = ((1 - soil.Campbell_n) *
            np.log(np.pi * rl * rl * L) / (2 * np.pi * L * (z[i+1]
                - z[i-1])))
        else:
            Rr[i] = 1E+20
            bz[i] = 0
    return

def PlantWaterUptake(tp, funcType, soil, theta, psi, z):
    pc = -1500
    RL = 3000000
    sp = 10
    pl = 0
    pb = 0
```

```
rb = 0
for i in range(1, n+1):
    rs[i] = bz[i] / hydraulicConductivityFromTheta(funcType, soil,
                                                    theta[i])
    pb = pb + psi[i] / (Rr[i] + rs[i])
    rb = rb + 1 / (Rr[i] + rs[i])
pb = pb / rb
rb = 1 / rb
if (pl > pb):
    pl = pb - tp * (RL + rb)
f = 11
while(abs(f)>10):
    xp = np.power(pl / pc,sp)
    sl = tp * (RL + rb) * xp * sp / (pl * (1 + xp) * (1 + xp)) - 1
    f = pb - pl - tp * (RL + rb) / (1 + xp)
    pl = pl - f / sl
tr = tp / (1 + xp)
for i in range(1, n+1):
    t[i] = (psi[i] - pl - RL * tr) / (Rr[i] + rs[i])
return tr,pl,pb
```

Output from this program is shown in Fig. 14.5 Note the shape of the water content profile with depth in Fig. 14.5(a). This shape is very different from that obtained from simulation of infiltration and redistribution only, since the plant takes up water at different rates from different soil layers. Figure 14.5(b) depicts the evapotranspiration rate as a function of time for a 10-day period, with a progressive decrease in evapotranspiration rate due to the water-limited conditions of the soil profile. Figure 14.5(c) shows the profile of the root water uptake. The negative values in the upper part of the profile are due to the well-known process of hydraulic lift, where the root acts like a connection between dry soil and wet soil, and in the night, when the plant does not transpire, water flows from wet soil into the root and out of the root into dry soil (Corak *et al.*, 1987). Therefore the area where root water uptake is negative means that the root has lost water to the soil. Indeed, Corak *et al.* (1987) maintained that in their experiment, the data supported the contention that the magnitude of water loss by roots to soil is controlled, at least in part, by soil factors.

This model can be used to give insight into several aspects of soil–plant water relations. The following observations can be made:

1. When soil water potential is uniform, water is taken from the layers of soil with the highest rooting densities.

2. As soil dries and water is taken from layers with fewer roots, the mean soil potential decreases, even though some roots are still in moist soil. This reduction in soil water potential decreases plant water potential, closes stomates and decreases transpiration and production.

3. When rooting density varies in a root system, soil hydraulic resistance appears to be relatively unimportant in all cases.

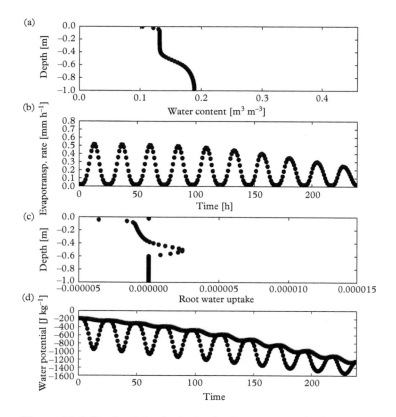

Figure 14.5 *Results of the simulation for the evapotranspiration program: (a) water content as a function of depth; (b) transpiration rate as a function of time; (c) plant water uptake as a function of depth; (d) leaf and soil water potential as a function of time. The dots represent the leaf water potential and the triangles the soil water potential.*

4. It should be possible to specify a value of the weighted mean soil water potential at which irrigation should occur for maximum production.

5. The process of water loss by roots to soil is determined by gradients in water potential and therefore the model confirms experimental evidence, where roots can redistribute water in the root zone.

....

14.7 EXERCISES

14.1. Use the model PSP_transpiration to find soil water distributions when T/Tp = 0.9 and T/Tp = 0.5 for a typical transpiration rate and root distribution. What causes the reduction in transpiration rate? How important is soil resistance?

14.2. Compare water extraction rates, water extraction patterns and fraction of available soil water used at three potential transpiration rates. What effect does transpiration rate have on available water? Available soil water is the water held between field capacity and permanent wilting water content to the depth of root.

14.3. Compare different levels of evapotranspiration by changing the initial soil water content conditions.

15

Atmospheric Boundary Conditions

All of the models developed so far have required specification of a surface boundary condition. In the case of evaporation, soil heat flow and plant water uptake, this boundary condition took the form of a specified potential or flux. In the case of infiltration into a bare soil, a flux boundary condition or a constant potential was set. In the case of evaporation and transpiration, a maximum flux can be specified, which could be reduced by the soil or crop, depending on water supply. In each case, information for determining the boundary condition was assumed to be available.

In working with the models, it was apparent that the actual values chosen for the atmospheric boundary conditions had a profound effect on model behaviour and that meaningful modelling results depend as heavily on using the correct boundary conditions as on the correct behaviour of the model itself. With this in mind, it is the objective of this chapter to develop the equations for heat and vapour exchange at soil and crop surfaces and show how these are used with the models derived in earlier chapters. The subject will not be developed in detail here, however, and the reader is referred to Campbell and Norman (1998) for a more complete treatment.

We are interested in considering the energy balance of a soil or a crop surface. The surface energy balance is usually defined with respect to an active layer of infinitesimal thickness. For convention, when the energy is directed towards the surface layer (energy sources), the sign is positive, and when the layer is losing energy, the sign is negative (energy sinks). Figure 15.1 depicts the exchange of energy at a surface. The indicated values refer to global annual averages (Ohmura and Raschke, 2005).

The general equation for the surface energy balance is written as

$$S_t(1 - a_l) \downarrow + L \downarrow + L \uparrow + H + L + G = 0 \tag{15.1}$$

where $S_t \downarrow$ is global short-wave radiation on a horizontal surface, a_l is the albedo, $L \downarrow$ is the long-wave downward radiation, $L \uparrow$ is the long-wave upward radiation, H is the sensible heat, L is the latent heat and G is the soil sensible heat flux. This last term G is not indicated in Fig. 15.1. When the radiation fluxes are summarized, eqn (15.1) is written as

$$R_n + H + L + G = 0 \tag{15.2}$$

Soil Physics with Python. First Edition. Marco Bittelli, Gaylon S. Campbell and Fausto Tomei.
© Marco Bittelli, Gaylon S. Campbell and Fausto Tomei 2015. Published in 2015 by Oxford University Press.

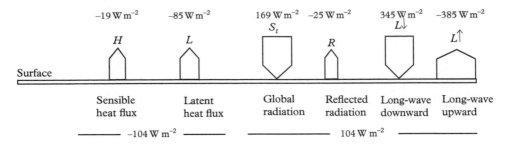

$-19\,\mathrm{W\,m^{-2}}$ $-85\,\mathrm{W\,m^{-2}}$ $169\,\mathrm{W\,m^{-2}}$ $-25\,\mathrm{W\,m^{-2}}$ $345\,\mathrm{W\,m^{-2}}$ $-385\,\mathrm{W\,m^{-2}}$

Sensible heat flux	Latent heat flux	Global radiation	Reflected radiation	Long-wave downward	Long-wave upward

——— $-104\,\mathrm{W\,m^{-2}}$ ———   ——— $104\,\mathrm{W\,m^{-2}}$ ———

Figure 15.1 *Schematic of the surface energy budget.*

where R_n is the net radiation. Let us provide a general description of each term:

- R_n: The net radiation is commonly positive during the day (the surface is gaining energy) and negative during the night (the surface is cooling off). The cooling is determined by net loss of long-wave (thermal) radiation.
- H: At mid-latitudes, the sensible heat is usually negative during the day, since the surface is usually at higher temperatures than the air and therefore the surface is losing heat. During the night, H is positive, since the soil is cooling off faster than the atmosphere and therefore the soil surface is cooler than the air and sensible heat is transferred from the atmosphere to the soil.
- L: In general, the latent heat term is negative during the day because of soil and plant evaporation. During the night, it is possible to have vapour condensation on the surface (due to vapour pressure gradients) and the latent heat can be positive.
- G: In general, the soil heat flux is negative during the day, since this energy flow moves into the soil and away from the reference plane into the deeper soil, so it is negative. At night, the flow of thermal energy is from the soil up to the plane and is then positive.

15.1 Radiation Balance at the Exchange Surface

The energy that heats crop and soil surfaces and evaporates water from them comes from the sun. A secondary source of energy is thermal radiation from the atmosphere. The first step in deriving equations describing the upper boundary is to specify the amount of energy that is available from these two sources. The term R_{abs} is the radiation absorbed at the surface and is calculated from

$$R_{\mathrm{abs}} = a_s S_t + \epsilon_a \sigma\, T_{aK}^4 \tag{15.3}$$

where S_t is global solar radiation, a_s is short-wave absorptivity, T_{aK} is the air temperature (in kelvin), σ is the Stefan–Boltzmann constant and ϵ_a is the atmospheric emissivity. For clear skies, ϵ_a can be calculated from the vapour concentration of the air (Brutsaert, 1975):

$$\epsilon_a = 0.58 c_{va}^{1/7} \tag{15.4}$$

where c_{va} is the vapour concentration of the air [g m^{-3}]. The atmospheric vapour concentration is normally an input variable, but it can also be approximated as the saturation vapour concentration at minimum temperature. When skies are cloudy, the atmospheric emissivity becomes (Unsworth and Monteith, 1975)

$$\epsilon_{ac} = (1 - 0.84 c_l)\epsilon_a + 0.84 c_l \tag{15.5}$$

where c_l is the fractional cloud cover. A simple relationship exists between c_l and the atmospheric transmission coefficient for solar radiation, T_t (the ratio of measured to potential daily global solar radiation). It is

$$c_l = 2.33 - 3.33 T_t \tag{15.6}$$

which is, of course, only valid for values of c_l between 0 and 1. If statements must be included in the computer program to set c_l to 0 if it is less than 0 and to 1 if it is greater than 1.

15.1.1   Potential Solar Radiation

To compute the global solar radiation, the first variable to be computed is the potential solar radiation. To compute the potential solar radiation at a given latitude and time of the day, it is necessary to know the position of the sun in the sky, which is given by its zenith angle θ. The zenith angle is calculated from

$$\cos \theta = \sin \lambda \sin \delta + \cos \lambda \cos \delta \cos[0.2618(t - t_0)] \tag{15.7}$$

where λ is the latitude, δ is the solar declination, t is time and t_0 is the time of solar noon. The earth turns at a rate of 0.2618 rad h^{-1}, so the factor 0.2618 is used to convert hours to radians. Time t is standard local time, ranging from 0 to 24. The solar declination δ can be calculated by

$$\sin \delta = 0.3985 \sin[4.869 + 0.0172 \mathcal{J} + 0.03345 \sin(6.224 + 0.0172 \mathcal{J})] \tag{15.8}$$

where $\mathcal{J}$ is the day of the year. This equation can also be written as

$$\sin \delta = 0.3985 \sin \left[2\pi \frac{\mathcal{J}}{365} - 1.414 + 0.03345 \sin \left(6.224 + \pi \frac{\mathcal{J}}{365} \right) \right] \tag{15.9}$$

The radiant flux density at a horizontal surface outside the earth's atmosphere at any time of the day is

$$Q_p = 1360 \cos \theta \tag{15.10}$$

where $1360 \, W \, m^{-2}$ is the solar constant and θ is the solar zenith angle.

By integration of the radiant flux density, we obtain the daily input of solar radiation on a horizontal surface:

$$Q_0 = 117.5 \frac{h_s \sin \lambda \sin \delta + \cos \lambda \cos \delta \sin h_s}{\pi} \quad (15.11)$$

where $\cos h_s = -\tan \lambda / \tan \delta$ and $117.5 \, MJ \, m^{-2}$ is the amount of energy on a $1 \, m^2$ surface that is perpendicular to the sun for the whole day. The rest of the equation adjusts this value for the length of the day and the angle that the sun makes with a horizontal surface.

15.1.2   Global Solar Radiation

Normally the daily global solar radiation is a measured input variable, but in cases where it is not available, it can be calculated from maximum and minimum temperature data as proposed by Bristow and Campbell (1984). When atmospheric absorption and scattering are considered, the potential solar radiation is reduced by a parameter τ_T to obtain the global short-wave radiation:

$$S_t = Q_0 \tau_T \quad (15.12)$$

Bristow and Campbell (1984) obtained the value of τ_T from daily temperatures variations on the basis of the assumption that sunny days with clear skies would have larger differences between maximum and minimum temperatures. However, it was found that average monthly or seasonal values of τ_T obtained from temperatures showed good correlations, but the correlation between radiation and temperature for a specific day was quite poor. Therefore these methods should not be used for simulations requiring daily values of global solar radiation, and experimental data are needed. When incoming short-wave radiation is measured, the coefficient τ_T can be obtained from the equation

$$\tau_T = \frac{S_t}{Q_0} \quad (15.13)$$

The short-wave absorptivity a_s in eqn (15.3), depends on the nature of the surface and, for soil, on the water content. For crops, a_s varies from 0.75 to 0.85. For bare soil surfaces that are dry, a_s is often around 0.7 and increases linearly with water content of the surface layer to around 0.88 at field capacity. The surface absorptivity can be calculated using data from the soil moisture model, and the atmospheric vapor concentration can usually be assumed to be constant over a day.

Below, two functions are shown in which the equations described above are implemented to compute the atmospheric emissivity ϵ_a by computation of potential radiation and from knowledge of measured radiation. Note that the saturation vapour pressure is computed from eqn (15.4). This code was used when fluxes were coupled in Chapter 12, where boundary conditions for heat, vapour and liquid water were specified.

```
from __future__ import division
import numpy as np

albedo = 0.2
sigma = 5.670e-8  #[W/m2K4]
latitude = 44.43/360.*2.*3.1415
M_w = 0.018
R = 8.31

def vaporConcentrationAir(T,relHum):  #[kg/m3]
    c_vsat = 0.611*1.e3*np.exp(17.27*T/(T+237.3)) * M_w/(R*(T+273.15))
    c_v = relHum*c_vsat
    return c_v

def atm_emissivity(measuredRadiation,day,T,relHum):
  sin_sD = 0.3985 * np.sin(4.869+0.0172*day+0.03345*
                            np.sin(6.224+0.0172*day))
  cos_sD = np.sqrt(1-sin_sD*sin_sD)
  h_s = np.arccos(-np.tan(latitude)*(sin_sD/cos_sD))
  potentialRadiation = 117.5 *(1.e6 *(h_s*np.sin(latitude)*
                                sin_sD+np.cos(latitude)*
                                cos_sD*np.sin(h_s))/3.1415)
  T_t = measuredRadiation/potentialRadiation
  c_l = 2.33-3.33*T_t
  if(c_l < 0):
    c_l = 0.
  elif(c_l > 1):
    c_l = 1.
  c_va = vaporConcentrationAir(T,relHum)*1.e3  # [g/m3]
  epsilon_a = 0.58*np.power(c_va,1./7.)
  emissivity = (1.-0.84*c_l)*epsilon_a + 0.84*c_l
  return emissivity
```

15.1.3 Thermal Radiation from the Surface

The other component of the radiation balance at the surface is the thermal radiation emitted from the surface. This is proportional to the fourth power of the surface temperature. Because the surface temperature is generally not known, it is convenient to write the surface emittance as the sum of two components. One component is proportional to the difference between surface and air temperature and can therefore be combined with the convective heat transfer term. The other term is proportional to the fourth power of the air temperature and is therefore known. The binomial expansion gives

$$T_{sK}^4 = (T_{aK} + \Delta T)^4 \approx T_{aK}^4 + 4T_{aK}^3 \Delta T \tag{15.14}$$

where only the first two terms of the expansion are retained, the others being negligible for $\Delta T \ll T_K$. The $\epsilon_a \sigma T_K^4$ term can now be subtracted from R_{abs} to give an approximation of the energy absorbed at the surface. This is often termed the net isothermal radiation. The term $4\epsilon_a \sigma T^3$, which multiplies the temperature difference between surface and air, can be thought of as a radiative conductance:

$$K_r = 4\epsilon_a \sigma T_K^3 \tag{15.15}$$

This is added to the boundary-layer conductance, which will be considered in Section 15.2, to form a combined conductance for heat transfer in the atmospheric boundary layer.

15.2   Boundary-Layer Conductance for Heat and Water Vapour

The models in Chapters 4 and 12 used boundary-layer conductances that were assumed to be known. Equations for computing these conductances will now be developed. Heat and water vapour are transported by eddies in the turbulent atmosphere above the crop. Boundary-layer conductance would therefore be expected to vary depending on the wind speed and level of turbulence above the crop. The level of turbulence, in turn, is determined by the roughness of the surface, the distance from the surface and the thermal stratification of the boundary layer. An equation that combines these terms in the appropriate way to give the boundary-layer conductance is

$$K_h = \frac{kC_g u^*}{\ln\left(\dfrac{z - d + z_h}{z_h}\right) + \Psi_h} \tag{15.16}$$

Here C_g is the volumetric specific heat of air ($1200\,\mathrm{J\,m^{-3}\,K^{-1}}$ at 20 °C and sea level), u^* is the friction velocity, k is von Kármán's constant (which is generally assumed to be 0.4), z is the height above the surface at which the temperature is measured, d is the zero-plane displacement of the surface, z_h is a surface roughness parameter for heat and Ψ_h is a stability correction factor for heat. This equation assumes that the resistances to heat and vapour flow are roughly equal (Flerchinger *et al.*, 1996). The friction velocity is defined as

$$u^* = \frac{ku}{\ln\left(\dfrac{z - d + z_m}{z_m}\right) + \Psi_m} \tag{15.17}$$

where z_m and Ψ_m are the roughness factor and stability correction for momentum and u is the mean wind speed measured at height z. The zero-plane displacement and roughness parameters depend on the height, density and shape of surface roughness elements. For typical crop surfaces, the following empirical correlations have been obtained:

$$d = 0.77h \tag{15.18}$$

$$z_m = 0.13h \tag{15.19}$$

$$z_h = 0.2z_m \tag{15.20}$$

where h is the height of the roughness elements (height of the crop). The stability correction parameters Ψ_h and Ψ_m correct the boundary-layer conductance for the effects of buoyancy in the atmosphere. When the air near the surface is hotter than the air above, the atmosphere becomes unstable and mixing at a given wind speed is greater than would occur in a neutral atmosphere. If the air near the surface is colder than the air above, the atmosphere is stable and mixing is suppressed. A stability parameter ζ has been defined, which is an index of the relative importance of thermal and mechanical turbulence in boundary-layer transport:

$$\zeta = \frac{-kzgH}{C_h T_K u^{*3}} \tag{15.21}$$

where H is the sensible heat flux in the boundary layer and g is the gravitational acceleration. The stability correction factors Ψ_h and Ψ_m are functions of ζ and depend on the difference between air and soil temperature. For stable conditions, when surface temperature is lower than air, the atmosphere is stable, but when surface temperature is higher than air temperature, the upward-moving warm air leads to turbulence. The stability correction factors can therefore be calculated depending on the magnitude of the sensible heat flux H:

$$\Phi_h = \begin{cases} 4.7\zeta & \text{if } H < 0 \\ -2\ln\left[\frac{1}{2}\left(1 + [1 - 16\zeta]^{\frac{1}{2}}\right)\right] & \text{if } H \geq 0 \end{cases} \tag{15.22}$$

and

$$\Phi_m = \begin{cases} \Phi_h & \text{if } H < 0 \\ 0.6\,\Phi_h & \text{if } H \geq 0 \end{cases} \tag{15.23}$$

The sensible heat flux density in eqn (15.21) is the product of the boundary-layer conductance and the difference between surface and air temperatures. Since the boundary-layer conductance is a function of the heat flux density, an iterative method must be used to find the boundary-layer conductance.

Boundary-layer conductances were used in Chapter 4 for soil heat flow models, in Chapter 11 for soil evaporation models and in Chapter 12 for evapotranspiration models. Of these, the evapotranspiration model is the most complex, since it involves exchange of heat and vapour from sources that are distributed throughout the plant canopy. A detailed plant canopy model is beyond the scope of this book. Interested readers should consult Campbell and Norman (1998) for more information.

Simple approximations for calculating evapotranspiration from crops will be discussed later in this chapter. For these, a boundary-layer conductance calculated assuming $\Phi_m = \Phi_h = 0$ is adequate. In Chapter 11, evaporation from the soil surface was modelled as an isothermal process. Stability corrections are not likely to be important in that model. During first-stage drying, evaporation prevents the surface from becoming hot, so stability corrections are small. Once the surface dries and becomes hot, boundary-layer conductance is relatively unimportant in determining evaporation rate anyway.

The use of stability corrections is important for simulating the temperature of a bare soil, since a dry soil, surface reaches temperatures well above air temperature during the day and can be well below air temperature on a clear night. Thermal stratification on a clear night can be strong enough to reduce sensible heat exchange between the soil surface and the air to almost nothing. If stability corrections are not made, predicted soil temperature profiles can have large errors. The implementation of the boundary layer conductance is shown below. In using the following program, it is important to iterate the entire soil heat flow calculation at least once, since surface temperature depends on the heat flux to the atmosphere, but the latter is determined, in part, by the surface temperature.

The code imports the module PSP_RadiativeConductance to compute atmospheric pressure. The parameters described above are then defined. The boundary-layer conductance is computed by the function **Kh**, which has wind speed, air temperature, surface temperature in kelvins and altitude as arguments. Note that these two temperatures are needed to iteratively compute H. At every iteration, the sensible heat H is computed from knowledge of the boundary-layer conductance multiplied by the temperature gradient between air and surface. The variables z, zh and zm are the height above the surface at which temperature is measured, the surface roughness parameter for heat and the surface roughness parameter for momentum, respectively. The variable vk is the von Kármán constant. The friction velocity is defined by the variable ustar, d is the zero-plane displacement and zeta is the stability parameter.

```
#PSP_boundaryLayerConductance.py
from __future__ import division, print_function
import numpy as np
import PSP_RadiativeConductance

g = 9.81

def Kh(u, T, Tk_0, altitude):
    cp = 29.3
    h = 0.01
    d = 0.77 * h
    zm = 0.13 * h
    zh = 0.2 * zm
    z = 1.5
    Psi_m = 0; Psi_h = 0
```

```
vk = 0.4
pa = PSP_RadiativeConductance.atm_pressure(altitude)
ro = 44.6 * (pa / 101.3) * (293.15 / Tk_0)   # molar density
                                                 of the gas
Ch = ro * cp            # volumetric heat of air (= 1200 J/m^3*K
                                              at 20C e sea level)
for i in range(3):
    ustar = vk * u / (np.log((z - d + zm) / zm) + Psi_m)
    K = vk * ustar / (np.log((z - d + zh) / zh) + Psi_h)
    H = K * (T - (Tk_0-273.15))
    zeta = -vk * z * g * H / (Ch * Tk_0 * np.power(ustar,3))
    if (zeta > 0):
        Psi_h = 4.7 * zeta
        Psi_m = Psi_h
    else:
        Psi_h = -2 * np.log((1 + np.sqrt(1 - 16 * zeta)) / 2)
        Psi_m = 0.6 * Psi_h
return K
```

15.3 Evapotranspiration and the Penman–Monteith Equation

The transpiration rate for a crop can be computed from

$$E = \frac{c_{vs} - c_{va}}{r_{vc} + r_{va}} \tag{15.24}$$

where c_{vs} and c_{va} are the vapour concentrations at the evaporating surface and in the air above the crop, respectively, and r_{vc} and r_{va} are the resistances to vapour diffusion for the canopy and atmospheric boundary layer, respectively. The canopy resistance r_{vc} is an equivalent resistance for all of the leaves in the canopy. If the total area of leaves in the canopy is assumed to consist of n groups of leaves, each having area index F_i and stomatal diffusion resistance r_{vsi}, then the canopy resistance can be calculated from

$$\frac{1}{r_{vc}} = \sum \frac{F_i}{r_{vsi}} \tag{15.25}$$

where the sum is taken over the n groups of leaves. If all of the leaves have the same diffusion resistance, then $r_{vc} = r_{vs}/F$, where F is the total leaf area index of the canopy. The boundary-layer resistance r_{va} is C_h/K_h, where K_h is the boundary-layer conductance from eqn (15.16).

The concentration of vapour at the evaporating surface is generally assumed to be the saturation concentration at the canopy temperature, since the humidity of the substo-matal cavities is always near unity. The surface concentration can therefore be calculated as previously shown from the saturation vapour pressure and canopy temperature.

The difficulty with eqn (15.24) is that canopy temperature must be known to predict the transpiration rate. The transpiration rate, however, influences the canopy temperature. One option would be to ignore differences in canopy and air temperatures and calculate the evaporation rate using the air temperature. This option can be evaluated by rewriting eqn (15.24) as

$$E = \frac{c'_{vs} - c'_{va}}{r_v} + \frac{c'_{va} - c_{va}}{r_v} \tag{15.26}$$

The second term on the right-hand side is the isothermal evaporation rate, or the evaporation that one would predict by assuming that the canopy and air temperatures are the same. The vapour concentration difference in the first term on the right-hand side of eqn (15.26) can be approximated as $s(T_s - T_a)$, where s is the slope of the saturation vapour concentration function at T_a (or preferably at the average of T_s and T_a). Values for s can be calculated from eqn (12.12).

The error that results from using the isothermal approximation can be assessed by taking the ratio of the first to the second terms on the right-hand side of eqn (15.26):

$$W = s\frac{T_s - T_a}{c'_{va} - c_{va}} \tag{15.27}$$

Order-of-magnitude estimates of W can be obtained by taking c_{va} as $10\,\mathrm{g\,m^{-3}}$ and T_a as 20 °C. Then $s = 1\,\mathrm{g\,m^{-3}}$, $c'_{va} = 17.3\,\mathrm{g\,m^{-3}}$ and $W = 0.13$ per degree temperature difference between canopy and air. If the vapour density remains constant but the temperature increases, then W decreases. It is 0.08 at 30 °C and 0.06 at 40 °C. Tall crops that are rapidly transpiring maintain surface temperatures that are generally within a few degrees of air temperature. For these conditions, the isothermal approximation may be adequate. In most cases, however, it appears that the temperature difference between crop and air must be accounted for.

Penman combined eqn (15.26) with the energy budget equation to obtain the well-known Penman formula for predicting potential evapotranspiration. A form of the equation more similar to the equation we will derive was obtained by Monteith (1964) and is known as the Penman–Monteith equation.

Similarly to eqn (15.2), the energy balance for the crop surface can be written as

$$R_{ni} - G - L - K_{hr}(T_s - T_a) = 0 \tag{15.28}$$

where R_{ni} is the isothermal net radiation discussed earlier, G is the ground heat flux density (calculated from equations in Chapter 5) and K_{hr} is the sum of the boundary-layer and radiative conductances (eqns (15.15) and (15.16)). Storage of heat in the canopy and energy used for photosynthesis have been ignored. Equations (15.26) and (15.28) can now be combined to give the Penman–Monteith equation:

$$E = \frac{s}{s + \gamma^*} \frac{R_{ni} - G}{L} + \frac{s}{s + \gamma^*} \frac{c'_{va} - c_{va}}{r_{vc} + r_{va}} \tag{15.29}$$

The apparent psychrometer constant γ^* is calculated from

$$\gamma^* = \frac{r_v K_{hr}}{L} \tag{15.30}$$

where $r_v = r_{vc} + r_{va}$. Equation (15.29) can be used to calculate evapotranspiration when R_{ni}, air temperature, vapour concentration and resistances are known. It is difficult to know the canopy resistance, however, since it depends on leaf water potential and therefore on transpiration rate. It is useful to calculate a potential transpiration rate using eqn (15.29) with a minimum canopy resistance. This potential transpiration rate can then be adjusted, through iteration, to find the correct actual transpiration for any given leaf water potential.

In eqn (15.29), only γ^* and r_{vc} change when stomates close. If E_p is defined as the potential transpiration rate when stomatal resistance is minimum, and the ratio E/E_p is taken using eqn (15.29), one obtains

$$E = E_p \frac{s + \gamma_p^*}{s + \gamma^*} \tag{15.31}$$

where γ_p^* represents the value from eqn (15.30) when crop resistance is minimum. Using eqn (15.31) and a relationship between canopy resistance and leaf water potential, it is then possible to correctly calculate the actual transpiration rate for any canopy water status and potential transpiration rate.

In most species, the leaf water potential has little effect on stomatal resistance until some critical value of potential is approached. At potentials below the critical value, stomatal resistance increases rapidly. An empirical equation that fits this behaviour well is

$$r_{vs} = r_{vs}^0 \left[1 + \left(\frac{\psi_L}{\psi_c} \right)^n \right] \tag{15.32}$$

Here r_{vs}^0 represents the stomatal resistance with no water stress, ψ_c is the critical leaf water potential (the water potential at which stomatal resistance reaches twice its minimum value) and n is an empirical constant that determines how steeply resistance increases with decreasing potential. Tongyai (1977) found values of n around 3 for Ponderosa pine seedlings, but data from Cline and Campbell (1976) show much higher values for broad-leaf species, with some values as high as 20.

Combining eqns (15.32), (15.30) and (15.31) gives

$$E = E_p \frac{Ls + K_{hr} \left(r_{va} + r_{vc}^0 \right)}{Ls + K_{hr} \left\{ r_{va} + r_{vc}^0 \left[1 + \left(\frac{\psi_L}{\psi_c} \right)^n \right] \right\}} \tag{15.33}$$

In eqn (15.33), the canopy resistance is assumed to have the same water potential dependence as the leaf stomatal resistance.

In the limit as the boundary-layer resistance becomes negligibly small, eqn (15.33) becomes

$$E = \frac{E_p}{\left[1 + \left(\dfrac{\psi_L}{\psi_c}\right)^n\right]} \tag{15.34}$$

which was used in Chapter 14 to find the leaf water potential. Equations (15.33) and (14.18) can be solved simultaneously using the Newton–Raphson procedure. Values for s, K_{hr}, r_{va} and r_{vs}^0 must be supplied to the subroutine.

15.4   Partitioning of Evapotranspiration

In the program PSP_transpiration, potential evapotranspiration was arbitrarily partitioned into 10 % evaporation and 90 % transpiration. The actual partitioning depends on the energy supplied to the crop and soil and on the resistances to transport. A complete model that correctly partitions evapotranspiration is given by Norman and Campbell (1983), but it is too complex for the present analysis. A simpler approach is to assume that the ratio of transpiration to evapotranspiration is the same as the ratio of radiation intercepted by the crop to total incident radiation. This is determined by the leaf area index of the crop and, averaged over a day, can be approximated by

$$T = ET\, e^{-KF} \tag{15.35}$$

where F is the leaf area index of the crop and K is the extinction coefficient. A typical value of K for a closed canopy is 0.5.

....

15.5 EXERCISE

15.1. Using the program presented, compute the boundary-layer conductance by providing data for wind speed, altitude, air temperature and surface temperature. Check the values of the conductance for conditions where $H > 0$ or $H < 0$.

Appendix A
Basic Concepts and Examples of Python Programming

A.1   Basic Python

The open source Python language interpreter was used for the programs written in this book. We chose Python because it is an object-oriented, easy-to-use language that is equipped with many powerful features and is very well suited for both teaching and applications. Some specific reasons for choosing Python are as follows:

1. The language and the applications are freeware and open source.
2. The language is multiplatform. The applications, shells and visual applications run on Microsoft, Macintosh and Linux.
3. It has a clean syntax.
4. There is a large and growing community of developers, who are dedicated to the growth of this language.
5. The language is object-oriented and has high-level data structures, combined with dynamic typing and dynamic binding.
6. It is equipped with a numerical package (numpy) that has a powerful n-dimensional array object, sophisticated functions, tools for integrating C++ and Fortran code, linear algebra, Fourier transform, complex number algebra and other capabilities.
7. The syntax is easy to learn and therefore it is well suited for teaching purposes.
8. The language can be used with a visual application (Visual Python) allowing easy visualization of objects, vectors, graphs and simulations.

Throughout this book, Python programs are presented to solve a variety of problems that are often encountered in soil physics. We name the specific file according to its specific purpose. The programs are organized in projects. The projects and file names begin with the acronym PSP, which stands for Python Soil Physics.

The modules `matplotlib`, `scipy` and `numpy` should also be installed, since they are used in several programs. Recent versions of Python install `scipy` and `numpy` automatically during the installation of the Python program. `matplotlib` is mainly used for plotting graphs, while `scipy` and `numpy` are used to include a variety of mathematical packages. Variables are described by long names, facilitating code readability. For symbolic mathematics, the package `sympy` is available, where algebra, differential calculus and matrices in symbolic format are available.

A.1.1   Programs and Modules Needed for this Book

To run the programs written in this book, the following programs and modules must be downloaded:

1. Python version 2.7
2. Tkinter for Python 2.7
3. Matplotlib for Python 2.7
4. Python imaging library (PIL) for Python 2.7
5. Scipy for Python 2.7
6. Numpy for Python 2.7
7. Visual Python 6.1 for Python 2.7

The programs are written using Python 2.7, but it is possible to incorporate the features of version 3 by writing at the beginning of the program the following line:

```
from  __future__  import print_function, division
```

For Python 2.x, this statement invokes the new Python 3.x print format and division scheme, while this statement is ignored by Python 3.x, thereby allowing the reader to utilize the programs written in this book with previous versions.

A.1.2   Python Documentation

The Python language and its visual application, Visual Python, are constantly updated with new features. Therefore the descriptions provided below have the aim of providing a general overview of the language and its main features. Some of these features may change in future versions. The primary reference for Python is the official Python website <http:www.python.org>. Within the documentation page, there are a variety of links to tutorials, references, forums and community development.

There are a variety of good books on the subject. *Learning Python* (Lutz, 2009) is a good book for both beginners and expert programmers. *Python Essential Reference* (Beazley, 2006) is a highly recommend read. *Python in a Nutshell* (Martelli, 2006) is also an excellent and comprehensive book. A recommended book on using Python in computational science is *Python Scripting for Computational Science* (Langtangen, 2009).

A.1.3   Running Programs in Python

As with other languages, programs are written using a text editor. When saving the file created by the text editor, use the extension `.py`. After writing the program, it can be run by using an *integrated development environment* (IDE) or by a command prompt. A useful IDE for Python is *Eclipse* (<http://www.eclipse.org/>) with its integrated `pydev` environment, which is a freeware environment for developing Python programs and projects. It is a well-developed IDE to write, debug and create applications in Python. The programs written in this book were written and debugged using `pydev` included in Eclipse.

Since Python is compiled to byte code and then interpreted in a virtual machine, when the program is executed, Python compiles the source code into a byte code and the Python virtual machine (PVM) runs the program. Throughout this book, and in accord with this usage in many Python books, the prefix for interactive prompt commands is >>>, while the output has no prefix. Moreover, functions (both internal and user-defined) are indicated in **bold**, while programs, variables and constructors are written in `teletype`. Constants are written in capital letters.

For project names, file names and variables, we use the first word with an initial lower case letter and the consequent words with initial capital letters. For instance, the file `PSP_basicProperties.py` is a file where a program is written to compute basic soil properties. Analogously, the variable `waterDensity` has the first word with a lower case initial letter and the second with a capitalized initial. Each file begins with the acronym PSP, which stands for Python Soil Physics, except the file containing the **main** function, which is called `main`. To make programs readable, we use variables with self-explanatory names and we also present a list of variables used in the Python programs.

A.1.4 Plotting and Visualization

In this book, we use the program `matplotlib` (<http://matplotlib.org/>) for plotting the results of the computations. `matplotlib` allows the plotting of time-changing variables, and is therefore very powerful for visualization of dynamic processes. `matplotlib` is a powerful library for plotting scientific data. It also allows the plot to be exported in many different formats. Recent versions of `matplotlib` may require the installation of the dependences, which are `numpy`, `dateutil`, `six`, `pyparsing`; for Windows, they are downloadable at <http://www.lfd.uci.edu/~gohlke/pythonlibs/> .

For visualization of numerical solutions in three dimensions or for mesh generation algorithms, the program Visual Python, version 6 (<http://www.vpython.org/>) is used. Further details are provided in Chapter 10. A stable version of the Python Imaging Library (PIL), version 1.1.7, Python 2.7 for Macintosh, can be downloaded from <http://www.astro.washington.edu/users/rowen/python/>. The user should download and install the following file for the PIL module:

```
PIL-1.1.7-py2.7-python.org-macosx10.6.dmg
```

A.1.5 Extensions for other languages

One of the advantage of Python is that it is flexible and easy to program. However, for certain types of calculations, Python (and any other interpreted language) can be slow. Usually, iterations over large arrays are difficult to do efficiently. Such calculations may be implemented in a compiled language such as C or Fortran. In this book, for the three-dimensional solution of water flow, we present two versions of the program. One version is written entirely in Python, while the second presents some parts of the program (arrays and computationally expensive functions) written in C and used with the program Cython. The latter version implemented in Cython is much faster than the former. Cython is an optimizing compiler for both the Python programming language and the extended Cython programming language. It makes writing C extensions for Python as easy as Python itself. The Cython language supports calls of C functions and declaration of C types on variables and class attributes. This allows the compiler to generate very efficient C code from Cython code. The language can be downloaded at <http://cython.org/> .

A.2   Basic Concepts of Computer Programming

In this section, to describe some basic concepts of computer programming, we present a simple example of a program written to solve a problem. A computer program consists of a series of instructions written to perform a specific task or solve a problem. For instance, it is very easy for a person to figure out if the black point labelled a in Fig. A.1 is inside the circle A or not. However, we need to devise a way to tell the computer if the point is inside or outside the circle. Visual applications in computer programming often deal with similar problems, such as how to determine if the user clicked inside or outside a specific icon or symbol on the computer screen to activate a specific function such as opening a file. A first way to check if the black point a is inside the circle A is the following: if the point is inside circle A, a straight line from the point and passing through the centre of the circle will cross the circle perimeter A only once. If the point a is outside the circle (in this case, let us consider the circle B), a straight line from the point a and passing through the centre of the circle B would cross the perimeter of the circle B twice.

Let us draw a line that, starting from the point, crosses the centre of the circle. If the line crosses the circle only once, then the point is inside the circle; otherwise, if the line crosses the circle twice, then the point is outside. From a geometrical standpoint, if the point has coordinates x_1 and y_1 and the centre of the circle has coordinates X_1 and Y_1, the line that starts in (x_1, y_1) and goes through (X_1, Y_1) will have the equation

$$\frac{x - X_1}{y - Y_1} = \frac{x_1 - X_1}{y_1 - Y_1}$$

where $(x_1 - X_1)(x - X_1) > 0$, while the circle of radius r of centre X_1, Y_1 will have the equation

$$(x - X_1)^2 + (y - Y_1)^2 = r^2$$

To determine the intersection point between the line and circle, a solution of the system of equations is needed. In this way, we determine if the point is inside or outside the circle based on the number of solutions of the system of equations.

An alternative and simpler approach would be to measure the distance between the centre of the point and the centre of the circle. If this distance is less than or equal to the radius of the circle, then the point would be considered inside the circle; otherwise, if the distance is larger than the

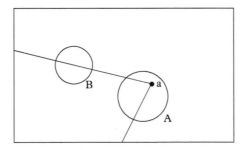

Figure A.1 *Example for determining if a point is inside or outside a circle by using a computer.*

radius, the point would be outside. To achieve this very simple task, we need to write a program, where the user is asked to click with the mouse on the screen. The computer calculates the position on the screen where the user clicked, it determines if the position is inside or outside the circle and then provides an output message, based on the latter method.

A.2.1 Flow Diagrams

A flow diagram shows the sequence of operations in a computer program. Figure A.2 depicts a flow diagram for the problem just discussed. To write this program, we need to assume that (a) we have a function that draws a circle on the screen if we provide the (X, Y) coordinates for the circle center and its radius, (b) we have a function that checks if the point where the user is clicking is inside or outside the circle by using the distance formula and (c) we have a function able to write the results on the screen. We also need to import a graphical user interface (GUI) to draw a canvas for the circle. One GUI implemented in Python is `tkinter`. A module to perform mathematical operations is also needed, called `math`. Let us call the first function **drawCircle** with arguments (myCanvas, x, y, r), where myCanvas defines the space on the screen where to draw the circle, x and y are the coordinates and r is the radius. We are not going to analyse the object `myCanvas.create_oval()` at this stage, since it is an internal function of the program, but we know that if we pass the correct numbers, the function will draw an oval on the screen of the desired size.

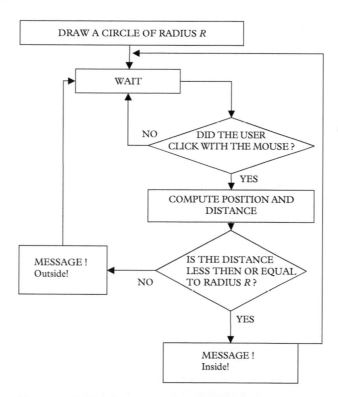

Figure A.2 *Flow diagram for the circle program.*

According to the program structure, a circle is just one form of an oval. Since ovals are defined by the size of the side of a rectangle with the oval inscribed, a circle is just an oval inscribed into a square. The program is as follows:

```
# PSP_circle.py
from __future__ import print_function, division
from math import sqrt
import sys
if sys.version_info >= (3,0):
    from tkinter import *
else:
    from Tkinter import *

myCenterX = 150
myCenterY = 150
myRadius = 50
myRoot = Tk()
myText = StringVar()

def drawCircle(myCanvas,x,y,r):
    myCanvas.create_oval(x-r,y-r,x+r,y+r,width=1,fill='')

def ButtonLeftHandler(myEvent):
    dX = myEvent.x - myCenterX
    dY = myEvent.y - myCenterY
    distance = sqrt(dX*dX + dY*dY)
    if (distance <= myRadius):
        myText.set("Inside!")
    else:
        myText.set("Outside!")

def main():
    myCanvas = Canvas(width=300, height=300, bg='white')
    myCanvas.pack(expand=YES, fill=BOTH)
    myLabel = Label(myRoot, textvariable = myText, font=("Helvetica",
                                                         12))
    myLabel.pack()
    myText.set("Try to click the mouse button in the canvas...")
    myCircle = drawCircle(myCanvas, myCenterX, myCenterY, myRadius)
    myRoot.bind_class ("Canvas", "<Button-1>", ButtonLeftHandler)
    myRoot.mainloop()
main()
```

The first line allows the program to be used in both Python versions 2.x and 3.x, as described above. Then, the square root function (sqrt) from the module math is imported. The following lines are written to run the program Tkinter. The if statement is for both Python versions 2.x and 3.x, since Tkinter was modified from version 2.x to version 3.x. Then the variables identifying the circle coordinates and the radius are defined.

Try to click the mouse button in the canvas...

Inside!

Figure A.3 *Program output.*

The widget `Canvas` is a rectangular area intended for drawing pictures or other complex layouts, implemented in `tkinter`. Its definition is written in the **main**( ) function. The `Canvas` object is created and called `myCanvas`. A series of `tkinter` objects such as `Label` and `set` are then created. The function **drawCircle**(`myCanvas,x,y,r`) is created to draw the circle by using the command `myCanvas.create_oval()` to generate the circle on the canvas.

The function **ButtonLeftHandler**(`myEvent`) is written to read the event of clicking with the left button with the mouse on the canvas and use that point (of specific coordinates) to compute the distance from the centre of the circle. The distance is computed by first determining the difference between the coordinate where the user clicked on the canvas and the centre of the circle. Then these values (`dX` and `dY`) are used in Pythagoras' theorem (line 16) to compute the distance. Finally, the `if` statement between lines 17 and 20 checks if the distance is less than or equal to the radius and prints the answer on the canvas. The output is shown in Fig. A.3.

A.2.2   Algorithm

When we attempt to perform a sequence of calculations using a computer, we need to write an *algorithm*. An algorithm is a method that defines, in an unambiguous manner, a finite sequence of steps to be performed in a specified order. The object of the algorithm is to implement a procedure to solve a problem or approximate a solution to a problem. When designing a program, it is useful to keep in mind the following steps:

1. Clearly state the purpose of the program and the necessary calculations.
2. Write a flow diagram of the program.
3. Write a pseudo-code.
4. Write the code.

5. Compile and debug the program.
6. Run the program.
7. Check the results.

When designing a program, we need to think about its components (objects), the relationships occurring among them, the input, the output and so on. Depending on the level of complexity and the size of the program, we would need to create different levels of organization or hierarchies. Designing well-organized flow diagrams is an important part of this process. A pseudo-code is used for describing the algorithm. It specifies the form of the input–output and the calculations. We can consider a pseudo-code as a code written in a language (English or Italian, say). For example we can simply write: define the variables `myCenterX`, `myCenterY` and `myRadius`. Draw a circle of given size by calling a function. Wait until user clicks on a chosen point on the screen with the mouse. Calculate the distance of the selected point from the circle center. If the distance is smaller than the radius, print inside else print outside. End.

After writing the pseudo-code, it is necessary to translate it into a code written in the selected language (Python, Visual BASIC, Fortran, C, C++, Java, etc.). When an algorithm is written (using an editor), it is not understood by the computer unless it is compiled. To compile is to decode a series of instructions written in a higher-order language and produce a machine language program. If the compiler finds errors in the program, it generates a series of error messages indicating the error type (usually a numerical code and its description) and its position (line number) in the program. After compilation, the computer generates an executable file. Since Python is a pre-compiled language, compilation is not necessary since it is done already while running the program. By executing this file, the program will run. It is always important to keep in mind that the absence of errors in the program does not imply that the program performs the computation without errors. Therefore a series of tests needs to be performed to verify the robustness, stability and convergence properties of the algorithms.

A.3   Data Representation: Variables

A program works with data, the material used to perform computations and operations. Organization of data, labelling and storage in a computer are fundamental elements of computer programming. In computer programming, a data type is created and variables are assigned to that type. The variable identifies an area in the computer memory where the data are stored. In most of the available programming languages (C, C++, Visual BASIC, Java, Fortran), a variable has a specific type. For instance, *integer* means that the variables listed are integers, in contrast to *single* or *double*, which are floating-point variables, i.e. numbers that have a fractional part. Different languages use different words to declare variables. For instance in Visual BASIC the declaration of a variable is done via the *Dim* statement, which is short for Dimension.

In Python, there is no need to declare a variable, since the variable takes the type of the data assigned to it, for instance:

```
>>> waterDensity = 1000
```

types the variable `waterDensity` as an integer. We can change the type by reassigning the variable:

```
>>> waterDensity = "one thousand"
```

In this case, the variable `waterDensity` is now a string. In Python, strings can be represented with either single or double quotes. Therefore 'one thousand' and "one thousand" are both string assignments. If we want the variable `waterDensity` to be a number with decimals (a floating-point number), in Python it is sufficient to include a point after the number, implying that the number can have a fractional part:

```
>>> waterDensity = 1000.
```

This feature is one of the ways Python differs from other languages such as C or BASIC and is one of the reasons for Python's flexibility. It is called *dynamic typing*, implying that types are determined at runtime and not as a response of a prior declaration in the program. While this feature can be very useful in making the program shorter, easier to read and flexible, the programmer must still be aware of what those types are and where they were first used. Moreover, it is possible to change the variables within the program. For instance, if we write

```
>>> 10/3
```

This is an operation between two integers and the result will be 3. However, it is possible to change it to a floating-point operation:

```
>>> float(10) / 3
```

The result will be now 3.333333333. In Python 2.x, since an operation between two integers such as 10/3 = 3 is missing, only the integer part of the number is considered. From Python 3.x, such an operation converts the numbers to floating-point and returns the value 10/3 = 3.33333333. To import this feature into Python 2.x, it is necessary to include the line

```
from __future__ import division
```

at the beginning of the program to import the module `division`. The main built-in types of variables used in Python are listed below.

A.3.1 Numeric Types

In Python, there are four numeric types:

- integers
- long integers
- double-precision
- double-precision complex numbers

Integers are whole numbers from negative to positive values and are of type `int`, for instance 0, 2, 9. The upper and lower dimensions depends on the number of bits used on a given computer to represent a number.

Double-precision numbers are real numbers of type `float` such as 0.1, 14.5, 3.8E+8. Floating-point variables hold real numbers characterized by decimal places. The size and numbers of digits of the floating-point number depend on the machine on which the program is running and in Python this information is contained in the sys.float_info, which is a module containing information about the precision and internal representation of numbers. In Python, a floating-point variable is created by simply assigning a number with a decimal point in it.

Double-precision complex numbers are characterized by a real and an imaginary part. In Python, a complex number is represented as (real+imaginary j), where j is $\sqrt{-1}$. The real and imaginary parts of a number are obtained by typing r.real and r.imag. Imaginary numbers are written with a suffix of j. Therefore a complex number A is assigned as $A = (1 + 3j)$ and $B = (2 + 5j)$ and their sum $A + B = (3 + 8j)$. Python supports the common operations for complex numbers (addition, subtraction, multiplication, division and modulus). The following is an example of complex addition and multiplication:

```
>>> (1+3j)+(2+6j)
(3+9j)
>>> 2j * complex(0,1)
(-2+0j)
>>> (2+4j)*(3-7j)
(34-2j)
```

As described above, in Python, a variable automatically takes the type of the data assigned to it. However, the constructors int(), float(), and complex() can be used to produce numbers of a specific type as a result of a specific arithmetic operation.

A.3.2   Boolean

Python has a specific type for Boolean, called bool, of values True and False, and these can be interchanged with 1 and 0, respectively.

A.4   Comments Rules and Indendation

A hash mask is used in front of a comment, such as

```
# This line is commented
```

For commenting many lines and avoiding repeating the sign, a *docstring* command is used by typing three double quotes at the beginning and end of the selected part of the code:

```
"""
This line is commented
this one too...
and also this one...

"""
```

Indentation rules are an important aspect of Python programming. Python identifies block boundaries by line indentations, which are empty spaces to the left of the code. For instance, statements indented to the same block line are part of the same block of code. The end of the block is determined by a lesser-indented line. This feature makes Python code easily readable because it uses many fewer parenthesis than other languages such as C or C++. Python functions use no explicit begin or end, nor do they use curly braces to mark where the function code starts and stops. The only delimiter is a colon (:) and the indentation of the code itself.

A.5   Arithmetic Expression

Arithmetic expressions in Python follow similar rules as other common programming languages. The principal operations are

```
x + y                    sum of x and y
x - y                    difference of x and y
x * y                    product of x and y
x / y                    quotient of x and y
x // y                   floored quotient of x and y
x % y                    remainder of x / y
-x  x                    negated
+x  x                    unchanged
abs(x)                   absolute value or magnitude of x
int(x)                   x converted to integer
float(x)                 x converted to floating point
complex(re, im)          a complex number with real part re,
                             imaginary part im
c.conjugate()            conjugate of the complex number c
divmod(x, y)             the pair (x // y, x % y)
pow(x, y)                x to the power y
x ** y                   x to the power y
sqrt(x)                  the square root of x
```

Operations such as the absolute number (**abs**), complex numbers (**complex**), the square root (**sqrt**) and others require that the module math be imported. When the math operators are used from in the Python shell or within a program, it is necessary to import the **math** module:

```
>>> from math import *
```

The asterisk indicates that all the packages in the math module are imported. If only one operator is needed, it is possible to import only that one:

```
>>> from math import sqrt
```

To decrease computational time, it is advisable to import only those packages that are needed. Similarly to other languages, the simple arithmetic operations and their computational order are

```
()   (parenthesis)       1) first
^    (exponential)       2) from right to left
* /  (mult/division)     3) third from left to right
+ -  (add/subtraction)   4) fourth from left to right
```

This means that parentheses are considered first, then exponentials are calculated, divisions, and so forth. The following is an example that describes the importance of a correct understanding of this idea. The following simple calculation computes the void ratio in soils, which mathematically is written as

$$e = \frac{\text{porosity}}{1 - \text{porosity}}$$

Computers need to know which operation we want to compute first; therefore, if we write $e = poros/1 - poros$, the division will be performed first ($poros/1$) and then $poros$ will be subtracted from the result of the first calculation. This is *wrong* because the result of this calculation will always be zero ($poros - poros = 0$). It is necessary to add parentheses to tell the computer to first compute ($1 - poros$) and then divide $poros$ by the result of the first calculation: $e = poros/(1 - poros)$.

It is important to be precise when coding a mathematical equation into a programming language. Here are three rules of thumb:

(a) Understand the rules of the programming language you are using.

(b) do not be afraid of using parentheses.

(c) Break the equation in different components, but only if the equation is really large.

Another important arithmetic issue is *mixed-mode arithmetic*. When we have a real and an integer number in the same expression the integer is converted into a real before the operation is performed:

```
3 + 8.0/5 -> 3 + 8.0/5.0 -> 3 + 1.6 -> 3.0 + 1.6 = 4.6
```

If the operation is performed between two integers, the results is always an integer. If the result has a decimal, the value is truncated:

```
3.0 + 8/5 -> 3.0 + 1 -> 3.0 + 1.0 =   4.0
```

As described already, Python version 2.x truncates the division between integers, while in version 3.x the result is not truncated.

A.6 Functions

A function provides a convenient way to include some computation that can be called several times within a program. Indeed, functions are the basic structures of Python programming. When a function is properly written, it is not necessary to know *how* the computation is performed, but it is sufficient to know *what* is performed. There are many functions in Python that have been previously implemented and are not explicitly in view (*built-in*), such as the mathematical operations described above. For specific computations and for expert programmers, it may be necessary sometimes to read and understand the built-in functions; however, in many cases, it is sufficient to know what computation is performed.

Functions can also be written by the programmer. In Python, a function is written with a statement called **def**, which generates a new function object and assigns it a name. Within the function, there is a `return` statement to provide the results of the function call.

A.6.1   Open, Read and Analyse Experimental Data

Below is an example of a program that reads data from an ASCII file. An ASCII file (also called text file) is a standard file format that can be freely interchanged and is readable from different operating systems. The following example shows how a function can be used to open and read a data file.

Compute hourly average of air temperature

The first program where functions are employed is a simple one, where experimental data stored in a *text* file must be read and the hourly average temperature computed. The *text* file contains experimental data on air temperature, collected every ten minutes by a datalogger. The program shows how to open, and read the file and compute an hourly average of air temperature. Two functions are written to solve this problem. The first (**read3VarFile**) is a function that reads the experimental data. The second (**computeMeanT**) computes the hourly average temperature. Below is the text for the file PSP_read3VarFile.py that contains the program written to read the experimental data. The program was named after the fact that it reads a file with three variables.

```
#PSP_read3VarFile.py
from csv import reader

def read3VarFile(myFile, nRowHeader, myDelimiter, x1, x2, x3,
                 printOnScreen):

        myReader = reader(open(myFile, "rt"), delimiter=myDelimiter)
        nRow = 0
        for row in myReader:
            nRow = nRow + 1
            # Header
            if (nRow <= nRowHeader):
                if (printOnScreen): print (row[0], row[1], row[2])
            # Values
            else:
                i = nRow - nRowHeader - 1
                x1.append(float(row[0]))
                x2.append(float(row[1]))
                x3.append(float(row[2]))
                if (printOnScreen):  print (x1[i], x2[i], x3[i])
```

The first line is commented and contains the file name. In the second line, the built-in function called **csv** used to read the files, is imported. The function is used to open comma-separated files (*csv*). The arguments of the function **read3VarFile**( ) are defined within the parentheses and are the file name, the number of rows occupied by the file header, the delimiter, the three variables and the option of printing on screen the data read in the file. The variable myReader is defined to use the reader instruction. The for loop is used to increment the row number and read the file row by row, while the if statement determines if the current row is a header row. The program then reads the actual data and appends them in arrays of floating points for the three variables x1, x2 and x3. The last line is used to print the results on screen if the option was selected as True in the program that calls this function.

The program `PSP_averageTair.py` is as follows:

```
#PSP_averageTair.py
from PSP_read3VarFile import *
import matplotlib.pyplot as plt

def  computeMeanT(hourMinute, airT, meanT):
    mySum = 0.
    myHour = 0
    nValues = 0
    for i in range(0, len(airT)):
        if (int(hourMinute[i] / 100) == myHour):
            mySum += airT[i]
            nValues += 1
        else:
            meanT.append(mySum / nValues)
            nValues = 1
            mySum = airT[i]
            myHour += 1
            if (myHour == 24): myHour = 0

def main():
    doy = []
    hourMinute = []
    airT  = []
    read3VarFile("tenMinutesTemp.txt", 1, '\t', doy, hourMinute,
                 airT, False)
    meanT = []
    computeMeanT(hourMinute, airT, meanT)

    for h in range(0, len(meanT)):
        plt.plot(h,meanT[h],'ro')
    plt.title('')
    plt.xlabel('Time [hour]',fontsize=16)
    plt.ylabel('Temperature [C]',fontsize=16)
    plt.tick_params(axis='both', which='major', labelsize=14)
    plt.tick_params(axis='both', which='minor', labelsize=14)

    plt.show()

main()
```

In this program, the file `PSP_read3VarFile.py` is imported to read the data file (as described above) and the program `matplotlib` to plot the results. The function **computeMean** is written to calculate the hourly mean temperature. The `if` statement is used to discriminate the end of an hour, based on the data format. Specifically, the output data format for the Campbell Scientific datalogger for the minutes has 10, 20, 30, 40, 50 and 100 corresponding to the first ten, twenty, thirty, forty and fifty minutes after midnight, with the first hour (1 a.m.) written as 100.

To discriminate between these incremental time steps, the number is divided by 100 and only the integer part of the number is taken, by using the operator **int**(x). If the integer part is equal to the variable myHour, then the variable mySum is incremented by adding values of airT. When the end of the hour is reached (for instance when HM=100), the condition is no longer met, since the integer is not equal to myHour, and the program goes to the else branch of the if condition, where mean temperature is computed. Below is an example of the first ten rows of the file tenMinutesTemp.txt. The first column holds the day of the year (Doy), the second the hour minutes as described above, the third the air temperatures and the fourth the percentage of relative humidity.

Doy	HM	Tair	RH
183	00	25.64	47.01
183	10	25.30	48.37
183	20	25.10	47.63
183	30	25.11	50.81
183	40	25.31	47.29
183	50	25.31	44.51
183	100	25.99	37.53
183	110	25.85	39.23
183	120	25.45	41.87

The function **main** is used to define the variables, call the reader and plot the data using the **plt.plot** command. The results of the computation are plotted in Fig. A.4.

Compute average temperature and cumulative precipitation

A similar example is now presented, in which hourly meteorological data on air temperature and precipitation are stored. The following is an example of the first ten rows of the file weather.txt. The first column holds the date, the second the day of the year (Doy), the third the hour minutes,

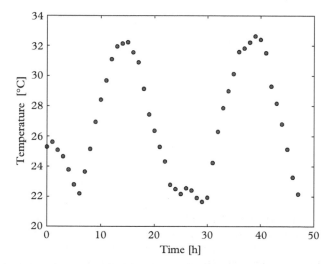

Figure A.4 *Plot of mean hourly temperatures versus time.*

the fourth the precipitation [mm], the fourth air temperature [°C] and the fifth the percentage relative humidity.

```
Year    Doy    HM  P Tair RH
2005    1      100 0 3.02 55.47
2005    1      101 0 2.33 54.74
2005    1      201 0 1.57 57.27
2005    1      301 0 2.65 54.22
2005    1      401 0 2.83 52.65
2005    1      501 0 3.03 48.72
2005    1      601 0 2.54 50.31
2005    1      701 0 2.57 49.38
2005    1      801 0 3.65 43.65
2005    1      901 0 4.29 45.40
```

The program imports a function contained in the file PSP_read6VarFile.py to open and read the file. The function implemented is very similar to **read3VarFile** presented above, but it reads six variables instead of three.

```
#PSP_read6VarFile.py
from csv import reader

def read6VarFile(myFile, nRowHeader, myDelimiter, x1, x2, x3,x4,x5,x6,
                printOnScreen):

        myReader = reader(open(myFile, "rt"), delimiter=myDelimiter)
        nRow = 0
        for row in myReader:
            nRow = nRow + 1
            # Header
            if (nRow <= nRowHeader):
                if (printOnScreen): print (row[0], row[1], row[2],
                                            row[3],row[4],row[5])
            # Values
            else:
                i = nRow - nRowHeader - 1
                x1.append(float(row[0]))
                x2.append(float(row[1]))
                x3.append(float(row[2]))
                x4.append(float(row[3]))
                x5.append(float(row[4]))
                x6.append(float(row[5]))
                if (printOnScreen):  print (x1[i], x2[i], x3[i],x4[i]
                                            ,x5[i],x6[i])
```

After reading the input data, the program PSP_weatherData.py computes daily average air temperature and cumulative precipitation and plots the results as shown in Fig. A.5.

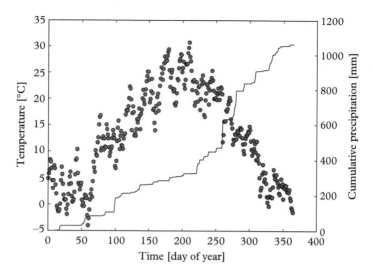

Figure A.5 *Plot of cumulative annual precipitation and mean daily air temperatures.*

```
#PSP_weatherData.py
from PSP_read6VarFile import read6VarFile
import matplotlib.pyplot as plt

def computeDaily(year, doy, hourMinute, prec, airT, rh, meanT,
                 cumPrec):
    mySumT = airT[0]
    mySumP = prec[0]
    nValues = 1
    for i in range(1, len(airT)):
        if (int(hourMinute[i] / 100) != 0):
            mySumT += airT[i]
            mySumP += prec[i]
            nValues += 1
        else:
            meanT.append(mySumT / float(nValues))
            cumPrec.append(mySumP)
            nValues = 1
            mySumT = airT[i]
            mySumP += prec[i]
    #last day
    meanT.append(mySumT / float(nValues))
    cumPrec.append(mySumP)

def main():
    year=[]
    doy = []
```

```
hourMinute = []
prec=[]
airT   = []
rh=[]

read6VarFile("weather.txt", 1, '\t', year, doy, hourMinute, prec,
             airT, rh, False)
meanT = []
cumPrec=[]
computeDaily(year, doy, hourMinute, prec,airT, rh, meanT, cumPrec)

myGraph = plt.figure()
myAxisLeft = myGraph.add_subplot(111)
myAxisRight = myAxisLeft.twinx()
plt.title('')
myAxisLeft.set_xlabel('Time [doy]',fontsize=16)
myAxisLeft.set_ylabel('Temperature [C]',fontsize=16)
myAxisRight.set_ylabel('Cumulative precipitation [mm]',fontsize=16)

#========================DAILY AVERAGE TEMP PLOT====================
for h in range(0, len(meanT)):
    myAxisLeft.plot(h, meanT[h], 'ro')

#========================CUMULATIVE PRECIPITATION====================
h = range(0, len(cumPrec))
myAxisRight.plot(h, cumPrec,'b-')

plt.show()

main()
```

The function **computeDaily** is written to calculate the daily mean temperature and cumulative precipitation. The `if` statement is used to discriminate the end of a day, based on the data format as described before. In *main*, the variables are defined and the program `read6VarFile` is called to open the file and import the data. Note that in this case we named it `read6VarFile` since it reads six variables. The function **computeDaily** is called and the plot is created, on two different *y* axes (Left and Right).

A.6.2   Call by Reference and by Value

Commonly in computer languages, there are two ways of passing a variable to a function: `call by value` and `call by reference`. The `call by value` is such that the function cannot modify the variable in the calling function; it can only alter its private, temporary copy. The `call by reference` allows the function to alter the original argument. In Python, it is not straightforward to pass a variable to a function and allow modification of the variable inside. Python's way of transferring arguments is based on application of an assignment operator between the argument and the value in the call. Langtangen (2009) calls this way of passing variables to functions `call by assignment`. If an argument x passed to a function can be changed inside the function and noticed in the calling function, then x must be a mutable object. A mutable object in Python is a list, a class instance, a dictionary or a Numerical Python array. Immutable objects such as numbers, strings and tuplets cannot be changed inside the function.

A.7   Flow Control

Flow control statements are the core of a programming language, and they control the order that computations are performed.

A.7.1   Loops: While and For

The `while` statement executes a block of code, based on the evaluation of a test:

```
while (test):
        (statement1)
else:
        (statement2)
```

The expression is evaluated, and if it is true, then the block of code is performed; otherwise the control exits the loop and an additional else part may be executed. Python executes directly from the Python shell. To use the shell, write each instruction separately on a different prompt row. To experiment with these short code segments, type them into the shell and see what happens. Here is an example:

```
>>> a= 0; b= 5
>>> while a < b:
>>>      print(a, end=" ")
>>>      a+=1
0 1 2 3 4
```

The end=" " is used to place all outputs on the same line separated by a space. The $a+ = 1$ is short for $a = a + 1$.

The `For` statement is used to repeat a block of statements a specific number of times. For loops, a counter variable is used, whose value is incremented or decremented with each repetition of the loop. The following is an example of `for` statements:

```
>>> for i in range(10):
>>>      print(i,end=" ")
0 1 2 3 4 5 6 7 8 9

>>> for i in range(1, 11):
>>>      print(i,end=" ")
1 2 3 4 5 6 7 8 9 10
```

Note that Python begins counting from zero in the first example (default), while if we want to print from 1 to 10, we need to write from 1 to 11, because Python counts up to $n-1$.

Whether to use While or For depends on how data are organized and the purpose of the loop. Usually the For loop is preferred when there is a clear initialization and a known increment, because this control flow keeps the control statements close together and visible at the top of the command. When, for instance, the number of data points is known, it is convenient to use a For loop. The While loop is used more often when the increment is unknown a priori or it changes depending on the file type.

Numerical Python has an option of doing mathematical operations with arrays as arguments for the operation, but it can also perform mathematical operations by using for loops. Although numpy is powerful with direct operations with arrays, it is rather slow with for loops. Here we present an example where we add two arrays x and y, each having 10 million entries. Two functions are defined: **func** and **func2**. In the first, the addition is done using numpy arrays (without a for loop), while **func2** performs the addition within a for loop. The results of the computation and the time necessary for the computation are then printed on screen.

```python
import numpy as np
from time import time

def func(x,y):
    return x+y

def func2(x,y,z):
    for i in range(np.size(x)):
        z[i] = x[i]+y[i]
    return x

x = np.ones(1e7)*5
y = np.ones(1e7)*7
print('x= ' + str(x))
print('y= ' + str(y))


t0 = time()
z= func(x,y)
print('The sum of x and y is: z=  '+ str(z))
t1= time()
print('Calculation based on array-operations took: '+ str(t1-t0)
      + ' seconds')

t2 = time()
y=np.ones(1e7)*7
func2(x,y,z)
t3 = time()
print('The sum of x and y is: z=  '+ str(z))
print('Calculation based on for-loop took: '+ str(t3-t2) + ' seconds')
```

The results of this computation are

```
>>>
x= [ 5.   5.   5.  ...,   5.   5.   5.]
y= [ 7.   7.   7.  ...,   7.   7.   7.]
The sum of x and y is: z=  [ 12.   12.   12.  ...,   12.   12.   12.]
Calculation based on array-operations took: 0.0448551177979 seconds
The sum of x and y is: z=  [ 12.   12.   12.  ...,   12.   12.   12.]
Calculation based on for-loop took: 6.82873296738 seconds
>>>
```

A.7.2   If-Else

The if-else statement is written to perform a choice (or multiple choices) or express a decision. The general statement can be written as

```
if <test1>:
        <statements1>
elif <test2>:
        <statements2>
else:
        <statements3>
```

The statement elif stands for else if, which allows for multiple statements. Remember that Python identifies block boundaries by line indentation. Statements indented to the same block line are part of the same block of code. The end of the block is determined when a lesser-indented line is encountered.

An example of how to use if statements is given below, in which Celsius temperatures are converted into Fahrenheit and vice versa. The user is asked to select between four options: to print the options, to convert from Celsius to Fahrenheit, to convert from Fahrenheit to Celsius or to quit the program. The if statement and two elseif statements are used to quit the program, compute the conversion or print the options again to allow the user to quit the program after the computation has been performed. Note that since the input() instruction returns a float, the instruction is wrapped into a **float()** to convert the input temperature from string to float. The program PSP_celsiusFahrenheit.py is as follows:

```
#PSP_celsiusFahrenheit.py
def print_options():
    print ("Options:")
    print (" 'p' print options")
    print (" 'c' convert from celsius")
    print (" 'f' convert from fahrenheit")
    print (" 'q' quit the program")

def celsius_to_fahrenheit(c_temp):
    return (9.0 / 5.0) * c_temp + 32.0
```

```
def fahrenheit_to_celsius(f_temp):
    return (f_temp - 32.0) * 5.0 / 9.0

choice = "p"
while choice != "q":
    if choice == "c":
        temp = float(input("Celsius temperature: "))
        print ("Fahrenheit:", celsius_to_fahrenheit(temp))
        print_options()
        choice=input()
    elif choice == "f":
        temp = float(input("Fahrenheit temperature: "))
        print ("Celsius:", fahrenheit_to_celsius(temp))
        print_options()
        choice=input()
    elif choice != "q":
        print_options()
        choice=input()
```

A.8   File Input and Output

The file I/O in Python depends on the type of file that the program is dealing with. The major distinction is between *text* files and *binary* files. The content of files that are opened in the text mode will be converted automatically as string (**str**). Files that are opened in binary mode are not converted into any format and are treated as raw binary files. To open a file in binary mode, it is necessary to add a lowercase **b** to the built-in *open* statement. The choice depends on the type of files the programmer is dealing with. If the programmer is dealing with a large dataset already in binary format, or with images in binary format, the best choice may be to treat the files as **bytes** or binary files.

If the files are data or test files, such as text files (.txt), csv, xml or html, it may be convenient to treat the file as text files. The open statement can be used to create a file or open an existing file depending on the processing mode. The following is an example of the open statement used to create a file:

```
>>> f= open('data.txt','w')
>>> f.write('1 2')
>>> f.close
```

This statement creates a new file called *data.txt* and writes the numbers 1 and 2 separated by a space. The processing mode w stands for write. The file is then closed. Details are provided by technical books on the various options available for file I/O.

A.9   Arrays

In Python, arrays are not data types like floats or strings. They need to be created as array type by importing a built-in standard module named `array`. An array of values can include characters, integers or floating-point numbers. Arrays are sequence types and behave very much like lists, except that the type of objects stored in them is constrained. The type is specified at object creation time by using a type code, which is a single character:

```
>>> from array import *
>>> x =array('i',[1,2,3,4,5])
```

In this case, the type code is an integer (i) and the members of the array are 1, 2, 3, 4 and 5. If we wanted to define an array of floating points, we would write

```
>>> from array import *
>>> x =array('f',[0.1,1.1,1.5,3.4,4.6])
```

The array can also be indexed and the elements printed by a `for` loop:

```
>>> from array import *
>>> a=array('i',[1,2,3,4,5])
>>> for i in a:
>>>      print(i)
1
2
3
4
5
```

A.9.1   Arrays in numpy

Another option to create, manipulate and perform computations with arrays is to use the module numpy. The numerical Python module has three implementations: `Numeric`, `numarray` and numpy. numpy is the latest version and contains all the implementations of the first two, plus some additional features. It is therefore recommended to utilize numpy. numpy is a Python package for scientific computing, with many important features such as array creation and management, linear algebra, random numbers, Fourier transform, among others.

A useful application of numpy is its array management. Here we present a program used to open, scan and read a data file. In the previous examples, we presented a method to open and read a data file having a given number of variables. Here, a program is shown where a larger data file is opened, scanned and read, without having to prescribe a priori the number of variables contained in the file. To perform this task, a general reader was created using the array functions implemented in numpy.

Below is an example of experimental data from a soil profile, with data collected on an hourly basis (`weather_soil.text`). The data were collected with a Campbell Scientific datalogger (Campbell Scientific, 2006) and the data represented year (`year`), day of the year (`doy`), hour and minute (`HM`), precipitation (`P`) [mm], air temperature (`Tair`) [°C], relative humidity (`RH`) [%], soil water content (`Wc1`,`Wc2`)[$m^3\ m^{-3}$], soil matric potential (`Mp1`, `Mp2`) [J kg^{-1}] and soil temperature (`Ts1`, `Ts2`) [°C] at −10 cm and −30 cm below the soil surface.

Year	Doy	HM	P	Tair	RH	Wc1	Wc2	Mp1	Mp2	Ts1	Ts2
2011	321	1148	0	10.57	58.19	0.223	0.322	-1269	-1711	9.12	9.00
2011	321	1150	0	10.21	60.83	0.225	0.308	-1059	-1684	9.08	9.22
2011	321	1152	0	9.96	62.1	0.241	0.31	-1064	-1598	8.72	8.71
2011	321	1154	0	9.52	63.71	0.232	0.32	-1398	-1517	9.46	9.27
2011	321	1156	0	9.42	63.71	0.232	0.322	-1301	-1170	8.87	8.6
2011	321	1158	0	9.17	65.14	0.227	0.325	-1630	-1193	9.93	9.48
2011	321	1200	0	8.93	65.37	0.23	0.322	-1587	-1241	7.37	8.9

When data are collected for long periods of time and many variables are collected, output files are large and data must be scanned, read and processed in efficient ways. Moreover, we may need to perform various operations, for instance computing the daily average temperature or the daily average soil water content. Here we have written a program that has been used in many programs throughout the book. It is an efficient, general file reader that allows us to read data organized in rows and columns.

The program #PSP_readDataFile.py implements two functions: **scanDataFile** and **readDataFile**. The first is used to read the file using the function **reader**, which is implemented in Python.

The function **scanDataFile** takes the file name and the delimiter as input and returns the number of rows, the number of columns and a Boolean used to warrant if the file has the same number of data for each column or otherwise if there are missing data. Then it checks if it has the same number of rows and columns (to identify possible errors in the file); otherwise, when the function is called, an attribute False is passed to the function call.

A matrix A is created and is filled with zeros by using the instructions of **numpy**. The data are then loaded in the matrix A with a for loop.

```
#PSP_readDataFile.py
from __future__ import print_function
import csv
import numpy as np

def scanDataFile(file, delimiter):
    reader = csv.reader(open(file, "rt"), delimiter=delimiter)
    nrRows = 0
    for row in reader:
        if (nrRows == 0):
            nrCols = len(row)
        else:
            if (len(row) != nrCols):
                #wrong file: nr fields is not uniform
                return (nrRows, nrCols, False)
        nrRows += 1
    return (nrRows, nrCols, True)

def readDataFile(file, nrHeaderFields, delimiter, printOnScreen):
    nrRows, nrCols, isFileOk = scanDataFile(file, delimiter)
    if (isFileOk == False): return (nrRows, False)
```

```
    if (nrRows == 1):
        A = np.zeros((nrRows, nrCols-nrHeaderFields))
    else:
        A = np.zeros((nrRows-nrHeaderFields, nrCols))

    reader = csv.reader(open(file, "r"), delimiter=delimiter)
    i = 0
    for row in reader:
        if (printOnScreen): print(row)
        if (nrRows == 1):
            #for rows
            for j in range(nrHeaderFields, len(row)):
                A[i, j-nrHeaderFields] = float(row[j])
        else:
            #for columns
            if (i >= nrHeaderFields):
                for j in range(0, len(row)):
                    A[(i-nrHeaderFields), j] = float(row[j])
        i += 1

    return(A, True)
```

An efficient way to incorporate these functions into different programs is to create a file called `PSP_readDataFile`. This file is imported into the programs where it is needed and then its functions can be called.

A.10   Reading Date Time

A modification of the program `#PSP_readDataFile.py` was written to read date format by adding the function `readGenericDataFile`, which reads strings by using lists. Since dates are strings and `numpy` arrays do not read strings, the strings must be converted into numbers. This can be done by using the function `date2num` included in the standard Python module `date time`. The following is an example of soil temperature data at four different depths, collected on an hourly basis and comma-separated:

```
date hour ,T1    ,T2    ,T3    ,T4
01-09-04 00.00,23.8,23.9,22.4,20.7
01-09-04 01.00,23.6,23.8,22.4,20.7
01-09-04 02.00,23.4,23.7,22.4,20.7
01-09-04 03.00,23.1,23.5,22.5,20.7
01-09-04 04.00,23,23.5,22.5,20.7
01-09-04 05.00,22.8,23.3,22.5,20.7
01-09-04 06.00,22.6,23.3,22.5,20.7
01-09-04 07.00,22.5,23.2,22.5,20.7
01-09-04 08.00,22.4,23.1,22.4,20.7
01-09-04 09.00,22.4,22.9,22.4,20.7
01-09-04 10.00,22.5,22.9,22.5,20.7
01-09-04 11.00,22.8,22.7,22.4,20.7
```

The file #PSP_readDataFile.py now contains the following function, which uses the list A=[] instead of the A = np.zeros arrays:

```
def readGenericDataFile(file, nrHeaderRows, delimiter, isPrintScreen):
    nrRows, nrCols, isFileOK = scanDataFile(file, delimiter)
    if (isPrintScreen): print ('nrRows =', nrRows, ' nrCols =',
                               nrCols)

    if (isFileOK == False): return (nrRows, False)
    myReader = csv.reader(open(file, "rt"), delimiter=delimiter)

    A = []
    i = 0
    for myRow in myReader:
        if (isPrintScreen): print(myRow)
        if (i >= nrHeaderRows):
            A.append(myRow)
        i += 1
    return(A, True)
```

This function is called by the program #PSP_readSoilTempData.py, which is written to read, scan and plot the soil temperature data. Conversion of the list is performed by defining a variable date = [] and appending the values within the for loop over the nrValues. The date is then converted to a number by using the function d=**date2num**(**date**). The variable d is now a number that can be plotted by matplotlib using the function plt.plot_date(). The program is as follows (the output is shown in Fig. 4.1 in Chapter 4):

```
#PSP_soiltemp.py
from __future__ import print_function, division
from PSP_readDataFile import readGenericDataFile
import numpy as np
import matplotlib.pyplot as plt
from matplotlib.dates import date2num
from datetime import datetime

NODATA = -9999

def main():
    A, isFileOk = readGenericDataFile("data/soilTemp_2days.txt", 1,
                                      ',', False)
    if (isFileOk == False):
        print ("Incorrect format in row: ", A)
        return()

    nrValues = len(A)

    # date array
    date = []
```

```
        for i in range(nrValues):
            strDate = A[i][0]
            date.append(datetime.strptime(strDate,"%d-%m-%y %H.%M"))
        d = date2num(date)

        #values (with simple error check)
        soilTemp = np.zeros((4, nrValues))
        for col in range(4):
            previous = NODATA
            for i in range(nrValues):
                isOk = False
                # missing data
                if (A[i][col+1] != ""):
                    value = float(A[i][col+1])
                    # void data > 1 day
                    if (i == 0) or ((d[i]-d[i-1]) < 1):
                        # wrong values
                        if (value > -50) and (value < 100):
                            # spikes
                            if (previous == NODATA) or (abs(value - previous)
                                                                    < 10):
                                isOk = True
                if isOk:
                    soilTemp[col,i] = value
                    previous = value
                else:
                    soilTemp[col,i] = np.nan
                    previous = NODATA

    print ("first date:", date[0])
    print ("last date: ", date[len(date)-1])

    plt.title('')
    plt.ylabel('Temperature [C]',fontsize=16)
    plt.xlabel('Date',fontsize=16)
    plt.plot_date(d, soilTemp[0],'r')
    plt.plot_date(d, soilTemp[1],'g')
    plt.plot_date(d, soilTemp[2],'b')
    plt.plot_date(d, soilTemp[3],'k')
    plt.tick_params(axis='both', which='major', labelsize=14)
    plt.tick_params(axis='both', which='minor', labelsize=14)
    plt.show()

main()
```

A.11   Object-Oriented Programming in Python

Python has powerful object-oriented programming (OOP) structures. Classes are defined using the following statements:

```
>>> class soil_data:
>>>     texture="silt"
>>>     sand = 30
>>>     silt = 30
>>>     clay = 40
>>>     def thisMethod(self)
>>>         return "these are soil data"
```

A class has *attributes* (the variables to store data) and *methods* (the functions that the class can provide). Here the class `soil_data` has an attribute named texture, which is a string, and three attributes named sand, silt and clay, which are integers. The statement **def** is used to define a method (which is similar to a function, but is called a method when it is written within a class). To access the data contained in the class, an object is needed. To create an object, the following statement is used: `read_data_object=soil_data()`. Here the object `read_data_object` is created. With this object, it is possible to access the data contained in the class `soil_data`. For instance, by typing in the Python console the statement

```
>>>   read_data_object.sand
 30
```

the value of sand (30) is read from the class and returned on the screen. It is also possible to execute the method contained in the class:

```
>>>   read_data_object.thisMethod
  "these are soil data"
```

A.12   Output and Visualization

The output of a Python program can be generated by importing a variety of available packages used to plot variables in various formats, such as the package Tkinter. However, packages like Tkinter are used mainly as 2D plotting packages.

For visualization of model output, it is necessary to use or implement visualization packages that are often difficult to code, complicated or not available. In this book, we use a very powerful graphics module, Visual Python. Visual Python is an easy-to-use 3D graphics module for Python. As is shown throughout this book, the programmer can create 3D objects (such as spheres, and curves) and position them in 3D space. Visual Python can automatically update the position of objects or curves in 3D and update the position of the objects, depending on the specific features of the written program. It is very advantageous for visualizing physical processes, since the programmer does not have to spend time on the display management, but can focus on the computational aspects of the program. The package is freeware and can be downloaded at <http://www.vpython.org>. To run the visual programs written in this book, the user must install Visual Python, version 6.

A.13 EXERCISES

A.1. Compute by hand the complex multiplication $(5-3j)*(4+2j)$. Reform the same operation using Python.

A.2. Read the code of the program `PSP_averageTair.py,` understand the purpose of each program line and identify the flow of instructions as described above. Run the program and test the results.

Appendix B
Computational Tools

This appendix provides examples of computational tools that are used in the book. Soil physics often deals with problems that must be solved by employing a specific computational tool. Numerical derivatives, integrals, linear and nonlinear interpolations, linear algebra, Laplace and Fourier transforms, and ordinary and partial differential equations are commonly used to solve soil physics problems. In the following sections, we present examples of solutions that are used throughout this book and that can also be utilized in a variety of different problems.

B.1 Numerical Differentiation

Numerical differentiation is based on the use of Lagrange polynomials. The theoretical basis for the determination of numerical derivatives from polynomials (and the associated proofs), on which the computation of numerical derivatives is based, is beyond the scope of this book. However, a description of the use of Lagrange polynomials for numerical differentiation is given by Burden and Faires (1997).

The derivative of a function is defined as

$$f'(x) = \lim_{h \to 0} \frac{f(x_0 + h) - f(x_0)}{h} \tag{B.1}$$

A numerical approximation of this number can be obtained by first assuming that $x_0 \in (a, b)$, and that $x_1 = x_0 + h$, with h small enough to ensure that $x_1 \in (a, b)$. We can write eqn (B.1) as

$$f'(x) \approx \frac{f(x_0 + h) - f(x_0)}{h} \tag{B.2}$$

Equation (B.2) is such that there is no information about the error term and therefore the truncation error cannot be estimated. Analysis and estimation of the error terms is described in detail by Burden and Faires (1997). When h is small, the simple difference quotient can be used to approximate eqn (B.2):

$$f'(x_0) = \frac{f(x_0 + h) - f(x_0)}{h} \tag{B.3}$$

This formula is called a *forward difference* for $h > 0$. If $h < 0$, it is called a *backward difference*. To overcome the problem of the truncation error, more evaluation points should be used. These generalizations are referred as *(n + 1)-point* formulas. Usually a higher number of evaluation points produces greater accuracy; however, issues concerning truncation error and computational time lead to the use of three- or five-point formulas. In some cases, however, the use of a higher number of points may lead to larger truncation errors.

Three points can be used with the application of the *three-point formula*

$$f'(x_0) = \frac{1}{h}\left[-\frac{3}{2}f(x_0) + 2f(x_0 + h) - \frac{1}{2}f(x_0 + 2h)\right] \tag{B.4}$$

Similarly, five Lagrange coefficients can be used to derive the *five-point formula*. With a similar derivation, the following equation is obtained:

$$f'(x_0) = \frac{1}{12h}[f(x_0 - 2h) - 8f(x_0 - h) + 8f(x_0 + h) - f(x_0 + 2h)] \tag{B.5}$$

Usually, the five-point formula gives superior results in terms of a smaller truncation error. Below we present a Python code, where a five-point derivative is implemented to compute the derivative of measured soil temperature, to investigate the rate of change of temperature with respect to time. The data are collected by a CR10X (Campbell Scientific) every 10 minutes. The first six hours of data from the file SoilTemp.txt are as follows:

```
#Hour Temp(C)
1    12.48893452
2    9.769722939
3    11.34481716
4    11.65844917
5    14.50185585
6    16.84011841
```

The data files are opened with the file PSP_readDataFile.py described in Appendix A. PSP_numericalDerivation.py is then used, where the function for computing the five-point derivative is implemented as follows:

```python
from PSP_readDataFile import readDataFile
import matplotlib.pyplot as plt

def firstDerivative5Points(y):
    myDerivative = []
    for i in range(0,2):
        myDerivative.append(0)
    for i in range(2, len(y)-2):
        dY = (1./(12.)) * (y[i-2] - 8.*y[i-1] + 8.*y[i+1] - y[i+2])
        myDerivative.append(dY)
    for i in range(len(y)-2, len(y)):
        myDerivative.append(0)
    return(myDerivative)

def main():
    myHour = []
    myTemp = []
    A, isFileOk = readDataFile("airTemp.dat", 1, '\t', True)
    if (isFileOk == False): print ("Incorrect format in row: ", A)
```

```
myHour = A[:,0]
myTemp = A[:,1]
myDerivative = firstDerivative5Points(myTemp)

fig = plt.figure(figsize=(10,8))
plt.plot(myHour, myTemp, 'ko')
plt.plot(myHour, myDerivative,'k')
plt.xlim(xmin=0,xmax=25)
plt.ylim(ymin=-5,ymax=40)
plt.xlabel('Time [hour]',fontsize=20,labelpad=8)
plt.ylabel('Temperature [C]',fontsize=20,labelpad=8)
plt.tick_params(axis='both', which='major', labelsize=20,pad=8)
plt.show()
```

```
main()
```

The third line of the **main** function is used to read the data, as described in Appendix A. The function **firstDerivative5Points(y)** implements eqn (B.5) for computation of the five-point derivative. The function takes an array y (which is the variable to be differentiated) and returns an array myDerivative. Figure B.1 depicts hourly temperature and its numerical derivative. This algorithm is used in Chapter 5, in the project PSP_travelTimeAnalysis, for computation of the first derivative of the TDR wave form.

B.2   Numerical Integration

In integration, which is the reverse process of differentiation, the task is to find a function whose derivative is known. If the derivative of a function is not defined or the integration has to be performed on discrete experimental data, analytical integration is not possible and so numerical

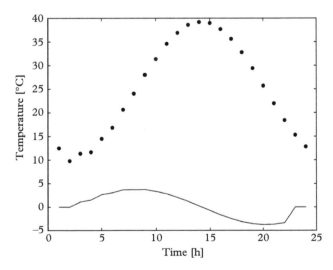

Figure B.1 *Plot of hourly temperature (black dots) and its numerical derivative (solid line) as a function of time.*

integration must be performed. For numerical integration, the methods are also based on inter-polation using Lagrange polynomials. The use of Lagrange polynomials for finding numerical integrals is described by Burden and Faires (1997).

In a similar fashion to what was done in the determination of a two-point derivative, we obtain the *trapezoidal rule*:

$$\int_a^b f(x)\, dx = \frac{h}{2}\left[f(x_0) + f(x_1)\right] - \frac{h^3}{12}f''(\xi)$$ (B.6)

where a and b are the integration limits. Equation (B.6) is called trapezoidal because the integral $\int_a^b f(x)\, dx$ is approximated by the area of a trapezoid. The error term involves the second derivative of the function (f''); therefore the trapezoidal rule returns exact results for functions whose second derivatives are zero. Further analysis on the evaluation of errors for numerical integration can be found in Burden and Faires (1997).

Applying the second Lagrange polynomial with nodes $x_0 = a$, $x_2 = b$ and $x_1 = a + h$, where $h = (b - a)/2$, *Simpson's rule* is obtained:

$$\int_{x_0}^{x_2} f(x)\, dx = \frac{h}{3}\left[f(x_0) + 4f(x_1) + f(x_2)\right] - \frac{h^5}{90}f^4(\xi)$$ (B.7)

The performance of a numerical integration technique can be evaluated by applying it to simple functions whose integrals over a given interval are known exactly. For instance, Table B.1 shows the results for the trapezoidal rule and Simpson's rule for simple functions, together with the analytical (exact) values.

It is worth noting that Simpson's rule performed much better here than the trapezoidal rule. A useful extension of Simpson's rule is the *three-eighths Simpson's rule*, which is obtained by adding one order to the Lagrange polynomial, thereby adding an additional evaluation point (x_0, x_1, x_2 and x_3):

$$\int_{x_0}^{x_2} f(x)\, dx = \frac{3h}{8}\left[f(x_0) + 3f(x_1) + 3f(x_2) + f(x_3)\right] - \frac{h^5}{90}f^4(\xi)$$ (B.8)

This equation is called *three-eighths Simpson's rule*, because the increment is $\left(\frac{3}{8} \cdot h\right)$.

B.2.1 Extended Methods

As discussed above, there are different approaches to computing numerical integrals. Press *et al.* (1992) present a numerical code that combines the trapezoidal and Simpson's equations, with the idea of increasing the number of evaluated values until a certain precision is obtained.

Table B.1 *Approximation of integration over the interval [0, 2] of simple functions $f(x)$ using the trapezoidal rule and the Simpson's rule*

$f(x)$	x^2	x^4	$\sin x$	e^x
Exact value	2.667	6.400	1.416	6.389
Trapezoidal rule	4.000	16.000	0.909	8.389
Simpson's rule	2.667	6.667	1.425	6.421

Usually, it is not known at the beginning how many points at which the function must be evaluated to obtain the required precision. Therefore the program must be able to add stepwise more points to be used for evaluating the integral. Here we have translated the original Fortran code into Python and adapted the code presented by Press *et al.* (1992) for numerical integration. The code has also been modified to include plotting options.

Extended trapezoidal

This function, called **trapzd**, calculates an integral stepwise. The calculation of the integral is based on the equation

$$\int_{x_1}^{x_N} f(x)\, dx = h\left[\frac{1}{2}f_1 + f_2 + f_3 + \cdots + f_{N-1} + \frac{1}{2}f_N\right] + O\left(\frac{(b-a)^3 f''}{N^2}\right) \tag{B.9}$$

where N is the number of points, $f(x)$ is the function to be integrated, the term $O(.)$ is the error in the interval $(b-a)$ and f'' is the second derivative of the function. We present here an example of the application of the extended trapezoidal rule to the equation

$$y = (x-1)^2 + 0.6x\sin(6x) + 1.4 \tag{B.10}$$

Equation (B.10) is integrated using the trapezoidal rule for different numbers of points N, as shown below.

Simpson

The function **qsimp**(func,a,b) is a function that calls the function **trapdz** sequentially with increasing n until a certain accuracy for the integral is obtained. The computation stops when the relative difference of two consecutive values is smaller than a certain threshold (EPS).

Simpson's rule (described above) is included in the calculation of the integral: when comparing the coefficients ($\frac{4}{3}$) and considering that the values st have been obtained by the points that are in the middle of the points that were used for calculating ost in the previous run (compared with Simpson's rule, where the coefficient of the middle value is $\frac{4}{3}$), where $x_0 = a$, $x_{2N} = b$ and $x_1, x_2, x_3, \ldots, x_{2N-1}$ are equally spaced between a and b with distance h. Simpson's rule is applied at the approximation symbol in the following equation:

$$\int_a^b f(x)\, dx = \int_{x_0}^{x_2} f(x)\, dx + \int_{x_2}^{x_4} f(x)\, dx + \int_{x_4}^{x_6} f(x)\, dx + \ldots + \int_{x_{2(N-1)}}^{x_{2N}} f(x)\, dx$$

$$\approx \frac{h}{3}[f(x_0) + 4f(x_1) + f(x_2)] + \frac{h}{3}[f(x_2) + 4f(x_3) + f(x_4)]$$

$$+ \frac{h}{3}[f(x_4) + 4f(x_5) + f(x_6)] + \ldots + \frac{h}{3}[f(x_{2N-2}) + 4f(x_{2N-1}) + f(x_{2N})]$$

$$= \frac{1}{3}\left\{4 \cdot \left[\frac{1}{2}f(x_0) + f(x_1) + f(x_2) + f(x_3) + \ldots + f(x_{2N-1}) + \frac{1}{2}f(x_{2N})\right]\right.$$

$$\left. -2 \cdot \left[\frac{1}{2}f(x_0) + f(x_2) + f(x_4) + f(x_6) + \ldots + f(x_{2(N-1)}) + \frac{1}{2}f(x_{2N})\right]\right\}$$

$$= (4 \cdot \text{st} - \text{ost})/3$$

$$= \text{s}$$

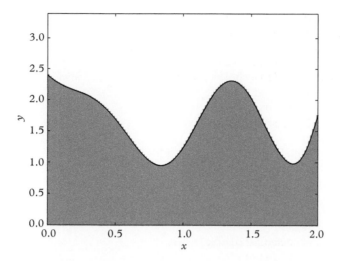

Figure B.2 *Integration of eqn (B.10) using the extended method, combining the trapezoidal rule and Simpson's rule.*

where st and ost are respectively the new and the old return values of **trapzd** when called from **simpy**. The project PSP_numericalIntegration contains the file PSP_Simpson.py with the program for the final integration. The function **plottrapzd**() is written to plot the results of the integration as shown in Fig. B.2.

```
#PSP_simpson.py
import numpy as np
import matplotlib.pyplot as plt
from math import *

def plottrapzd(func,a,b,n):
  it = 1;
  for j in range(1,n):
    it <<= 1;
  x = np.zeros(2+2*it);
  y = np.zeros(2+2*it);
  d = (b-a)/float(it);
  x[0]=a;
  y[0]=func(a);
  x[1]=a+0.5*d;
  y[1]=func(a);
  for i in range(1,it):
    x[2*i]=a+(float(i)-0.5)*d;
    y[2*i]=func(a+float(i)*d);
    x[2*i+1]=a+(float(i)+0.5)*d;
    y[2*i+1]=func(a+float(i)*d);
  x[2*it]=b-0.5*d;
```

```
        y[2*it]=func(b);
        x[2*it+1]=b;
        y[2*it+1]=func(b);
        x_p = np.zeros(1+it);
        y_p = np.zeros(1+it);
        for i in range(0,1+it):
          x_p[i] = x[2*i]+0.5*d;
          y_p[i] = y[2*i];
        x_p[0] = x[0];
        plt.fill_between(x,y,0,color='0.5')
        x2 = np.zeros(1000);
        y2 = np.zeros(1000);
        for i in range(0,1000):
          x2[i]=a+(b-a)/1000.*float(i+1);
          y2[i]=func(x2[i]);
        plt.plot(x2,y2,color='black')
        plt.ylim(0,max(y2)+1.);
        plt.xlim(a,b);
        plt.xlabel('x',fontsize=16,labelpad=4)
        plt.ylabel('y',fontsize=16,labelpad=4)
        plt.tick_params(axis='both', which='major', labelsize=16,pad=4)
        plt.tick_params(axis='both', which='minor', labelsize=16,pad=4)

def trapzd(func, a, b, n):
    if (n == 1):
        trapzd.s=0.5*(b-a)*(func(a)+func(b));
        return trapzd.s;
    else:
        it = 1;
        for j in range(1,n-1):
            it <<= 1;
        tnm=float(it);
        del_=(b-a)/tnm;
        x=a+0.5*del_;
        sum_ = 0.0;
        for j in range(1,it+1):
            sum_ += func(x);
            x+=del_;
        trapzd.s=0.5*(trapzd.s+(b-a)*sum_/tnm);
        return trapzd.s;

def qsimp(func, a, b):
    EPS = 1.0e-6;
    JMAX = 20;
    ost=0.0;
    os=0.0;
    for j in range(1,JMAX+1):
      st=trapzd(func,a,b,j);
```

```
    s=(4.0*st-ost)/3.0;
    if (j > 5):
      if (fabs(s-os) < EPS*fabs(os) or (s == 0.0 and os == 0.0)):
          plottrapzd(func,a,b,j);
          plt.show()
          return s;
    os=s;
    ost=st;
  return 10.0;

def func(x):
    return (x-1.)*(x-1.)+0.6*x*sin(6.*x)+1.4;
integral = qsimp(func,0,2);
print ("Numerical solution:")
print (integral)
```

The output of the program PSP_Simpon.py is the result of the numerical integration

$$\int_0^2 [(x-1)^2 + 0.6x\sin(6x) + 1.4]\, dx \simeq 3.288 \tag{B.11}$$

Since this is a numerical solution, the integration does not provide an exact value, but an approximate one.

B.3 Linear Algebra

The numpy module presents many applications for numerical computation. The classes that represent matrices and basic operations such as matrix multiplications and transpose are a part of numpy. The module linalg contains functions to perform linear algebra operations.

B.3.1 Matrix Multiplication

Matrix multiplication is easily performed in Python. Let us consider the following matrix multiplication:

$$\begin{vmatrix} 0 & -1 & 2 \\ 4 & 11 & 2 \end{vmatrix} \times \begin{vmatrix} 3 & -1 \\ 1 & 2 \\ 6 & 1 \end{vmatrix} = \begin{vmatrix} 11 & 0 \\ 35 & 20 \end{vmatrix} \tag{B.12}$$

To perform this multiplication in Python, we first need to define the two matrices:

```
>>> from numpy import *
>>> A = [[0, -1, 2],
         [4, 11, 2]]
>>> B = [[3, -1],
         [1, 2],
         [6, 1]]
```

```
>>>dot(A,B)
array([[11,   0],
       [35,  20]])
```

Note that A is a 2×3 matrix, B is a 3×2 matrix and the resulting matrix is 2×2. The square brackets inside are used to separate the rows from each other. Using the notation for a numpy array, matrix multiplication is done by using the dot() notation.

B.3.2   Inverse of a Matrix

The inverse of a matrix A is the matrix B such that $AB = I$, where I is the identity matrix. The identity matrix is a matrix having 1s down the main diagonal. The matrix B is denoted by $B = A^{-1}$.

Let us consider the matrix

$$A = \begin{vmatrix} 0 & -1 \\ 4 & 11 \end{vmatrix} \tag{B.13}$$

The inverse of the matrix A is

$$\begin{vmatrix} 2.75 & 0.25 \\ -1 & 0 \end{vmatrix} \tag{B.14}$$

To find the inverse in Python, we first need to define the matrix A and then compute the inverse:

```
>>> from numpy import *
>>> A = [[0, -1],
            [4, 11]]
>>> linalg.inv(A)
array([[2.75,   0.25],
       [-1., -0.        ]])
>>>
```

To check that this computation is correct, we can multiply the matrices A and A^{-1} to see if we obtain the identity matrix. We first define the matrix A^{-1} as B:

```
>>> from numpy import *
>>> A = [[0, -1],[4, 11]]
>>> B= [[2.75,0.25],[-1,0.]]
>>>dot(A,B)
>>>array([[1.,0.],
            [0.,1.]])
```

The results are correct, since the result of the **dot** product between A and B is the identity matrix I.

B.3.3   Gaussian Elimination

Gaussian elimination is a direct method for solving a linear system

$$A \cdot x = b \tag{B.15}$$

This system can be written in matrix form and we can then employ linear algebra techniques for solving it. Details of solutions of systems of linear equations can be found in linear algebra books (e.g. Lai, 2009). The following is an example of a system of equations, expressed in matrix form (the raised dot denotes matrix multiplication):

$$\begin{vmatrix} 5 & 2 & 0 & 0 \\ 2 & 5 & 2 & 0 \\ 0 & 2 & 5 & 2 \\ 0 & 0 & 2 & 5 \end{vmatrix} \cdot \vec{x} = \begin{vmatrix} 5 \\ -4 \\ 4 \\ 5 \end{vmatrix} \tag{B.16}$$

The objective of Gaussian elimination is to solve this system of equations for the vector $\vec{x}$. This algorithm is based on inversion of the matrix A. To store the matrix, numpy arrays are employed. Below, the Gaussian algorithm is implemented, with the file `PSP_readDataFile` being imported to read the data as described above. The numpy package is then imported and the function **GaussElimination**(A, b) is defined. The first part of the algorithm is used to divide the elements in i and m by the matrix $A[m, m]$. The following lines are then used for the backward insertion. The **main** file contains the lines to read the matrices A and b from files, call the function `Gauss_elimination(A,b)` and return the unknown vector $\vec{x}$.

```
#PSP_GaussEliminationm.py
from PSP_readDataFile import readDataFile
from numpy import *

def GaussElimination(A,b):
    n=len(b)
    for m in range(0,n-1):
        for i in range(m+1,n):
            dummy=A[i,m]/A[m,m]
            A[i,m]=0.
            for j in range(m+1,n):
                A[i,j]-=(dummy*A[m,j])
            b[i]-=(dummy*b[m])
    # backward insertion
    x=zeros(n,float)
    x[n-1]=b[n-1]/A[n-1,n-1]
    for i in range(n-2,-1,-1):
        x[i]=b[i]
        for j in range(i+1,n):
            x[i]-=(A[i,j]*x[j])
        x[i]/=A[i,i]
    return (x)

def main():
    A, isFileOk = readDataFile("A.txt", 0, ',', False)
    if (isFileOk == False): print ("Incorrect format in row: ", A)
    x, isFileOk = readDataFile("b.txt", 0, ',', False)
    b = x[0,:]
    print("matrix A\n",A)
```

```
    print("vector b\n",b)

    x= GaussElimination(A,b)
    print("unknown vector x\n",x)

main()
```

The result of the above computation is as follows:

```
>>>
matrix A
 [[ 2. -1.  0.  0.  0.]
  [-1.  2. -1.  0.  0.]
  [ 0. -1.  2. -1.  0.]
  [ 0.  0. -1.  2. -1.]
  [ 0.  0.  0. -1.  2.]]
vector b
 [ 5. -4.  4. -4.  5.]
unknown vector x
 [ 3.  1.  3.  1.  3.]
>>>
```

B.3.4   Thomas Algorithm

The tridiagonal matrix algorithm (TDMA), also known as the Thomas algorithm, is a simplified form of Gaussian elimination that can be used to solve tridiagonal systems of equations. It is based on LU decomposition, in which the matrix system $Ax = b$ is rewritten as $LUx = b$, where L is a lower-triangular matrix and U is an upper-triangular matrix. A tridiagonal system may be written as

$$a_i x_{i-1} + b_i x_i + c_i x_{i+1} = d_i \tag{B.17}$$

where $a_1 = 0$ and $c_n = 0$. In matrix form,

$$\begin{vmatrix} b(1) & c(1) & 0 & 0 \\ a(2) & b(2) & c(2) & 0 \\ 0 & a(3) & b(3) & c(3) \\ 0 & 0 & a(4) & b(4) \end{vmatrix} \begin{vmatrix} x(1) \\ x(2) \\ x(3) \\ x(4) \end{vmatrix} = \begin{vmatrix} d(1) \\ d(2) \\ d(3) \\ d(4) \end{vmatrix} \tag{B.18}$$

For such systems, the solution can be obtained in $O(n)$ operations instead of the $O(n^3)$ required by Gaussian elimination. A first sweep eliminates the a's, and then an (abbreviated) backward substitution produces the solution. Examples of such matrices commonly arise from the discretization of 1D problems (e.g. the 1D Poisson problem). Because of its numerical efficiency and simplicity, the Thomas algorithm is extensively used in this book to solve linear system of equations and for numerical solution of partial differential equations (PDEs).

The forward elimination phase is given by (the modified coefficients are indicated by primes')

$$c_i' = \frac{c_i}{b_i} \tag{B.19}$$

$$d_i' = \frac{d_i}{b_i} \tag{B.20}$$

$$b_{i+1}' = b_{i+1} - a_{i+1} c_i \tag{B.21}$$

$$d_{i+1}' = d_{i+1} - a_{i+1} d_i \tag{B.22}$$

The backward substitution is given by

$$x_n = \frac{d_n}{b_n} \tag{B.23}$$

$$x_i = d_i - c_i x_{i+1} \tag{B.24}$$

For further details and derivation of the Thomas algorithm, see Conte and deBoor (1972) or Morton and Mayers (1994).

The Python code is shown below. The **main** file is used to load the matrices and call the program PSP_ThomasAlgorithm that contains the Thomas algorithm. The data are read by the program PSP_readDataFile described in Appendix A. Note that in this program there are two functions: Thomas and ThomasBoundaryCondition. The first function solves the whole matrix, while the second solves from the row of index first to the row of index last. The rows before the first and after the last are used to set the boundary conditions.

In the program PSP_main_ThomasAlgorithm.py, lines 8–12 are used to read the file A.txt, which contains the matrix A, while lines 15–17 are used to read the file b.txt, which contains the matrix of the known terms. Lines 20–27 are written to create the arrays (using numpy) and to assign the values to the diagonals of the matrix (lines 24–27). Finally, the unknowns (x) are obtained by calling the function Thomas, which takes as arguments the diagonal elements a, b, c, d.

```python
from __future__ import print_function, division
from PSP_readDataFile import readDataFile
from PSP_ThomasAlgorithm import Thomas
import numpy as np

def main():
    myOutput, isFileOk = readDataFile("A.txt", 0, ',', False)
    if (isFileOk == False):
        print ("Incorrect matrix format in row: ", myOutput)
        return (False)
    A = myOutput
    print("matrix A\n",A)

    myOutput, isFileOk = readDataFile("b.txt", 0, ',', False)
    d = myOutput[:,0]
    print("vector d\n",d)
```

```
    # initialize vectors
    n = len(d)
    a = np.zeros(n, float)
    b = np.zeros(n, float)
    c = np.zeros(n, float)
    for i in range(n):
        if (i > 0): a[i] = A[i,i-1]
        b[i] = A[i,i]
        if (i < (n-1)): c[i] = A[i,i+1]

    x = Thomas(a, b, c, d)
    print("unknown vector x\n",x)

main()
```

PSP_main_ThomasAlgorithm.py is the core program in which the Thomas algorithm is implemented. In lines 9–13, the forward substitution scheme is written, while lines 16–18 contain the backward substitution.

When a solution is obtained by imposing specific boundary conditions, the algorithm is modified. The function **ThomasBoundaryCondition**( ), between lines 22 and 33, is written to solve a matrix where the first and the last elements are specified.

```
#PSP_ThomasAlgorithm.py
from __future__ import division
import numpy as np

def Thomas(a, b, c, d):
    n=len(d)
    x =np.zeros(n, float)

    for i in range(0, n-1):
        c[i] /= b[i]
        d[i] /= b[i]
        b[i + 1] -= a[i + 1] * c[i]
        d[i + 1] -= a[i + 1] * d[i]

    # back substitution
    x[n-1] = d[n-1] / b[n-1]
    for i in range(n-2, -1, -1):
        x[i] = d[i] - c[i] * x[i + 1]

    return(x)


def ThomasBoundaryCondition(a, b, c, d, x, first, last):

    for i in range(first, last):
        c[i] /= b[i]
```

```
      d[i] /=  b[i]
      b[i + 1] -=  a[i + 1] * c[i]
      d[i + 1] -=  a[i + 1] * d[i]

   # back substitution
   x[last] = d[last] / b[last]
   for i in range(last-1, first-1, -1):
       x[i] = d[i] - c[i] * x[i + 1]
```

The result of the above computation is as follows:

```
>>>
matrix A
 [[ 2. -1.  0.  0.  0.]
  [-1.  2. -1.  0.  0.]
  [ 0. -1.  2. -1.  0.]
  [ 0.  0. -1.  2. -1.]
  [ 0.  0.  0. -1.  2.]]
vector d
 [ 5. -4.  4. -4.  5.]
unknown vector x
 [ 3.  1.  3.  1.  3.]
>>>
```

Clearly the results are the same as for the Gaussian algorithm (note that the known vector is called `vector d`, instead of `vector b`). However, as mentioned above, the solution can be obtained in $O(n)$ operations instead of $O(n^3)$; therefore this algorithm is much faster than Gaussian elimination (Morton and Mayers, 1994).

....

B.4 EXERCISES

B.1. Run the code of the program `PSP_derivativeTair`, understand the purpose of each program line and analyse the results plotted by the program.

B.2. Utilize the program `PSP_Simpson.py` to compute the integral of the function $\sin(x) \times e^{(-x \times 0.5)}$

List of Symbols

Lower Case Latin Symbols

a	pore radius [m]
a_l	albedo [–]
a_s	total surface area [m^2]
	short-wave absorptivity [–]
b	Campbell water retention curve parameter [–]
b_g	diffusivity parameter for soil air-filled porosity [–]
c	speed of light [m s^{-1}]
	concentration of gas or solute [kg m^{-3} or mol kg^{-1} or g L^{-1}]
c_l	fractional cloud cover [–]
c_{va}	vapour concentration of air [kg m^{-3}]
c_{vs}	soil surface vapour concentration [kg m^{-3}]
c_v	vapour concentration of soil [kg m^{-3} or g m^{-3}]
c_v'	saturation vapour concentration [kg m^{-3} or g m^{-3}]
$co(i)$	nodal CO$_2$ concentrations [kg m^{-3}]
d	particle diameter [m]
	characteristic length [m]
	zero-plane displacement [m]
d_d	geometric mean particle diameter for sand fraction [m]
d_g	geometric mean particle diameter [m]
d_t	geometric mean particle diameter for silt fraction [m]
d_y	geometric mean particle diameter for clay fraction [m]
$df(i)$	gas diffusivity in element i
e	vapour pressure [Pa]
	void ratio [m^3 m^{-3}]
e_a	actual vapour pressure [Pa]
e_s	vapour pressure at saturation [Pa]
$\vec{e}_x$	unit vector [–]
f	logarithm of scaling factor [–]
f_g	flux density of gas [g m^{-2} s^{-1}]
f_h	flux density of heat [W m^{-2}]
f_i	flux density of component i [g m^{-2} s^{-1}]
f_l	flux density of liquid phase [kg m^{-2} s^{-1}]
f_s	flux density of solute [kg m^{-2} s^{-1}]
f_v	flux density of water vapour [kg m^{-2} s^{-1}]
f_w	flux density of water [kg m^{-2} s^{-1}]
	empirical parameter for thermal conductivity [–]
f_{wi}	infiltration rate [kg m^{-2} s^{-1}] or [m^3 s^{-1}]
	water flux density per pore [kg m^{-2} s^{-1}]
	empirical weighting factor for thermal conductivity [–]

g	gravitational acceleration [m s^{-2}]
$\bar{g}$	geometric mean [–]
g_a	shape factor for thermal conductivity [–]
g_c	shape factor for thermal conductivity [–]
h_r	relative humidity [–]
h_a	atmospheric relative humidity [–]
h_{pond}	immobile water [m]
h'_s	mobile water [m]
h_s	surface hydraulic depth [m]
	relative humidity at soil surface [–]
	sunset hour angle [–]
h^*	thickness of water film [m]
i	iteration counter [–]
	array index [–]
	spatial element for numerical solution [–]
k	von Kármán constant [–]
$k(i)$	element conductances [–]
k_d	linear partition coefficient [m^3 kg^{-1}]
k_g	gas weighting factor [–]
k_l	Langmuir constant [–]
k_s	solids weighting factor [–]
k_w	water weighting factor [–]
l	tortuosity parameter [–]
m	van Genuchten parameter [–]
m_d	fractional mass of sand fraction [–]
m_g	mass of gas phase [kg]
	diffusivity parameter for soil air-filled porosity [–]
m_l	mass of liquid phase [kg]
m_s	mass of solid phase [kg]
m_t	total mass of solid, gas and liquid phases [kg]
	fractional mass of silt fraction [–]
m_y	fractional mass of clay fraction [–]
n	van Genuchten parameter [–]
	empirical parameter for surface tension dependence on temperature [–]
p	probability [–]
p_1	sampling point [–]
p_2	sampling point [–]
p_3	sampling point [–]
q	soil property for thermal conductivity [–]
r	radius [m]
	autocorrelation function [–]
r_c	capillary radius [m]
r_k	autocorrelation function [–]
r_s	soil surface resistance for water vapour transfer [s m^{-1}]
r_v	aerodynamic resistance for water vapour transfer [s m^{-1}]
r_{va}	boundary-layer resistance to vapour diffusion [s m^{-1}]
r_{vc}	canopy resistance to vapour diffusion [s m^{-1}]
r_{vsi}	stomatal diffusion resistance [s m^{-1}]
s	standard deviation [–]
	slope of saturation vapour concentration [kg m^{-3} °C]

t	time [s]
t_0	time of solar noon [s or day]
u	wind speed [m s^{-1}]
	sink or source term [s^{-1} or m^3 s^{-1}]
u^*	friction velocity [m s^{-1}]
v	velocity [m s^{-1}]
v_I	infiltration rate [m^{-2} s^{-1}]
w	gravimetric water content [kg kg^{-1}]
x	spatial coordinate [m]
y	spatial coordinate [m]
z	spatial coordinate [m]
z_d	damping depth [m]
z_h	surface roughness parameter for heat [m]
z_m	surface roughness parameter for momentum [m]
z_0	reference z-coordinate [m]

Upper Case Latin Symbols

A	area [m^2]
$\mathbf{A}$	stiffness matrix
A_m	surface area per unit mass [m^2 g^{-1} or m^2 kg^{-1}]
A_{svl}	Hamaker constant [J]
A_v	surface area per unit volume [m^2 cm^{-3} or m^2 m^{-3}]
$A(0)$	amplitude of temperature wave [°C]
C	solute present per unit mass of soil [kg kg^{-1}]
	water capacity [kg J^{-1}]
	Courant number [–]
	autocovariance [–]
$\mathbf{C}$	mass matrix
C_d	drag coefficient [–]
C_g	volumetric heat capacity of gas phase [J m^{-3} K^{-1}]
C_k	autocovariance function [–]
C_h	volumetric specific heat of soil [J m^{-3} K^{-1}]
C_l	volumetric specific heat of liquid [J m^{-3} K^{-1}]
C_o	volumetric specific heat of organic matter [J m^{-3} K^{-1}]
C_s	volumetric heat capacity of the solid matrix [J m^{-3} K^{-1}]
C_m	volumetric specific heat of minerals [J m^{-3} K^{-1}]
C_0	autocovariance at lag zero [–]
D	fractal dimension [m]
	computational domain for integration [–]
	water diffusivity [m^2 s^{-1}]
D_g	gas diffusivity [m^2 s^{-1}]
D_h	thermal diffusivity [m^2 s^{-1}]
D_m	molecular diffusion of ion in soil [m^2 s^{-1}]
D_o	diffusion coefficient of ion in water [m^2 s^{-1}]
D_v	vapor diffusivity [m^2 s^{-1}]
D_w	diffusivity of gas in water [m^2 s^{-1}]
D_0	binary diffusion coefficient for gas in air [m^2 s^{-1}]

E	evaporation rate [mm h^{-1}]
E_h	thermal energy [J]
E_p	potential evaporation rate [mm h^{-1}]
EC	electrical conductivity [S m^{-1}]
F	distribution function [–]
	total leaf area index of crop [–]
F_b	force of buoyancy [N]
F_d	force of drag [N]
F_g	force of gravity [N]
F_h	hydrostatic force [N]
F_{ij}	water flux between ith and jth nodes
G	soil heat flux [W m^{-2}]
H	sensible heat [W m^{-2}]
	hydraulic head [m]
I	cumulative infiltration [kg m^{-2}]
I_ζ	incomplete beta function [–]
$\mathcal{J}$	Julian day [–]
K	thermal conductance [m^2 s^{-1}]
	hydraulic conductivity [kg s m^{-3}]
	extinction coefficient [–]
$\overline{K}$	mean hydraulic conductivity [kg s m^{-3}]
K_g	conductance for gas flow [m^2 s^{-1}]
K_H	Henry's law constant [–]
K_h	boundary-layer conductance [mol m^{-2} s^{-1}]
K_{hr}	sum of boundary layer and radiative conductances [mol m^{-2} s^{-1}]
K_r	radiative conductance [mol m^{-2} s^{-1}]
K_s	saturated hydraulic conductivity [kg s m^{-3}]
K_v	vapour conductivity [kg s m^{-3}]
K_0	minimum value of saturated hydraulic conductivity [kg s m^{-3}]
K^*	reference state hydraulic conductivity for Miller–Miller media [kg s m^{-3}]
	conveyance function for Manning's equation [–]
L	length [m]
	latent heat flux [W m^{-2}]
$L\downarrow$	long-wave downward radiation [W m^{-2}]
$L\uparrow$	long-wave upward radiation [W m^{-2}]
L_f	latent heat of fusion [J kg^{-1}]
L_i	density of root system in layer i [m^{-2}]
L_v	latent heat of vaporization [J kg^{-1}]
M	mass of soil particles [kg]
	Manning roughness parameter [s m$^{-0.33}$]
M_i	molecular mass [kg mol^{-1} or g mol^{-1}]
M_T	total mass of particles with radius less than R
M_w	molecular mass of water [kg mol^{-1}]
N	number [–]
	solute absorbed by soil surface per unit mass of soil [kg kg^{-1}]
P	pressure [Pa]
	atmospheric pressure [Pa]
P_i	partial pressure of gas [Pa]

P_0	standard-state pressure [Pa]
Pe	Péclet number [–]
Q_g	gas flux [g s^{-1}]
Q_p	radiant flux density at horizontal surface [W m^{-2}]
Q_0	daily input of solar radiation on horizontal surface [MJ m^{-2}]
R	gas constant [J mol^{-1} K^{-1}]
	radius [m]
	resistance [m^4 kg^{-1} s^{-1}]
R_{abs}	radiation absorbed at soil surface [W m^{-2}]
Re	Reynolds number [–]
R_f	retardation factor for solute transport [–]
R_L	resistance in leaf [m^4 kg^{-1} s^{-1}]
$R_{L,upper}$	upper size limit for fractal behaviour
R_n	net radiation [W m^{-2}]
R_r	resistance in root [m^4 kg^{-1} s^{-1}]
R_s	resistance in soil [m^4 kg^{-1} s^{-1}]
R_x	resistance in xylem [m^4 kg^{-1} s^{-1}]
S	particles falling distance [m]
S_c	water saturation at air-entry potential [m^3 m^{-3}]
S_e	degree of saturation [m^3 m^{-3}]
S_I	sorptivity [–]
$S_t \downarrow$	global short wave on a horizontal surface [W m^{-2}]
T	Celsius temperature [°C]
$\overline{T}$	mean soil temperature [°C]
T_a	air temperature [°C or K]
T_C	critical temperature [°C or K]
T_{max}	maximum temperature [°C or K]
T_{min}	minimum temperature [°C or K]
T_K	kelvin temperature [°C or K]
T_{aK}	air temperature [°C or K]
T_{sK}	soil temperature [°C or K]
T_0	standard-state temperature [°C or K]
T_s	soil temperature [°C or K]
T_t	atmospheric transmission coefficient for solar radiation [–]
V	volume [m^3]
V_f	volume of fluids [m^3]
V_g	volume of gas [m^3]
V_m	molar volume of air [m^3]
V_l	volume of liquid phase [m^3]
V_s	volume of solid phase [m^3]
V_t	total volume of solid, gas and liquid phases [m^3]

Lower Case Greek Symbols

α	van Genuchten's parameter in soil moisture characteristic [m^{-1}] [m^{-1} or kg J^{-1}]
	geometric parameter for dielectric mixing model [–]
	osmotic coefficient [–]
α_s	source strength for sink–source term [g m^{-3}]

β contact angle [degrees or radians]

       parameter vector [–]

γ surface tension [N m^{-1} or J m^{-2}]

       semivariogram function [–]

γ_0 surface tension constant [N m^{-1} or J m^{-2}]

γ_k semivariance [–]

γ^* apparent psychrometer constant [kg m^{-3} °C]

δ solar declination [–]

ϵ_a atmospheric emissivity [–]

ϵ_{ac} atmospheric emissivity for cloudy skies [–]

ϵ_b bulk relative dielectric permittivity [–]

ϵ_g gas-phase relative dielectric permittivity [–]

ϵ_l liquid-phase relative dielectric permittivity [–]

ϵ_r relative dielectric permittivity [–]

ϵ_r' real part of relative permittivity [-]

ϵ_r'' imaginary part of relative permittivity [–]

ϵ_s solid-phase relative permittivity [–]

ϵ_0 permittivity of free space [F m^{-1}]

ζ variable depending on gas-filled porosity [–]

ζ_{ij} ratio of bulk exchange area between ith and jth nodes

η dynamic viscosity [kg m^{-1} s^{-1}]

       weighting factor for numerical solution [–]

θ volumetric water content [m^3 m^{-3}]

       angle [radians or degrees]

θ_d volumetric water content in the dry end [m^3 m^{-3}]

θ_r residual water content parameter for van Genuchten's equation [m^3 m^{-3}]

θ_s volumetric saturated water content [m^3 m^{-3}]

θ_0 volumetric cutoff water content for thermal conductivity model [m^3 m^{-3}]

θ^* reference-state water content for Miller–Miller media [m^3 m^{-3}]

ϑ dispersivity [m]

ι correlation length for autocovariance models [–]

κ scaling factor for Miller–Miller media [m]

λ thermal conductivity [W m^{-1} K^{-1}]

       latitude [degrees]

λ_a thermal conductivity of air [W m^{-1} K^{-1}]

λ_f thermal conductivity of fluid [W m^{-1} K^{-1}]

λ_g thermal conductivity of gas [W m^{-1} K^{-1}]

λ_s thermal conductivity of solid [W m^{-1} K^{-1}]

λ_w thermal conductivity of water [W m^{-1} K^{-1}]

μ magnetic permeability of material

       mean

μ_r relative magnetic permeability [–]

μ_r' real part of relative magnetic permeability [–]

μ_r'' imaginary part of relative magnetic permeability [–]

μ_0 magnetic permeability of free space [H m^{-1}]

ν exponent of fractal dimension [–]

       number of particles in solution per molecule of solute

$\hat{\rho}_a$ molar density of air [mol m^{-3}]

ρ_b bulk density [g cm^{-3} or kg m^{-3}]

ρ_l liquid water density [g cm^{-3} or kg m^{-3}]
ρ_s particle density [g cm^{-3} or kg m^{-3}]
σ electrical conductivity [S m^{-1}]
     standard deviation [–]
     Stefan–Boltzmann constant [W m^{-2} K^{-4}]
σ_{dc} zero-frequency (dual-current) electrical conductivity (S m^{-1})
σ_g geometric standard deviation [μ m]
ς exponent of Burdine (1953) model for hydraulic conductivity [–]
τ shear stress [N m^{-2}]
     tortuosity parameter [–]
τ_T parameter for atmospheric adsorption and scattering [–]
ϕ angle [radians or degrees]
     latitude [radians]
     random angle for random functions in Miller scaling [radians or degrees]
ϕ_f porosity [–]
ϕ_g volumetric fraction of gas phase (gas-filled porosity) [m^3 m^{-3}]
ϕ_o volumetric fraction of organic matter content [m^3 m^{-3}]
ϕ_s volumetric fraction of solid phase [m^3 m^{-3}]
χ transformation variable for Philip's infiltration model [–]
     correlation length [–]
ψ water potential [J kg^{-1}]
ψ_e Campbell's air-entry potential parameter [J kg^{-1}]
ψ_f soil water potential at wetting front [J kg^{-1}]
ψ_g gravitational soil water potential [J kg^{-1}]
ψ_h hydrostatic soil water potential [J kg^{-1}]
ψ_i soil water potential at upper boundary [J kg^{-1}]
ψ_L water potential in sub-stomatal cavities [J kg^{-1}]
ψ_m matric soil water potential [J kg^{-1}]
ψ_o osmotic soil water potential [J kg^{-1}]
     oven dry matric potential [J kg^{-1}]
ψ_{os} osmotic potential of saturation extract [J kg^{-1}]
ψ_t total soil water potential [J kg^{-1}]
ψ_Ω overburden pressure soil water potential [J kg^{-1}]
ψ^* reference-state matric potential for Miller–Miller media [J kg^{-1}]
ψ_{xL} water potential in mesophyll and cell walls [J kg^{-1}]
ψ_x water potential in root [J kg^{-1}]
ψ_{xr} water potential in xylem [J kg^{-1}]
ω angular frequency [s^{-1}]
ζ gas diffusion parameter for porous media [–]
     stability parameter [–]

Upper Case Greek Symbols

Δ difference operator
     slope of saturation vapour pressure function [Pa °C^{-1}]
Λ characteristic length for Miller–Miller media [m]
Λ^* reference characteristic length for Miller–Miller media [m]
Φ objective function for least squares curve fitting [-]
     matric flux potential [–]

Φ_e air-entry value for matric flux potential [–]
Ψ_h stability correction factor for heat [–]
Ψ_m stability correction factor for momentum [–]
Σ sum operator [–]
Θ dispersivity [m]
Ω spatial domain [m^3]
      random variable [radians]

List of Python Variables

This is a list of the variables used in the Python programs. Where possible, a correspondence to the symbols used in the mathematical equations is included. Not all the variables are listed, but only the main ones described in the chapters.

Variables

A	array used to read a data file
a	slope of the line for travel time analysis
a	lower diagonal of tridiagonal matrix (a)
aerodynamicResistance	aerodynamic resistance
air_entry	air-entry value (ψ_e)
airPermittivity	gas phase dielectric permittivity (ϵ_g)
airT	air temperature (T_a)
airT0	mean air temperature ($\overline{T}$)
albedo	albedo
alpha	geometrical parameter for dielectric mixing model (α)
ampT	amplitude for temperature oscillation ($A(0)$)
area	node area (A)
	triangle area for three-dimensional solution (A)
	interface area for three-dimensional solution (A)
atmPressure	atmospheric pressure (P_a)
average	average of a vector
avgFirstValues	average of first entries of a vector
b	intercept of the line for travel-time analysis
b	solution of Levenberg–Marquardt
b	main diagonal of tridiagonal matrix (b)
b0	initial guess for Levenberg–Marquardt
bg	diffusivity parameter for soil air-filled porosity (b_g)
binaryDiffCoeff	binary diffusion coefficient (D_0)
bmin, bmax	lower and upper limit for Levenberg–Marquardt
boundaryLayerCond	boundary-layer conductance (K_h)
boundaryOxygenConc	oxygen concentration at boundary (c)
boundaryT	temperature at boundary
bulkDensity	bulk density (ρ_b)
bulkPermittivity	bulk dielectric permittivity (ϵ_b)
c	speed of light (c)
	upper diagonal of tridiagonal matrix (c)
C	water-holding capacity divided by time step
Ca	volumetric specific heat of air (C_g)
C_T	heat capacity (C_h)

CAMPBELL	flag for hydraulic functions
Campbell_b	Campbell water curve retention parameter (b)
Campbell_he	Campbell air-entry potential parameter (ψ_e)
CampbellMFP_he	air-entry potential in matric flux potential
capacity	water-holding capacity (C)
CELL_CENT_FIN_VOL	flag for numerical scheme: cell-centred finite volume
check	Boolean variable to check if a condition is true
clay	clay content (m_y)
color	three-entry array of an RGB colour
colorGrid	array for the colour values of a 2D image
computeOnlySurface	flag for choosing computation of surface runoff only
Conc	solute concentration per unit mass of soil at end of time step
concWat	solute concentration per unit mass of water
CourantNr	Courant number (Cr)
concentration, co	concentration (c)
conductivity	unsaturated hydraulic conductivity (K)
currentSe	degree of saturation (S_e)
cylinder	Visual Python cylinder object
currentDeltaT	current time step (Δt)
D	fractal dimension (D)
d	vector of constants (d)
	zero plane displacement (d)
degreeSaturation	degree of saturation (S_e)
delta	element dimension (d_x)
deltaSpace	delta space between points for TDR waveform
deltaT	difference in temperature
deltaT_max	maximum time step
deltaT_min	minimum time step
deltaTime	delta time between points TDR waveform
denom, denominator	denominator
density	density (ρ)
depth	soil depth (z)
dg	parameter for soil air-filled porosity (d_g)
diameter	particle diameter (d)
distance	distance
dpsi	water potential difference ($d\psi$)
dT	temperature difference (T)
dt	time difference (δt)
dt0	travel time used in TDR analysis (d_x)
Dv	vapour diffusivity (D_v)
DX	direction of tangent
dx	element dimension (d_x)
dy	element dimension (d_y)
dz	element dimension (d_z)
conductivityHVRatio	ratio between horizontal and vertical hydraulic conductivities
endTime	time when simulation stops ( )
energyBalance	energy balance
E_p	potential evaporation rate (E)

EPS	acceptable error in approximations
epsilon	indicator of water flow direction (ϵ)
Etp	potential evapotranspiration (E_p)
EXPONENTIAL	flag for statistical function type: exponential
f	distribution of characteristic length in heterogeneous field (f)
f_c	convective component of solute flux
f_d	diffusive component of solute flux
factor	weighting factor for explicit and implicit
first	first node for Thomas algorithm
first	index for travel time analysis
firstPointTime	absolute time for travel time analysis
flatLine	line for travel time analysis
flow	water flow for three-dimensional solution (f_w)
	boundary water flow for three-dimensional solution (f_w)
flux	water flux at upper boundary during one time step ()
g	gravitational acceleration (g)
g	weighting factor for explicit and implicit (η)
gasPorosity	gas-filled porosity (ϕ_g)
GAUSSIAN	flag for statistical function type: Gaussian
gc	shape factor
	element conductance (g)
geomParameter	geometric parameter for dielectric mixing model (a)
geometricFactor	geometric factor for soil grid
globalRadiation	global short-wave radiation (S_t)
gravWaterContent	gravimetric water content (w)
H	total hydraulic head (H)
H0	previous total hydraulic head (H^t)
handlePermittivity	probe-handle dielectric permittivity ()
heatFlux	heat flux (f_h)
height	vertical extension of image in number of pixels
heightCylinder	height of Visual Python cylinder object
i	integer used in `for` loops
index	index for travel-time analysis
	index of linked cell for three-dimensional solution
index1	index for travel time analysis
index2	index for travel time analysis
indexFlatLine	index for regression in travel time analysis
indexMaxDerivative	index for derivative in travel time analysis
indexMaxDy	index for travel time analysis
indexMinDerivative	index for derivative in travel time analysis
indexP0	index for travel time analysis
indexP2	index for travel time analysis
indexRegr1	index for regression in travel time analysis
indexRegr2	index for regression in travel time analysis
indexRegr3	index for regression in travel time analysis
indexSecondMaxDerivative	index for travel time analysis
indexZeroDerivative	index for travel time analysis
initialWaterPotential	initial water potential (h)

isFreeDrainage	flag for lower boundary condition
IPPISCH_VG	flag for hydraulic function
j	integer used in `for` loops
K	hydraulic conductivity (K)
Kh	boundary-layer conductance (K_h)
k_mean	element dimension (d_x)
K11	conductivities for water-potential-driven water and vapour flow
K12	conductivities for water-potential-driven heat flow
K21	conductivities for temperature-driven vapour flow
K22	conductivities for temperature-driven heat energy flow
k	von Kármán constant (k)
	hydraulic conductivity (K)
ka	gas weighting factor (k_g)
klQ	constant for solution concentration ($k_l \cdot Q$)
Ks	saturated hydraulic conductivity (K_s)
ks	solids weighting factor (k_s)
kw	water weighting factor (k_w)
L	length (L)
L	length of root per unit volume of soil (L_i)
lambda_	mean thermal conductivity
last	past node for Thomas algorithm
last	first node for Thomas algorithm
latentHeat	latent heat exchange
line1, line2, line3	index for travel time analysis
liquidDensity	liquid water density (ρ_l)
liquidPermittivity	liquid phase relative dielectric permittivity (ϵ_l)
liquidViscosity	liquid water viscosity (η_l)
longWaveRadiation	long-wave radiation (η_l)
shortWaveAbsRadiation	long-wave absorbed radiation
lowerDepth	depth of lower boundary
Lv	latent heat of vaporization (L_v)
lx,ly	correlation length of heterogeneous field (τ)
massBalance	mass balance
MAXDELTAINDEX	constant used for calculation of tangent
maxApproximationsNr	maximum number of approximations
maxIterationsNr	maximum number of iterations
maxTimeStep	maximum time step
maxThickness	maximum soil layer thickness
MBRThreshold	mass balance ratio threshold
meanT	mean temperature ($\overline{T}$)
meanType	type of mean
MFP	matric flux potential (Ψ)
mg	diffusivity parameter for air-filled porosity (m_g)
minThickness	minimum soil layer thickness
Mualem_L	tortuosity parameter (l)
myBeta	value of incomplete beta function
myColor	three-entry array for an RGB Colour
myMax	variable used for maximum value
myMin	variable used for minimum value

myOutput	variable used to read data file
mySum	variable used for summations
myWC	vector of water contents
myWP	vector of water potentials
n	number of elements (n)
NetRadOnSoil	net radiation (R_n)
NetAbsRadiation	net absorbed radiation
newSum	variable used to sum the energy balance
NEWTON_RAPHSON_MFP	matric flux potential with Newton–Raphson
NEWTON_RAPHSON_MP	matric potential with Newton–Raphson
NODATA	
nrIterations	number of iterations during one time step
nrOccupiedBoxes	number of occupied boxes in box counting technique
nrPixels	number of pixels in image
nrPixelsThreshold	threshold for box counting technique
nrPores	number of pores
nrTemperatures	number of temperature values
nrValues	number of values
nrXBoxes	number of boxes in horizontal direction
nrYBoxes	number of boxes in vertical direction
numer, numerator	numerator
oldConc	solute concentration of previous time step
oldT	temperature distribution of previous time step
oldTheta	water content distribution of previous time step
Oldvapor	water vapour concentration
omega	angular frequency (ω)
oxygenDiff	diffusion coefficient for oxygen in air (D_0)
p0,p1,p2	points for travel time analysis
particle	Visual Python spherical object
particleDensity	particle density (ρ_s)
particleDiameter	particle diameter (d)
permittivity	permittivity for TDR analysis
picture	image object
pl	leaf water potential (ψ_L)
pond	pond depth (h_{pond})
poresPercentage	percentage of pore pixels
porosity	porosity (ϕ_f)
probeHandle	physical length of TDR probe
probeLength	TDR probe length
psi	water potential (ψ)
Psi_m	stability correction for momentum (Ψ_m)
Psi_h	stability correction for heat (Ψ_h)
q	power for liquid return flow cutoff
R	gas constant (R)
r	radius (r)
rs	soil resistance (R_s)
rw	resistance per unit length of root
rl	radius of root
RadEmitFromSoil	long-wave radiation emitted from soil

RL	resistance of leaf (R_L)
R_r	root resistance (R_r)
reflecCoeff	reflection coefficient
relativeHumidity	relative humidity (h_r)
residualTolerance	residual tolerance
respRate	respiration rate (R)
RESTRICTED_VG	flag for hydraulic function
rgbThreshold	threshold for colour value
ro	molar density of air
rootDepth	depth of deepest root
roughness	Manning roughness parameter (M)
satMassWetness	saturated gravimetric water content
satWaterContent	volumetric saturated water content (θ_s)
Se	degree of saturation (S_e)
sensibleHeat	sensible heat exchange
shortWaveAbsRadiation	short-wave absorbed radiation
sigma	standard deviation of Gaussian function (σ)
simulationLength	length of simulation
sinkSource	sink–source term (u)
size	size
slope	slope of saturation vapour pressure function (Δ) geometric slope
smooth	smoothed vector for TDR analsis
soil	class storing soil parameters
soilTemperature	soil temperature
solidContent	volumetric solid phase (θ_s)
solidPermittivity	solid-phase dielectric permittivity (ϵ_s)
solutionDensity	density of sodium HMP solution (ρ_{hmp})
solutionViscosity	dynamic viscosity of HMP solution (η_{hmp})
solver	type of solver
stcor	Stefan correction (*stcor*)
step	time-step size
success	Boolean variable for successful iteration
sumHeatFlux	sum of heat flux at boundary
sumInfiltration	sum of water infiltration at upper boundary
sumX, sumX2, sumY, sumY2	sum of vector used in TDR analysis
svp	saturation vapour pressure (e_s)
SX	used to define direction of tangent
T	temperature (u)
temperature	temperature in degrees Celsius (T)
temperatureK	temperature in degrees kelvin (K)
thermalConductivity	thermal conductivity (λ)
thermalConductivityfluid	thermal conductivity of fluid (λ_f)
thermalConductivitygas	thermal conductivity of gas (λ_g)
thermalConductivitysolid	thermal conductivity of solid (λ_s)
thermalConductivitywater	thermal conductivity of water(λ_w)
theta	volumetric water content (θ)
thetaIni	initial volumetric water content (θ)
thetaR	residual volumetric water content (θ_r)

thetaS	saturated volumetric water content (θ_s)
threshold	threshold
time	time (t)
timeShift	time shift for temperature oscillation ()
timeStepMax	maximum time step
timeVector	time vector for recorded TDR window (t)
TimeOfDay	hour of day
tolerance	tolerance for mass or energy balance equation (ϵ)
totalDepth	total soil depth (z)
totalIterationNr	total number of iterations (N)
tp	potential transpiration rate (E)
travelTime	travel time (t)
u	source strength (u)
ustar	friction velocity (u^*)
ubPotential	upper boundary potential (ψ_0)
upperDepth	depth of upper boundary ()
VAN_GENUCHTEN	flag for hydraulic functions: van Genuchten ()
Vapor	water vapour concentration at end of time step
VG_alpha	van Genuchten's parameter (α)
VG_he	Ippisch–van Genuchten parameter (ψ_e)
VG_m	van Genuchten parameter (m)
VG_n	van Genuchten's parameter (ns)
VG_thetaR	van Genuchten's parameter (θ_r)
viscosity	viscosity (η)
voidRatio	void ratio (e)
vol	element volumes (V)
volume	element volume for three-dimensional solution (V)
Vp	fraction of speed of light for travel time computation (v)
waterContent, wc	volumetric water content (θ)
waterDensity	liquid water density (ρ_l)
waterFluxDensity	water flux during current time step (f_l)
waterPotential	water potential (ψ)
waterRetentionCurve	type of model for water retention curve
waterTemperature	water temperature (T)
waterVapourDiff0	binary diffusion coefficient for gas in air (D_0)
wcMalicki	water content from Malicki model (θ)
wcMixModel	water content from Roth *et al.* model (θ)
wcTopp	water content from Topp *et al.* model (θ)
wf	liquid return flow cutoff (w_f)
width	horizontal extension of image in number of pixels
windSpeed	wind speed (u)
windowBegin	window begin for TDR data ()
windowWidth	window width for TDR data ()
x	space variable (x)
xwo	cutoff water content for liquid return flow
y	space variable (y)
z	space variable (z)
zeta	stability parameter (ζ)
zh	roughness parameter (z_h)
zm	roughness factor (z_m)

List of Python Projects

The following are the Python projects for each chapter:

PSP_boxCounting	Chapter 2
PSP_capillaries	
PSP_basicProperties	
PSP_sedimentation	
PSP_gasDiffusion	Chapter 3
PSP_thermalConductivity	Chapter 4
PSP_heat	
PSP_travelTimeAnalysis	Chapter 5
PSP_waterRetentionFitting	
PSP_columnWaterContent	
PSP_Poiseuille	Chapter 6
PSP_unsaturatedConductivity	
PSP_MillerMiller	Chapter 7
PSP_infiltrationRedistribution1D	Chapter 8
PSP_triangulation	Chapter 9
PSP_Criteria3D	Chapter 10
PSP_evaporation	Chapter 11
PSP_coupled	Chapter 12
PSP_soluteTransportAnalytical	Chapter 13
PSP_soluteTransportNumerical	
PSP_transpiration	Chapter 14
PSP_boundaryLayerConductance	Chapter 15
PSP_circle	Appendix A
PSP_averageTair	
PSP_weatherData	
PSP_CelsiusFahrenheit	
PSP_readDataFile	
PSP_soilTemp	
PSP_numericalDerivation	Appendix B
PSP_numericalIntegration	
PSP_GaussElimination	
PSP_ThomasAlgorithm	

References

Abramowitz, M. and Stegun, I. A. (1970). *Handbook of Mathematical Functions.* Dover Publications, New York.

Acutis, M. and Donatelli, M. (2003). SOILPAR 2.00: Software to estimate soil hydrological parameters and functions. *Eur. J. Agron.,* **18**, 373–377.

Allen, T. (1981). Interaction between particles and fluids in a gravitational field. In *Particle Size Measurement.* Chapman and Hall, London.

Ball, B. C. (1981). Pore characteristic of soils from two cultivation experiments as shown by gas diffusivities and permeabilities and airfilled porosities. *J. Soil Sci.,* **32**, 483–498.

Barone, V. (2009). Invited speech presented at the Italian Society for the Progress of Sciences, Bari, October 13, 1933. As described in *Enrico Fermi, Atomi, Nuclei Particelle, Scritti Divulgativi ed Espositivi, 1923–1952,* pp. 110–111. Universale Bollati Boringhieri, Torino.

Baveye, P., Parlange, J. Y., and Stewart, A. B. (1997). *Fractals in Soil Science.* CRC Press, Boca Raton, FL.

Bear, J. (1972). *Dynamics of Fluids in Porous Media.* Dover Publications, New York.

Beazley, D. (2006). *Python Essential Reference* (3rd edn). SAMS, Indianapolis, IN.

Bechtold, M., Vanderborght, J., Ippisch, O., and Vereecken, H. (2011). Efficient random walk particle tracking algorithm for advective-dispersive transport in media with discontinuous dispersion coefficients and water contents. *Water Resour. Res.,* **47**, W10526, doi:10.1029/2010WR010267.

Bird, R. B., Stewart, W. E., and Lightfoot, E. N. (1960). *Transport Phenomena.* Wiley, New York.

Bittelli, M., Campbell, G. S., and Flury, M. (1999). Characterization of particle-size distributions in soils with a fragmentation model. *Soil Sci. Soc. Am. J.,* **63**, 782–788.

Bittelli, M., Flury, M., and Roth, K. (2004). Use of dielectric spectroscopy to estimate ice content in frozen porous media. *Water Resour. Res.,* **40**, W04212, doi:10.1029/2003WR002343.

Bittelli, M., Salvatorelli, F., and P. Rossi (2008*b*). Correction of tdr-based soil water content measurements in conductive soil. *Geoderma,* **143**, 133–142.

Bittelli, M., F. Ventura, G. S. Campbell, Snyder, R. L., Gallegati, F., and P. Rossi (2008*a*). Coupling of heat, water vapor, and liquid water fluxes to compute evaporation in bare soils. *J. Hydrol.,* **362**, 191–205.

Bittelli, M., Tomei, F., Pistocchi, A., Flury, M., Boll, J., Brooks, E. S., and Antolini, G. (2010). Development and testing of a physically based, three-dimensional model of surface and subsurface hydrology. *Adv. Water. Resour.,* **33**, 106–122.

Bittelli, M., Valentino, R., Salvatorelli, F., and P. Rossi (2012). Monitoring soil-water and displacement conditions leading to landslide occurrence in partially saturated clays. *Geomorphology,* **173–174**, 161–173.

Bresler, E. (1973). Simultaneous transport of solutes and water under transient unsaturated flow conditions. *Water Resour. Res.,* **9**, 975–986.

Bresler, E., McNeal, B. L., and Carter, D. L. (1982). *Saline and Sodic Soils: Principles–Dynamics–Modeling.* Springer-Verlag, New York.

Bristow, K. L. and Campbell, G. S. (1984). On the relationship between incoming solar radiation and daily maximum and minimum temperature. *Agric. For. Meteorol.*, **31**, 159–166.

Bristow, K. L., Campbell, G. S., and Calissendorf, C. (1984). The effects of texture on the resistance to water movement within the rhizosphere. *Soil Sci. Soc. Am. J.*, **48**, 266–270.

Brutsaert, W. (1975). On a derivable formula for long-wave radiation from clear skies. *Water Resour. Res.*, **11**, 742–744.

Buchan, G. D., Grewal, K. S., and Robson, A. B. (1993). Improved models of particle-size distribution: an illustration of model comparison techniques. *Soil Sci. Soc. Am. J.*, **57**, 901–908.

Buck, A. L. (1981). New equations for computing vapor pressure and enhancement factor. *J. Appl. Metereol.*, **20**, 1527–1532.

Burden, R. L. and Faires, J. D. (1997). *Numerical Analysis* (6th edn). Brooks and Cole, Washington.

Burdine, N. T. (1953). Relative permeability calculation from size distribution data. *Trans. AIME*, **198**, 71–78.

Burgess, T. M. and Webster, R. (1980). Optimal interpolation and isarithmic mapping of soil properties: I The semivariogram and punctual kriging. *J. Soil Sci.*, **31**, 315–331.

Campbell, G. S. (1974). A simple method for determining unsaturated conductivity from moisture retention data. *Soil Sci.*, **117**, 311–387.

Campbell, G. S., Jr, Jungbauer, J. D., Bidlake, W. R., and Hungerford, R. D. (1994). Predicting the effect of temperature on soil thermal conductivity. *Soil Sci.*, **158**, 307–313.

Campbell, G. S. (1977). *An Introduction to Environmental Biophysics*. Springer-Verlag, New York.

Campbell, G. S. (1985). *Soil Physics with Basic: Transport Models for Soil–Plant Systems*. Elsevier, Amsterdam.

Campbell, G. S. and Norman, J. M. (1998). *An Introduction to Environmental Biophysics* (2nd edn). Springer-Verlag, New York.

Campbell, G. S. and Shiozawa, S. (1992). Prediction of hydraulic properties of soils using particle-size distribution and bulk density data. In *Proceedings of the International Workshop on Indirect Methods for Estimating Hydraulic Properties of Unsaturated Soils* (ed. M. van Genuchten et al.), pp. 317–328. University of California, Riverside.

Campbell Scientific (2006). *CR10X Measurement and Control Datalogger, Instruction Manual*. Campbell Scientific, Inc., Logan, UT.

Carlson, B. C. (1972). The logarithmic mean. *Am. Math. Monthly*, **79**, 615–618.

Carminati, A. (2012). A model of root water uptake coupled with rhizosphere dynamics. *Vadose Zone J.*, **11**, doi:10.2136/vzj2011.0106.

Carminati, A., Moradi, A. B., Vetterlein, D., Vontobel, P., Lehmann, E., Weller, U., Vogel, H. J., and Oswald, S. E. (2010). Dynamics of soil water content in the rhizosphere. *Plant Soil*, **332**, 163–176.

Carminati, A., Schneider, C. L., Moradi, A. B., Zarebanadkouki, M., Vogel, H. J., Hildebrandt, A., Weller, U., Schuler, L., and Oswald, S. E. (2011). How the rhizosphere may favor water availability to roots. *Vadose Zone J.*, **10**, 1–11.

Carslaw, H. S. and Jaeger, J. C. (1959). *Conduction of Heat in Solids*. Clarendon Press, Oxford.

Cavazza, L. (1981). *Fisica del terreno agrario*. UTET, Bologna, Italy.

Cheng, S. W., Dey, T. K., and Shewchuk, J. R. (2012). *Delauney Mesh Generation*. CRC Press, Boca Raton, FL.

Childs, E. C. and Collis-George, N. (1950). The permeability of porous materials. *Proc. R. Soc. Lond.*, **A201**, 392–405.

Cignoni, P., Montani, C., and Scopigno, R. (1998). DeWall: A fast divide and conquer Delaunay triangulation algorithm in E^d. *Computer-Aided Design*, **30**, 333–341.

Cline, R. G. and Campbell, G. S. (1976). Seasonal and diurnal water relations of selected forest species. *Ecology*, 57, 367–373.

Conte, S. D. and deBoor, C. (1972). *Elementary Numerical Analysis*. McGraw-Hill, New York.

Corak, S. J., Blevins, D. G., and Pallardy, S. G. (1987). Water transfer in an alfalfa/maize association. *Plant Physiol.*, 84, 582–586.

Cowan, I. R. (1965). Transport of water in the soil–plant–atmosphere system. *J. Appl. Ecol.*, 2, 221–239.

Currie, J. A. (1965). Diffusion within soil microstructure: a structural parameter for soil. *J. Soil Sci.*, 16, 279–289.

Cussler, E. L. (1997). *Diffusion: Mass Transfer in Fluid Systems* (2nd edn), Chaps. 3 and 7. Cambridge University Press, Cambridge.

Dagan, G. and Neuman, S. P. (2005). *Subsurface Flow and Transport: A Stochastic Approach*. Cambridge University Press, Cambridge.

Dal Ferro, N., Delmas, P., Duwig, C., Simonetti, G., and Morari, F. (2012). Coupling x–ray microtomography and mercury intrusion porosimetry to quantify aggregate structures of a cambisol under different fertilisation treatments. *Soil Till. Res.*, 199, 13–21.

Darcy, D. (1856). *Les Fontaines de la Ville de Dijon*. Dalmont, Paris.

de Vries, D. A. (1963). Thermal properties of soil. In: *Physics of Plant Environment*. (ed. W. R. Wijk), pp. 210–235. North-Holland, Amsterdam.

Deresiewicz, H. (1958). Mechanics of granular media. *Adv. Appl. Mech.*, 5, 233–306.

Di Giammarco, P., Todini, E., and Lamberti, P. (1996). A conservative finite elements approach to overland flow: the control volume finite element formulation. *J. Hydrol.*, 175, 267–291.

Douglas, D. and Peucker, T. (1973). Algorithms for the reduction of the number of points required to represent a digitized line or its caricature. *Canadian Cartogr.*, 10, 112–122.

ESRI (2011). Arcgis desktop: Release 10. Technical report, Environmental Systems Research Institute, Redlands, CA.

Flerchinger, G. N., Hudson, C. L., and Wight, J. R. (1996). Modeling evapotranspiration and surface energy budgets across a watershed. *Water Resour. Res.*, 32, 2539–2548.

Flury, M. and Gimmi, T. F. (2002). Solute diffusion. In *Methods of Soil Analysis. Part 4. Physical Methods* (ed. J. Dane and G. Topp), pp. 1323–1350. American Society of Agronomy, Madison, WI.

Gardner, W. R. (1958). Some steady-state solutions to the unsaturated flow equation with application to evaporation from a water table. *Soil Sci.*, 85, 228–232.

Gardner, W. R. (1960). Dynamic aspects of water availability to plants. *Soil Sci.*, 89, 63–73.

Gee, G. W. and Bauder, J. W. (1986). Particle-size analysis. In *Methods of Soil Analysis. Part 1. Physical and Mineralogical Methods* (2nd edn) (ed. A. Klute), pp. 383–411. American Society of Agronomy, Madison, WI.

Gee, G. W. and Or, D. (2002). Particle-size analysis. In *Methods of Soil Analysis. Part 1. Physical and Mineralogical Methods* (ed. A. Klute), pp. 255–289. American Society of Agronomy, Madison, WI.

Green, W. H. and Ampt, G. A. (1911). Studies in soil physics: the flow of air and water through soils. *J. Agr. Sci.*, 4, 1–24.

Grismer, M. E. (1987). Note: Water vapor adsorption and specific surface. *Soil Sci.*, 144, 233–236.

Guggenheim, E. A. (1945). The principle of corresponding states. *J. Chem. Phys.*, 13, 253–262.

Hallikainen, M. T., Ulaby, F. T., Dobson, M. C., El-Rayes, M. A., and Wu, L. K. (1985). Microwave dielectric behavior of wet soil–Part 1: Empirical models and experimental observations. *IEEE Trans. Geosci. Remote Sens.*, 23, 25–34.

Harrison, L. P. (1963). Fundamental concepts and definitions relating to humidity. In: *Humidity and Moisture* (ed. A. Wexler), Vol. 3. Reinhold, New York.

Hasted, J. B. (1973). *Aqueous Dielectrics*. Chapman and Hall, London.

Ippisch, O., Vogel, H.J., and Bastian, P. (2006). Validity limits for the van Genuchten–Mualem model and implications for parameter estimation and numerical simulation. *Adv. Water Resour.*, **29**, 1780–1789.

International Society of Soil Science, ISSS (1927). Commission One. International Congress of Soil Science, Washington, D.C.

IUPAC (1972). *Manual of Symbols and Terminology*. International Union of Pure and Applied Chemistry, London.

Iwamatsu, M. and Horii, K. (1996). Capillary condensation and adhesion of two wetter surfaces. *J. Colloid Interface Sci.*, **182**, 400–406.

Jay, L. (1991). Comparison of existing methods for building triangular irregular network, models of terrain from grid digital elevation models. *J. Geogr. Inf. Syst.*, **5**, 267–285.

Johnson, W. M., McClelland, J. E., McCaleb, S. B., Ulrich, R., Harper, W. G., and Hutchings, T. B. (1960). Classification and description of soil pores. *Soil Sci.*, **12**, 1–52.

Jury, W. A. and Roth, K. (1990). *Transfer Functions and Solute Movement through Soil*. Birkhäuser, Basel.

Kleinbaum, D. G., Kupper, L. L., Nizam, A., and Muller, K. E. (2008). *Applied Regression Analysis and Other Multivariable Methods*. Thomson Brooks Cole, Belmont, CA.

Kroener, E., Vallati, A., and Bittelli, M. (2014a). Numerical simulation of coupled heat, liquid water and water vapor in soils for heat dissipation of underground electrical power cables. *Appl. Therm. Eng.*, **70**, 510–523.

Kroener, E., Zarebanadkouki, M., Kaestner, A., and Carminati, A. (2014b). Nonequilibrium water dynamics in the rhizosphere: how mucilage affects water flow in soils. *Water Resour. Res.*, **50**, 6479–6495.

Lai, D. C. (2009). *Linear Algebra and its Applications*. Pearson, New York.

Lai, S. H., Tiedji, J. M., and Erickson, E. (1976). In situ measurements of gas diffusion coefficient in soils. *Soil Sci. Soc. Am. J.*, **40**, 3–6.

Langtangen, H. P. (2009). *Python Scripting for Computational Science* (3rd edn). Springer-Verlag, Berlin.

Ledieu, J., Ridder, P. De, Clerck, P. De, and Dautrebande, S. (1986). A method of measuring soil moisture by time-domain reflectometry. *J. Hydrol.*, **88**, 319–328.

Levenberg, K. (1944). A method for the solution of certain nonlinear problems in least squares. *Q. Appl Math.*, **2**, 164–168.

Lumley, J. L. and Panofsky, A. (1964). *The Structure of Atmospheric Turbulence*. Wiley, New York.

Lutz, M. (2009). *Learning Python*. O'Reilly, Sebastopol, CA.

Malicki, M. A., Plagge, R., and Roth, C. H. (1996). Improving the calibration of dielectric TDR soil moisture determination taking into account the solid soil. *Eur. J. Soil Sci.*, **47**, 357366.

Mandelbrot, B. (1975). *Les objects fractals: Forme, hasard et dimension*. Flammarion, Paris.

Marquardt, D. W. (1963). An algorithm for least-squares estimation of non-linear parameters. *J. Soc. Ind. Appl. Math.*, **11**, 431–441.

Marshall, T. J. (1958). A relation between permeability and size distribution of pores. *J. Soil Sci.*, **9**, 1–8.

Marshall, T. J. (1959). The diffusion of gases through porous media. *J. Soil Sci.*, **10**, 79–82.

Martelli, A. (2006). *Python in a Nutshell* (2nd edn). O'Reilly, Sebastopol, CA.

Miller, E. E. and Miller, R. D. (1956). Physical theory for capillary flow phenomena. *J. Appl. Phys.*, **27**, 324–332.

Monteith, J. L. (1964). Evaporation and environment. In *The State and Movement of Water in Living Organisms. Symposia of the Society for Experimental Biology*, No. 19 (ed. G. E. Fogy), p. 205. Cambridge University Press, Cambridge.

Morton, K. W. and Mayers, D. F. (1994). *Numerical Solution of Partial Differential Equations.* Cambridge University Press, Cambridge.

Mualem, Y. (1976). A new model for predicting the hydraulic conductivity of unsaturated porous media. *Water Resour. Res.*, **12**, 513–522.

Nellis, G. and Klein, S. (2009). *Heat Transfer.* Cambridge University Press, Cambridge.

Norman, J. N. and Campbell, G. S. (1983). Application of a plant–environment model to problems in irrigation. *Adv. Irrig.*, **2**, 155–188.

Ohmura, A. and Raschke, E. (2005). Energy budget at the earth's surface. In *Observed Global Climate* (ed. M. Hantel). Springer-Verlag, Berlin.

Ong, S. K. and Lion, L. W. (1991). Effects of soil properties and moisture on the adsorption of trichloethylene vapor. *Water Res.*, **25**, 29–36.

Or, D., Lehmann, P., Shahraeeni, E., and Shokri, N. (2013). Advances in soil evaporation physics—a review *Vadose Zone J.*, **12**, doi:10.2136/vzj2012.0163.

Pachepsky, Y., Crawford, J. W., and Rawls, W. J. (eds.) (2000). *Fractals in Soil Science.* Elsevier, Amsterdam.

Papendick, R. I. and Campbell, G. S. (1980). Theory and measurement of water potential. In *Relations in Microbiology*, pp. 1–22. American Society of Agronomy Special Publication 9, Madison, WI.

Passioura, J. B. (1972). The effect of root geometry on the yield of wheat growing on stored water. *Aust. J. Agr. Res.*, **23**, 745–752.

Patankar, S. V. (1992). *Numerical Heat Transfer and Fluid Flow.* Taylor & Francis, London.

Penman, H. L. (1940). Gas and vapor movements in the soil: I. The diffusion of vapors through porous solids. *J. Agr. Sci.*, **30**, 437–461.

Philip, J. R. (1957). The theory of infiltration. *Soil Sci.*, **83**, 345–357.

Philip, J. R. and de Vries, D. A. (1957). Moisture movement in porous materials under temperature gradients. *Trans. Am. Geophys. Union*, **38**, 222–231.

Pieri, L. M., Bittelli, M., and Pisa, P. Rossi (2006). Laser diffraction, transmission electron microscopy and image analysis to evaluate a bimodal Gaussian model for particle size distribution in soils. *Geoderma*, **135**, 118–132.

Pieri, L. M., Bittelli, M., Wu, J. Q., Dun, S., Flanagan, D. C., Pisa, P. Rossi, Ventura, F., and Salvatorelli, F. (2007). Using the water erosion prediction project (WEPP) model to simulate field-observed runoff and erosion in the Apennines mountain range, Italy. *J. Hydrol.*, **336**, 84–97.

Pistocchi, A. and Tomei, F. (2003). Implementation of a 3D coupled surface–subsurface numerical flow model within the framework of the criteria decision support system. Proceedings of Italian Meeting of Agrometeorology, Bologna, Italy.

Poling, B. E., Prausnitz, J. M., and Connell, J. P. O. (2000). *The Properties of Gases and Liquids.* McGraw-Hill, New York.

Posadas, A. N. D, Gimenez, D., Quiroz, R., and Protz, R. (2003). Multifractal characterization of soil pore systems. *Soil Sci. Soc. Am. J.*, **67**, 1361–1369.

Press, W. H., Teukolsky, S. A., Vetterling, W. T., and Flannery, B. P. (1992). *Numerical Recipes. The Art of Scientific Computing* (2nd edn). Cambridge University Press, Cambridge.

Pritchard, D. T. and Currie, J. A. (1982). Diffusion coefficients of carbon dioxide, nitrous oxide, ethylene and ethane in air and their measurement. *J. Soil. Sci.*, **33**, 175–184.

Purcell, W. R. (1949). Capillary pressures—their measurement with mercury and the calculation of permeability. *Trans. AIME*, **186**, 39–48.

Raju, G. G. (2003). *Dielectrics in Electric Fields*. Marcel Dekker, New York.

Ramer, U. (1972). An iterative procedure for the polygonal approximation of plane curves. *Comput. Graphics Image Process.*, **1**, 244–256.

Ramo, S., Whinnery, J. R., and Duzer, T. Van (1994). *Fields and Waves in Communication Electronics*. Wiley, New York.

Rawls, W. J., Ahuja, L. R., and Brakensiek, K. L. (1992). Estimating soil hydraulic properties from soils data. In *Proceedings of the International Workshop on Indirect Methods for Estimating Hydraulic Properties of Unsaturated Soils* (ed. M. van Genuchten et al.). University of California, Riverside.

Richards, L. A. (1948). Porous plate apparatus for measuring moisture retention and transmission by soils. *Soil Sci.*, **66**, 105–110.

Robin, M. J., Gutjahr, A. L., Sudicky, E. A., and Wilson, J. L. (1993). Cross-correlated random field generation with the direct fourier transform method. *Water Resour. Res.*, **29**, 2385–2397.

Robinson, R. A. and Stokes, R. H. (1965). *Electrolyte Solutions*. Butterworths, London.

Roth, K. (1995). Steady state flow in an unsaturated, two-dimensional, macroscopically homogeneous, Miller-similar medium. *Water Resour. Res.*, **31**, 2127–2140.

Roth, K., Schulin, R., Fluhler, H., and Attinger, W. (1990). Calibration of time domain reflectometry for water content measurement using a composite dielectric approach. *Water Resour. Res.*, **26**, 2267–2273.

Russo, D. and Bresler, E. (1981). Soil hydraulic properties as stochastic processes: I An analysis of field spatial variability. *Soil Sci. Soc. Am. J.*, **46**, 682–687.

Russo, D. and Bresler, E. (1982). Soil hydraulic properties as stochastic processes: II Errors of estimate in an heterogeneous field. *Soil Sci. Soc. Am. J.*, **46**, 20–26.

Sallam, A., Jury, W. A., and Letey, J. (1984). Measurement of gas diffusion coefficient under relatively low air-filled porosity. *Soil Sci. Soc. Am. J.*, **48**, 3–6.

Santamarina, J. C., Klein, K. A., and Fam, M. A. (2001). *Soils and Waves: Particulate Materials Behavior, Characterization and Process Monitoring*. Wiley, New York.

Schaap, M. G., Leij, F. J., and van Genuchten, M.Th. (2001). Rosetta: a computer program for estimating soil hydraulic parameters with hierarchical pedotransfer functions. *J. Hydrol.*, **251**, 163–176.

Scheidegger, A. E. (1960). *Physics of Flow through Porous Media*. Macmillan, New York.

Scholander, P., Bradstreet, E., Hemmingsen, E., and Hammel, H. (1965). Sap pressure in vascular plants: negative hydrostatic pressure can be measured in plants. *Science*, **148**, 339–346.

Seaman, J. C., Chang, H., Goldberg, S., and Simunek, J. (2012). Reactive transport modeling. *Vadose Zone J.*, 3–6.

Shiozawa, S. and Campbell, G. S. (1991). On the calculation of mean particle diameter and standard deviation from sand, silt, and clay fractions. *Soil Sci.*, **152**, 427–431.

Simunek, J. and van Genuchten, M. T. (1994). *The CHAIN2D Code for Simulating the Two-Dimensional Movement of Water, Heat, and Multiple Solutes in Variably-Saturated Porous Media*. Research Report No. 136, US Salinity Laboratory, Agricultural Research Service, US Department of Agriculture, Riverside, California.

Simunek, J., van Genuchten, M. T., and Sejna, M. (2005). *The HYDRUS-1D Software Package for Simulating the One-Dimensional Movement of Water, Heat and Multiple Solutes in Variably-Saturated Media. Version 3.0*. Department of Environmental Sciences. University of California, Riverside.

Slocum, T. A., McMaster, R. B., Kessler, F. C., and Howard, H. H. (2009). *Thematic Cartography and Geovisualization* (3rd edn). Prentice-Hall, Upper Saddle River, NJ.

Smith, O. L. (1991). An analytical model of the decomposition of soil organic matter. *Soil Biol. Biochem.*, **11**, 585–606.

Soil Survey Division Staff (1993). Soil survey manual. Soil Conservation Service. U.S. Department of Agriculture Handbook 18. Washington, D.C.

Taina, I. A., Heck, R. J., and Elliot, T.R. (2008). Application of X-ray computed tomography to soil science: a literature review. *Can. J. Soil Sci.*, **88**, 1–19.

Taylor, S. A. and Stewart, G. L. (1960). Some thermodynamic properties of soils. *Soil Sci. Am. Proc.*, **24**, 243–247.

Tomei, F. (2005). *Numerical Analysis of Hydrological Processes*. Master's Thesis, Faculty of Mathematics, Department of Computer Science, University of Bologna, Italy.

Tongyai, M. L. C. (1977). *Predicting Ponderosa Pine Selling Survival from Environmental and Plant Parameters*. PhD. Dissertation, Washington State University, Pullman.

Topp, G. C., Annan, J. L., and Davis, A. P. (1980). Electromagnetic determination of soil water content: measurements in coaxial transmission lines. *Water Resour. Res.*, **16**, 574–582.

Tuller, M. and Or, D. (2005). Water films and scaling of soil characteristic curves at low water contents. *Water Resour. Res.*, **41** W09403, doi:10.1029/2005WR004142.

Turcotte, D. L. (1997). *Fractals and Chaos in Geology and Geophysics*. Cambridge University Press, Cambridge.

Unsworth, M. H. and Monteith, J. L. (1975). Geometry of long-wave radiation at the ground. *Q. J. R. Meteorol. Soc.*, **101**, 13–24.

USDA (1975). *Soil Survey*. US Department of Agriculture.

van de Griend, A. and Owe, M. (1994). Bare soil surface resistance to evaporation by vapor diffusion under semiarid conditions. *Water Resour. Res.*, **30**, 181–188.

van de Pol, R. M., Wierenga, P. J., and Nielsen, D. R. (1977). Solute movement in a field soil. *Soil Sci. Soc. Am. J.*, **41**, 10–13.

van Genuchten, M. T. (1980). A closed-form equation for predicting the hydraulic conductivity of unsaturated soils. *Soil Sci. Soc. Am. J.*, **44**, 892–898.

van Genuchten, M. T., Leij, F. J., and Yates, S. R. (1991). The RETC code for quantifying the hydraulic functions of unsaturated soils. EPA/600/2-91/065, US Environmental Protection Agency, Ada, OK.

van Kampen, N. G. (1981). *Stochastic Approaches in Physics and Chemistry*. North-Holland, New York.

Vogel, H.-J. (2002). Topological characterization of porous media. In *Morphology of Condensed Matter. Lecture Notes in Physics*, Vol. 600 (ed. K. R. Mecke and D. Stoyan), pp. 75–92. Springer-Verlag, Berlin.

Vogel, H.-J., Weller, U., and Schluter, S. (2010). Quantification of soil structure based on minkowski functions. *Comput. Geosci.*, **36**, 1236–1245.

Warrick, A. W. and Nielsen, D. R. (1980). Spatial variability of soil physical properties. In *Applications of Soil Physics* (ed. D. Hillel), pp. 224–319. Academic Press, New York.

Webster, R. (1977). *Quantitative and Numerical Methods in Soil Classification and Survey*. Clarendon Press, Oxford.

Index